Socioeconomic, Sustainability, Health, and Human Factors 2018

Transportation Research Record: Journal of the Transportation Research Board Volume 2672, Issue 3

ISBN: 978-1-5108-8175-4

Printed from e-media with permission by:

Curran Associates, Inc.
57 Morehouse Lane
Red Hook, NY 12571

Some format issues inherent in the e-media version may also appear in this print version.

Published by Sage Publications USA

ISBN (Print) 978-1-5108-8175-4
ISBN 2018 Printed Set (All Issues) 978-1-5108-7735-1

Printed by Curran Associates, Inc. (2019)

Additional copies of this publication are available from:

Curran Associates, Inc.
57 Morehouse Lane
Red Hook, NY 12571 USA
Phone: 845-758-0400
Fax: 845-758-2633
Email: curran@proceedings.com
Web: www.proceedings.com

Volume 2672 Number 3 December 2018

TRR
JOURNAL OF THE TRANSPORTATION RESEARCH BOARD

Contents

Socioeconomic, Sustainability, Health, and Human Factors 2018

Articles

TRANSPORTATION RESEARCH BOARD

CORE SPONSORS

Transportation Departments of the 50 States and the District of Columbia

Federal Government

U.S. Department of Transportation
- Federal Aviation Administration (FAA)
- Federal Highway Administration
- Federal Motor Carrier Safety Administration (FMCSA)
- Federal Railroad Administration (FRA)
- Federal Transit Administration
- National Highway Traffic Safety Administration (NHTSA)
- Office of the Assistant Secretary of Transportation for Research and Technology (OST-R)

U.S. Air Force Civil Engineer Center (AFCEC)
U.S. Coast Guard (USCG)
U.S. Department of Energy
U.S. Department of the Interior (DOI)
U.S. Environmental Protection Agency (EPA)

Nongovernmental Organizations

American Public Transportation Association (APTA)
Association of American Railroads (AAR)
California Air Resources Board

The National Academies of

SCIENCES • ENGINEERING • MEDICINE

The **National Academy of Sciences** was established in 1863 by an Act of Congress, signed by President Lincoln, as a private, nongovernmental institution to advise the nation on issues related to science and technology. Members are elected by their peers for outstanding contributions to research. Dr. Marcia McNutt is president.

The **National Academy of Engineering** was established in 1964 under the charter of the National Academy of Sciences to bring the practices of engineering to advising the nation. Members are elected by their peers for extraordinary contributions to engineering. Dr. C. D. Mote, Jr., is president.

The **National Academy of Medicine** (formerly the Institute of Medicine) was established in 1970 under the charter of the National Academy of Sciences to advise the nation on medical and health issues. Members are elected by their peers for distinguished contributions to medicine and health. Dr. Victor J. Dzau is president.

The three Academies work together as the National Academies of Sciences, Engineering, and Medicine to provide independent, objective analysis and advice to the nation and conduct other activities to solve complex problems and inform public policy decisions. The Academies also encourage education and research, recognize outstanding contributions to knowledge, and increase public understanding in matters of science, engineering, and medicine.

Learn more about the National Academies of Sciences, Engineering, and Medicine at **www.national-academies.org.**

The **Transportation Research Board** is one of seven major programs of the National Academies of Sciences, Engineering, and Medicine. The mission of the Transportation Research Board is to increase the benefits that transportation contributes to society by providing leadership in transportation innovation and progress through research and information exchange, conducted within a setting that is objective, interdisciplinary, and multimodal. The Board's varied activities annually engage about 7,000 engineers, scientists, and other transportation researchers and practitioners from the public and private sectors and academia, all of whom contribute their expertise in the public interest. The program is supported by state transportation departments, federal agencies including the component administrations of the U.S. Department of Transportation, and other organizations and individuals interested in the development of transportation.

Learn more about the Transportation Research Board at **www.TRB.org**.

Article

TRR
JOURNAL OF THE TRANSPORTATION RESEARCH BOARD

Transportation Research Record
2018, Vol. 2672(3) 1–11

DOI: 10.1177/0361198118775868
journals.sagepub.com/home/trr
SAGE

Transportation and Land Property Rights: Economic Impacts on Agricultural Productivity

Lucas Bispo de Oliveira Alves[1] **and Hironori Kato**[1]

Abstract

This paper empirically investigates the relationship between the effectiveness of land property rights and the impacts of transportation on agricultural output. It is assumed that property rights are a pre-condition for the functioning of land markets. The hypothesis is that if land markets are functional then the impacts of transportation are higher because farmers can respond to improvements in transportation either by making productivity-enhancing investments or by engaging in efficient land transactions. An agricultural production function was estimated where both transportation and property rights were treated as constituents of total factor productivity, using panel data from 76 countries from 2007 to 2013. Ordinary least squares and two-stage least squares regressions revealed that, contrary to our expectations, the impacts of transportation are higher in countries with less effective property rights systems. Three possible explanations are raised. First, it is possible that better property rights are correlated with higher government intervention in agriculture, which negatively impacts productivity. Second, possibly larger benefits from improvements in property rights are to be made in the earlier stages of development. Third, it is possible that high productivity is associated with large farms because of economies of scale. Since they are more likely to be found in land-rich countries, which tend to have less effective property rights, the present paper raises concerns that land transactions triggered by transportation projects may lead to outcomes that are not Pareto-optimal, with possible consequences for social indicators such as rural poverty.

Transportation infrastructure is critical for agricultural development. In the U.S.A., for example, agricultural goods accounted for 30% of all goods transported by all modes—measured in ton-miles—in 2007, which makes agriculture the largest user of the transportation system (*1*). Agricultural regions are often distant from population centers, which increases the need for transportation. Meanwhile, insufficient transportation networks, mostly in developing countries, keep the costs of access to markets high, which limits prospects for economic uplifting.

The introduction of an efficient transportation system benefits the agricultural sector directly by diminishing costs for farmers while also possibly generating indirect efficiency-enhancing effects where access to markets leads to specialization and agglomeration. Additionally, opportunities to explore comparative advantages boosts investments in land improvement, machinery, seeds, etc., which increases land values (*2*). One necessary condition for these indirect effects to take place is a functional land market. Ineffective property rights systems may lead to difficulties selling or renting land quickly and at low costs, which could inhibit and distort transactions. Farmers may also be unable to offer their land as guarantees to banks to obtain funding for their investments. The risk of expropriation may decrease incentives for investment (*3*).

Most of the lands expected to be claimed for agriculture in the next decades depend on new transportation systems and are located in countries with relatively low effectiveness of property rights (*4*). As the world population is expected to increase to over nine billion people by 2050, demand for food is forecast to rise by 70% compared with 2010 levels (*5*). Investments in transportation are expected to unlock unexplored agricultural potential and increase productivity worldwide. Consequently, demand for agricultural land is rising and effective property rights systems are required to allow for the desired effects from new transportation in terms of

[1]Department of Civil Engineering, The University of Tokyo, Tokyo, Japan

Corresponding Author:
Address correspondence to Lucas Bispo de Oliveira Alves: bispo-lucas@ip.civil.t.u-tokyo.ac.jp

specialization, agglomeration, and investment. In fact, the importance of effective property rights for agricultural development has caught the attention of governments and bilateral/multilateral aid agencies, which motivates initiatives for surveying, registering, and titling land worldwide as well as improving the performance of related public institutions (*6*).

This paper examines a hypothesis that the impacts of transportation on agricultural production are higher when property rights are more effective, using a panel dataset from 76 countries from 2007 to 2013. It treats both quality of transportation and effectiveness of property rights—measured by international institutions—as components of agricultural total factor productivity (TFP). Then it provides insights into which policies can maximize the impacts of transportation investment according to the characteristics of each country or area.

The paper is organized as follows. The next section presents the literature review. Then, the dataset characteristics and the methodology employed in an empirical analysis are presented, followed by the results and discussion. Finally, concluding remarks and further issues are presented.

Literature Review

Property Rights and Agricultural Land Markets

The property rights system in a given area consists of rules and/or regulations regarding who may own land and under what circumstances ownership can be transferred. It also comprises the institutions that process these transactions and safeguard ownership. An effective system is assumed to eliminate ambiguities regarding land ownership, diminish land conflicts, and support land transactions safely, quickly, and at low transaction costs.

Common use of land constitutes the rule in around four billion hectares or one third of the world's total land area, in contrast to private ownership, which is prevalent in the developed world (*7*). Where land is privately owned, it may be documented in formal ways, generally by the issuance of state-backed titles. In these settings, the performance of notaries, courts, and the police plays a role in the property rights system. In other places, private property is supported only by customary and tacit understandings between neighbors. Around 1.5 billion people are customary tenants (*7, 8*). Therefore, property rights systems can be classified across two dimensions: communal or private, and customary or formal systems. Most countries practice a mix of different systems across their territories. Private and formal systems are usually assumed to be the most effective (*9*).

Assuming the property rights system can support active land markets, there are three channels through which effective property rights enable the indirect impacts of transportation. First, in the case that property is secure—there is no risk of expropriation (e.g., by the government, amidst violent conflicts, or by corporations)—farmers have incentives to invest in land improvements or better technologies. Inversely, if the property rights system does not provide sufficient security, there may be not only lower investments but also conflicts involving land claims (*10*). The second channel is through access to credit. If property rights are effective, allowing for easy transactions, land can be used as collateral for obtaining bank loans (i.e., mortgages), which also favors investment (*3*). Third, if voluntary and mutually beneficial land transactions are possible at a low cost, land can be rented or sold to farmers or companies that have a higher value for it and can use it more productively. Through these three channels, territorial patterns of specialization and agglomeration emerge, increasing productivity (*11*).

Review of Empirical Literature

Past researchers have investigated the theoretical relationships between transportation, economic outcomes, and property rights. For example, Alston et al. (*12*) provided insights regarding the association between transportation costs and property rights. They presented an analytical framework with a demand curve for land titles, which is a function of the present value of land net of transportation costs, so that land located closer to markets is more likely to be titled than land located farther away. By comparing the benefits of titling with the costs involved in obtaining titles, such as accessing public institutions and lobbying, individuals decide whether or not to demand titles. This means that transportation could trigger the expansion of a formal property rights system. They also showed that their framework was consistent with experimental data in Brazil.

Many studies have investigated the relationship between effective property rights systems and economic outcomes through the channels described above. Early works attempted to evaluate the impact of property rights on economic performance by means of international comparisons. In order to measure the degree of effectiveness of the property rights system several indicators that provide country-level views have been proposed and used. Earlier studies relied mostly on a measure of risk of expropriation as a proxy for secure property rights, which is provided in the International Country Risk Guide (ICRG). More recently, new indicators have become available, such as the International Property Rights Index and The Heritage Foundation's Index of Economic Freedom. Acemoglu et al. (*13*) used a sample dataset of 64 countries and related their gross domestic product (GDP) per capita in 1995 to ICRG's measure of effectiveness of property rights. They employed two-stage least squares regressions in order to address potential problems of endogeneity stemming from reversed causality or omitted variables. Results indicate a positive impact of property rights on GDP per capita while ruling out other possible factors, such as geography or religion, as major determinants of economic performance. Kerekes and Williamson (*14*) confirmed the results of Acemoglu et al. (*13*) by utilizing indicators provided both by ICRG and by The Heritage Foundation. Using a range of model specifications, they found that more effective property rights result in improved capital formation and economic growth. Other studies suggest different possible relationships between the variables. Chong and Calderón (*15*) argued that there may be a two-way causality between economic growth and property rights. Bose et al. (*16*) suggested a non-linear relationship between the two variables, with property rights having a larger impact for poorer countries.

Despite these favorable results, studies conducted at the local level that attempted to investigate the channels through which property rights influence economic outcomes have found mixed results. Research designs vary substantially according to specific local contexts where questionnaire surveys are often used for data collection. Benin and Pender (*17*) studied locations in Ethiopia and found that only one out of 15 investment-related variables was affected by their proxy of tenure insecurity. Deininger and Jin (*18*), however, also investigated Ethiopian locations and found that expectation of future expropriation has a strong effect on investment. Similarly, Goldstein and Udry (*19*), in a study conducted in Ghana, found that lower risk of expropriation leads to longer fallowing of land. Place and Migot-Adholla (*20*) studied the effects of titling and land registration in four locations in Kenya and found that the effects on perceived land rights, credit use, crop yield, and land concentration were weak. They concluded that, should governments aim at agricultural development, investments in infrastructure would likely be more effective than titling alone. Jacoby and Minten (*21*), in a similar study conducted in Madagascar, found that having a title has little effect on farm-specific investment, productivity, and land values.

According to the UK Department for International Development (*8*), such inconsistency in the results is not only observable across countries but even within countries. Unresponsive investment-related variables may be constrained by underdeveloped financial markets, low educational attainment of farmers, prevalence of traditional techniques, etc. A diversity of context-specific variables may interfere with the impacts of property rights on economic performance.

The present paper follows the earlier studies in terms of approach, proposing a macro-level international comparison, but relies on an agricultural production function instead of focusing on economic growth. In doing so, it highlights land as a fundamental factor of production for the agricultural sector, and the interaction between transportation and property rights as constituents of agricultural productivity can be appreciated.

Method

This study sets up a hypothesis that the impacts of transportation are higher if the property rights system is more effective. This reflects an assumption that if property rights are not effective then land markets are dysfunctional, in which case efficiency-enhancing transactions and investments, identified as indirect impacts of transportation, are compromised. An econometric approach employed in this study treats a measure of the quality of transportation and an indicator of effectiveness of property rights as constituents of TFP across a sample of 76 countries, which were classified across two dimensions, following a methodology proposed by the World Bank (*22*). The sample is the largest possible selection of countries for which all variables were available for at least two consecutive years. This criterion was adopted in order to avoid measurement errors that yielded outliers in the dataset.

Countries are categorized into four types according to their availability of land for further agricultural expansion and the gap between their agricultural yield and a theoretical maximum. The first type contains countries, mostly in Latin America, that tend to be competitive global players in agricultural markets and have suitable land for expansion, low yield gaps, and relatively secure land tenure for investors, although their property rights systems may not be fully effective across their territories. The second type includes countries like Mongolia and Ukraine that have suitable land but display large yield gaps and may experience restrictions to agricultural production in infrastructure, education, or financial markets, or even in political situations. The third type covers countries with little land available and high yield gaps. Most are developing countries that can be constrained by a diversity of factors, ranging from the aforementioned to armed conflicts and corruption. Finally, countries with little land available and low yield gaps tend to be very populous countries that use land intensively and/or developed countries, which apply advanced agricultural technologies and face few or none of the aforementioned restrictions. The full sample and their classifications are illustrated in Figure 1.

The dependent variable—gross agricultural production value—displays considerable variability across

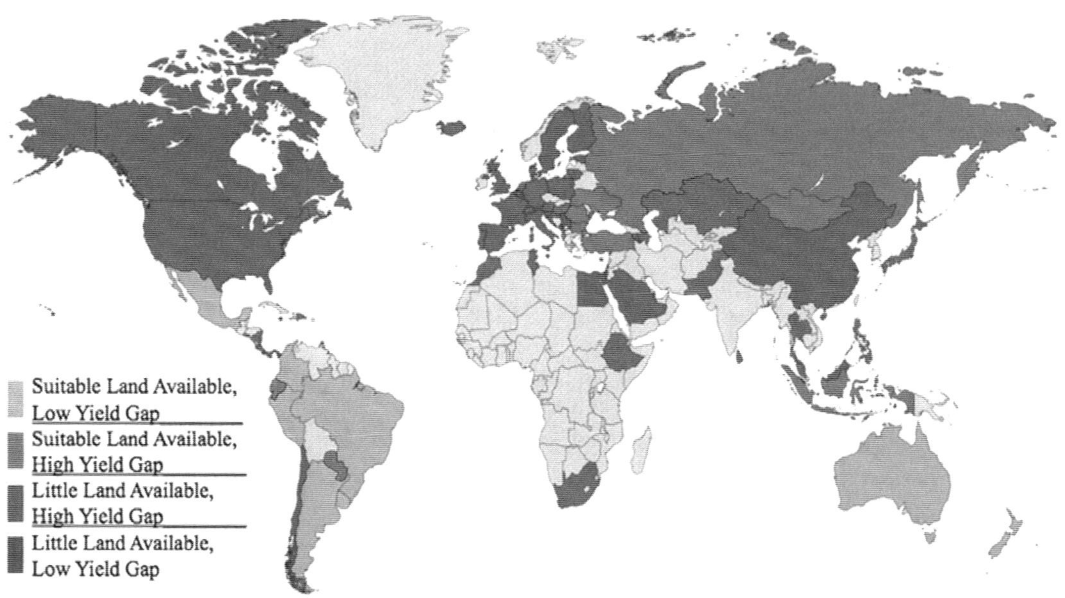

Countries not included: Afghanistan, Algeria, American Samoa, Andorra, Angola, Anguilla, Antigua and Barbuda, Aruba, Bahamas, Bahrain, Bangladesh, Belize, Benin, Bermuda, Bhutan, Bolivia, Bosnia and Herzegovina, Botswana, British Virgin Islands, Brunei Darussalam, Burkina Fasso, Burundi, Cape Verde, Cameroon, Cayman Islands, Central African Republic, Chad, Channel Islands, Hong Kong, Macau, Taiwan, Comoros, Congo, Cook Islands, Côte d'Ivoire, Cuba, Czechia, DPRK, Congo D.R., Dominica, Equatorial Guinea, Eritrea, Falkland Islands, Faroe Islands, Fiji, French Guiana, French Polynesia, Gabon, Georgia, Ghana, Gibraltar, Greenland, Granada, Guadeloupe, Guam, Guatemala, Guinea, Guinea-Bissau, Haiti, Holy See, Honduras, India, Iran, Iraq, isle of Man, Kenya, Kiribati, Kuwait, Kyrgyzstan, Lao P.D.R., Latvia, Lebanon, Lesotho, Liberia, Libya, Liechtenstein, Madagascar, Malawi, Maldives, Mali, Marshall Islands, Martinique, Mauritania, Mauritius, Mayotte, Micronesia, Monaco, Montserrat, Mozambique, Myanmar, Namibia, Nauru, Nepal, New Caledonia, Niger, Nigeria, Niue, Norfolk Island, Northern Mariana Island, Palestine, Oman, Pacific Islands Trust Territory, Palau, Papua New Guinea, Pitcairn Islands, Puerto Rico, Republic of Korea, Moldova, Reúnion, Rwanda, Saint Helena, Ascension and Tristan da Cunha, Saint Kitts and Nevis, Saint Lucia, Saint Pierre and Miquelon, Saint Vincent and the Grenadines, Samoa, San Marino, São Tomé and Príncipe, Senegal, Seychelles, Sierra Leone, Singapore, Slovakia, Solomon Islands, Somalia, South Sudan, Sudan, Suriname, Svalbard and Jan Mayen Islands, Swaziland, Syria, Tajikistan, Macedonia, Timor-Leste, Togo, Tokelau, Tonga, Turkmenistan, Turks and Caicos Islands, Tuvalu, Uganda, U.A.E., Tanzania, U.S. Virgin Islands, Uzbekistan, Vanuatu, Venezuela, Vietnam, Wallis and Futuna Islands, Western Sahara, Yemen, Zambia and Zimbabwe.

Figure 1. Countries in the sample dataset and their classifications.

countries, due to unobservable factors. This wide variability suggests the adequacy of the fixed effects method. By adopting different intercepts for each country, this procedure allows unobservable country-dependent effects to be captured. This choice is also supported by Haussmann's test.

A Cobb-Douglas agricultural production function is assumed as Equation 1 shows:

$$Y_{it} = K_{it}^{\beta_1} L_{it}^{\beta_2} G_{it}^{\beta_3} f(.) \tag{1}$$

where Y_{it} represents the gross agricultural production value in country i and year t; K_{it} represents the capital of country i and year t; L_{it} represents the labor of country i and year t; G_{it} represents the land of country i and year t; $f(.)$ represents the TFP function; and β_1, β_2, and β_3 are unknown coefficients. The TFP function is the portion of agricultural production that cannot be explained by the quantities of the factors of production.

Three types of models are employed in the model estimation using the above production function. The first model specifies the TFP function as

$$f(TQ_{it}, PR_{it}) = \exp(\mu + a_i + \beta_4 PR_{it} + \beta_5 TQ_{it} + \beta_6 PR_{it} \cdot TQ_{it}) \tag{2}$$

where TQ_{it} is the quality of transportation in country i and year t; PR_{it} is the effectiveness of property rights in country i and year t; μ is the error component; a_i is the country-specific constant in country i; and β_4, β_5, and β_6 are unknown coefficients. By plugging the TFP function in Equation 2 into the production function (1) and taking the natural logarithm for both sides of the whole equation, we obtain the expression for the base regression model as:

$$lnY_{it} = \mu + \alpha_i + \beta_1 lnK_{it} + \beta_2 lnL_{it} + \beta_3 lnG_{it} + \beta_4 PR_{it} + \beta_5 TQ_{it} + \beta_6 PR_{it} \cdot TQ_{it} + \varepsilon_{it} \quad (3)$$

where ε_{it} is the error component. The main coefficient we are interested in is β_6, which captures the interaction between effectiveness of property rights and the quality of transportation. Differentiating Equation 3 with respect to TQ_{it} yields the impact of quality of transportation on agricultural output as:

$$\frac{\partial lnY_{it}}{\partial TQ_{it}} = \beta_5 + \beta_6 PR_{it} \quad (4)$$

The estimated coefficient $\hat{\beta}_6$ is expected to have a positive sign, meaning that higher levels of effectiveness of property rights lead to increased impacts of transportation. The first model is estimated using the ordinary least squares (OLS) method.

Next, the second model introduces a proxy for quality of education into Equation 2 yielding an expanded model as:

$$f(TQ_{it}, PR_{it}, ED_{it}) = \exp(\mu + a_i + \beta_4 PR_{it} + \beta_5 TQ_{it} + \beta_6 PR_{it} \cdot TQ_{it} + \beta_7 ED_{it}) \quad (5)$$

where ED_{it} represents the quality of education in country i and year t. This procedure is adopted in order to tackle possible biases caused by the omission of relevant variables. Education is strongly correlated with each country's level of development and public institutions, which is also a relevant component of TFP. The second model is also estimated with the OLS method.

Finally, the third model incorporates instrumental variables (IVs) in order to address concerns of endogeneity related to quality of transportation, as described by Arman et al. (*23*). In this case, endogeneity stems from reversed causality. The mechanisms for this may include higher budget availability following growth in the agricultural sector or even increased lobbying from agricultural companies, both leading to investment in transportation. Endogeneity may bias the estimations. The model then incorporates the variables that can explain the quality of transportation without direct causality to agricultural output. For each country, its rate of savings, gross fixed capital formation, and GDP per capita were selected as instruments for our measure of the quality of transportation. Note that, even though agriculture may represent a low percentage of GDP in developed countries, there is no clear correlation between GDP per capita and agricultural output. In fact, countries with high GDP per capita, such as the U.S.A., Australia, or France, have comparatively high agricultural production, as do countries with a much lower GDP per capita, such as Brazil or China. As for savings and gross fixed capital formation, as macroeconomic variables they explain relatively little about agricultural output, especially in countries where agriculture is labor intensive.

A regression with quality of transportation as the dependent variable and the instruments as independent variables is specified. The fitted values of the first stage regressions (TQ^*_{it}) described in Equation 6 replace the original quality of transportation variable (TQ_{it}) in Equation 3, yielding a second stage regression. TQ^*_{it} consists of the original variables cleared from their correlation with the error term ε_{it}.

$$TQ^*_{it} = \pi_0 + \pi_1 RS_{it} + \pi_2 GFKF_{it} + \pi_3 GDPpc_{it} + v_{it} \quad (6)$$

where RS_{it} represents the rate of savings in country i and year t; $GFKF_{it}$ represents the gross fixed capital formation in country i and year t;$GDPpc_{it}$ represents the GDP per capita in country i and year t; v_{it} represents the error component; and π_0, π_1, π_2, and π_3 are unknown coefficients.

Data and Variables

Table 1 summarizes the potential variables used in the model estimation. Observations are annual and span from 2007 to 2013 for most countries, while some years were not available for all countries, yielding an unbalanced panel with a total of 479 entries. If all seven years of observations were available for all countries, the total number of observations would be 532, meaning that 90% of the observations are effectively available. Missing observations are most often either the first or the last years in the series for their countries.

Gross agricultural production value has been compiled by multiplying gross production in physical terms by output prices at the farm gate, thus leaving aside any industrial processes undertaken by companies other than the farms themselves. The sum of arable land and permanent crops approximates the extension of land effectively used for agricultural purposes for each given country and year. Gross agricultural capital stock and employment in agriculture are estimates produced by the Food and Agriculture Organization of the United Nations (FAO) based on several sources. The indicators of quality of transportation are available in the World Economic Forum's (WEF) Global Competitiveness Report and are based on annual surveys that assess executives' opinions about the business environment in each country. This indicator facilitates the comparison between countries with markedly different transportation needs, as executives confront them with each country's assets, thus providing a measure of the mismatch between them. Indicators of effectiveness of property rights are components of the International Property Rights Index,

Table 1. Potential Variables in the Dataset

Variable	Unit	Source	Comments
Gross Agricultural Production Value	US$ 2005 (Mil.)	FAOSTAT (24)	Value of production at farm gate
Agricultural Land	1,000 ha	FAOSTAT (24)	The sum of arable land and permanent crops
Gross Agricultural Capital Stock	US$ 2005 (Mil.)	FAOSTAT (24)	
Employment in Agriculture	1,000 People	FAOSTAT (24)	
Quality of Railways	Index (1–7)	WEF (25)	WEF's annual Executive Opinion Survey
Quality of Roads	Index (1–7)	WEF (25)	WEF's annual Executive Opinion Survey
Quality of Ports	Index (1–7)	WEF (25)	WEF's annual Executive Opinion Survey
Quality of Transportation	Index (1–7)	WEF (25)	Average of quality of roads, railways, and ports
Legal and Political Environment Index (LPEI)	Index (1–10)	IPRI (26)	Indicators provided by the World Bank and WEF
Physical Property Rights Index (PPRI)	Index (1–10)	IPRI (26)	Indicators provided by the World Bank and WEF
Land Property Rights Index	Index (1–10)	IPRI (26)	Average of LPEI and PPRI
Quality of Education	Percentage	World Development Indicators (27)	Rate of enrollment in tertiary education
Rate of Savings	Percentage	FAOSTAT (24)	
Gross Fixed Capital Formation	US$ 2005 (Mil.)	FAOSTAT (24)	
GDP *per Capita*	US$ 2005 (Mil.)	FAOSTAT (24)	

compiled by the Property Rights Alliance. The Legal and Political Environment Index (LPEI) is based on indicators of judicial independence, rule of law, political stability, and corruption. The Physical Property Rights Index (PPRI) accounts for the protection of physical property rights and ease of access to loans, and also considers the cost and time required to register property. The proxy for education, provided by the World Bank, consists of the percentage of the population with up to 5 years of age after the expected age of completion of secondary education that is enrolled in tertiary education. It is closely correlated with the quality and reach of the primary/secondary educational systems. Rate of savings, gross fixed capital formation, and GDP per capita are obtained from each country's national accounts.

Table 2 presents descriptive statistics for the indicators of quality of transportation and effectiveness of property rights for the dataset used in this study.

First, Table 2 shows a large variability in all macroeconomic variables, which is due to the disparity in terms of size of economy and economic performance across countries. Second, it shows that the mean in the quality of railways is significantly lower than the means in the qualities of roads and ports. This may reflect the difficulties of investment in and/or maintenance of railways. Third, the mean of LPEI is lower than that of PPRI. This may be because the emergence of basic protective mechanisms to property, perhaps diffused social values that are enforceable by community coercion, precedes the development of state institutions, such as representative assemblies and courts. Fourth, all indicators of quality of transportation and property rights display considerable variability across the sample, even though they tend to be relatively stable for each country over the seven-year timespan employed in this study.

Figure 2 displays a correlation plot between the Land Property Rights Index and Gross Production Value for the year 2013.

At first glance, this unconstrained analysis may show a negative relationship between the variables, expressed by the fitted line, which is consistent for every year. However, further examination of this relationship should be elaborated with the regression analysis since it seems variable across countries.

Estimation Results

Table 3 summarizes the estimation results for the base and the expanded models using an OLS method and the IV model using two-stage least squares. Note that *R2* for most of the regressions is in the 20% to 30% range while the adjusted *R2* tends to be below 5%. At first glance, these results may seem discouraging, however, it is common for these fitness measures in panel data regressions

Table 2. Descriptive Statistics

	Minimum	Maximum	First quartile	Third quartile	Mean	Median	Standard deviation
Gross. Agr. Prod. Value	4.51E + 04	5.86E + 08	2.07E + 06	2.11E + 07	2.39E + 07	6.20E + 06	7.02E + 07
Agricultural Land	9.3	515,400	784.06	12,104	43,796.86	5,101	98,973.88
Gross. Agr. Capital Stock	356	1,005,535	5,672	61,210.5	74,482.34	26,620	151,143.56
Employment	1.8	307,310	113.8	2,137.5	6,357.25	478.1	33,643.51
Quality of Railways	1.01	6.8	2.53	4.68	3.57	3.45	1.45
Quality of Roads	1.44	6.72	3.30	5.45	4.32	4.31	1.31
Quality of Ports	2.07	6.79	3.68	5.42	4.54	4.62	1.08
Quality of Transportation	1.93	6.54	3.31	5.28	4.24	4.22	1.16
LPEI	2.56	8.98	4.2	7.46	5.84	5.71	1.84
PPRI	3.16	8.79	5.67	7.2	6.45	6.33	1.11
Land Property Rights Index	3.60	8.87	4.95	7.25	6.15	6.05	1.40
Quality of Education	5.25	110.26	39	72.33	55.39	57.83	21.86
Rate of Savings	−9.99	60.78	17.05	26.76	22.64	21.44	9.98
Gross Fixed Capital Form.	542.10	3.01E + 06	7,885.08	96,559.22	150,147.44	32,462.50	386,523.99
GPD *per Capita*	212.33	88,748.23	4,093.26	36,183.27	19,112.98	8,875.05	19,500.09

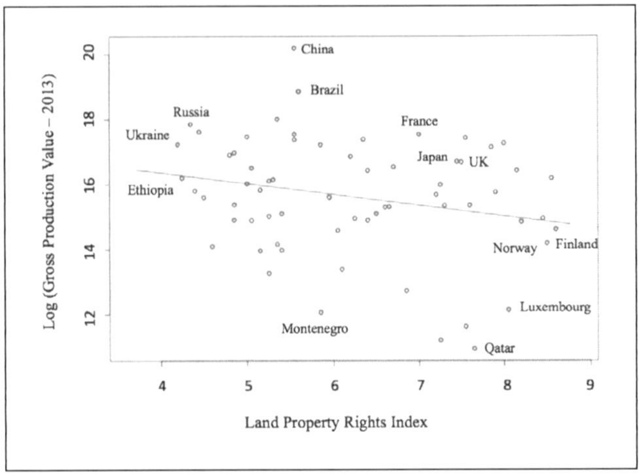

Figure 2. Land property rights index versus log (gross production value in 2013).

to be comparatively low. This is because *R2* and adjusted *R2* are unable to account for variations in the dependent variable that is explained by the intercepts, while using different intercepts for each cross-section unit is the core of the fixed effects methodology and essential in a dataset that displays stark variations between countries for the dependent variable (*28*).

First, the results show that the measure of agricultural capital stock has a significant influence on the agricultural production in all models, with elasticities varying from 22% to 38%. But the estimated coefficient of employment is significantly positive only in the basic model, with an elasticity of 2%, while that of agricultural land is significantly positive in the expanded model only, with an elasticity of 34%. These results can be understood in light of the fact that there are large cross-country differences in terms of production technology, with the predominance of capital stock, often captured as the level of mechanization in agriculture as the major determinant of productivity.

Second, the results show that the estimated coefficients of land property rights ($\hat{\beta}_4$) are significantly positive in all models while those of their cross-term ($\hat{\beta}_6$) are significantly negative in all models. The marginal effects of the land property rights (PR_{it}) are derived from Equations 1 and 2 as $\hat{\beta}_4 + \hat{\beta}_6 TQ_{it}$; thus, the results lead to the land property rights positively influencing the agricultural product when TQ_{it} is lower than 7 in the base model, 5.5 in the expanded model, and 6.5 in the IV model, respectively; otherwise they could have negative impacts. Note that the median of TQ_{it} is 4.22 as shown in Table 2. This means that the marginal effects of land property rights may be decreasing as the quality of transportation increases. This possibly reflects the fact that developed countries have advanced transportation systems in place and most of the gains from transportation

Table 3. Estimation Results of Pooled Models

	Method: Fixed effect panel data estimation						
	Dependent variable: ln(gross agricultural production value)						
		Base model		Expanded model		IV model (2SLS)	
	Independent variables	Coef.	*t*-values	Coef.	*t*-values	Coef.	*t*-values
β_1	ln (G. Agr. Capital Stock)	0.27**	6.28	0.22**	3.99	0.38**	6.70
β_2	ln (Employment)	0.02**	2.86	0.00	−0.30	0.00	0.41
β_3	ln (Agricultural Land)	0.04	1.62	0.34**	4.08	0.04	1.53
β_4	Land Property Rights	0.12**	4.69	0.10**	3.48	0.12**	4.44
β_5	Quality of Transportation	0.14**	3.80	0.11**	2.78	0.13**	3.59
β_6	Prop. Rights × Qual. Transp.	−0.02**	−3.77	−0.02**	−2.69	−0.02**	−3.53
β_7	Education	n.a.		0.002**	3.04	n.a.	
		n = 76, T = 2–7, N = 479		n = 76, T = 2–7, N = 479		n = 76, T = 2–7, N = 479	
	R^2	0.20		0.24		0.21	
	Adjusted R^2	0.03		0.05		0.04	
	F-statistic	16.82 on 6 and 394 DF. *p*-value: <0.01		14.21 on 7 and 393 DF. *p*-value: <0.01		16.57 on 6 and 394 DF *p*-value: <0.01	

Note: **: 1% significance; *: 5% significance.
Source: Author elaboration.

on agricultural productivity are likely to have been absorbed in the past.

Third, even though the estimated coefficient of education is statistically significant in the expanded model, it displays little economic relevance, which may be explained by the low contribution of labor as a factor of production compared with capital and land. Low economic significance for the proxy of education mitigates concerns of misspecification in the base model.

Finally, the results also show that the estimated coefficients with respect to the quality of transportation ($\hat{\beta}_5$) are significantly positive in all models. The marginal effects of the quality of transportation (TQ_{it}) are derived as $\hat{\beta}_5 + \hat{\beta}_6 PR_{it}$ as shown in Equation 4, thus the results lead to land property rights positively influencing agricultural production when PR_{it} is lower than 6 in the base model, 5 in the expanded model, and 6 in the IV model, respectively; otherwise they could have negative impacts. Note that the median of PR_{it} is 6.05 as shown in Table 2. This suggests that countries with poorer land property rights may enjoy positive effects from improved quality of transportation, but countries with higher land property rights may gain no benefits, or even experience negative impacts. This may be an unexpected result.

It should be noted that alternative specifications were tested, including a quadratic term for transportation, in order to account for decreasing impacts, as suggested by Adamopolous (*29*) and Canning and Pedroni (*30*). Year dummies were also tested in order to isolate potential short-term volatilities and the measures of quality and transportation were replaced by each of their modal components. None of these procedures, however, yielded significantly different results.

Next, Table 4 presents the estimation results of regressions for each of the four country categories separately using the base model.

Countries with suitable land for expansion and low yield gaps displayed the highest level of statistical significance for all explanatory variables. Brazil, Argentina, and Uruguay are examples of countries in this category. The highly capital-intensive sort of agriculture performed in these countries, as pointed out by Deininger et al. (*22*), may justify the negative coefficient calculated for "employment." The magnitude of the impact of transportation and negative marginal effects of property rights are higher for these countries than for those with little land for expansion and high yield gaps, even though results for this category of countries must be interpreted with caution, due to the fact that the factors of production were insignificant. Countries with little land for expansion and low yield gap—mostly developed countries—did not display statistical significance for most regressors. The impacts of transportation, in particular, must have been absorbed long before the timespan of the study. As for countries with suitable land for expansion and high gap, the insignificance of the coefficients may be explained by the interference of several other constraints. These are, in fact, the poorest countries in the sample, some of which experience armed conflicts, severe poverty, and corruption.

Table 4. Estimation Results of Regressions Using Subsamples

Method: Fixed effect panel data estimation: Base model

Dependent variable: ln(gross agricultural production value)

	Independent variables	Little land, low gap Coef.	Little land, low gap *t*-values	Suitable land, low gap Coef.	Suitable land, low gap *t*-values	Little land, high gap Coef.	Little land, high gap *t*-values	Suitable land, high gap Coef.	Suitable land, high gap *t*-values
β_1	ln (G. Agr. Capital Stock)	0.34**	6.96	0.38**	3.23	0.14	1.59	0.57.	1.69
β_2	ln (Employment)	0.03	1.40	−0.02.	−2.17	0.04	1.82	−0.03	−0.73
β_3	ln (Agricultural Land)	0.00	−0.30	0.47**	4.59	0.34	1.57	0.59	1.72
β_4	Land Property Rights	−0.01	−0.30	0.32**	4.87	0.43**	4.05	0.09	0.44
β_5	Quality of Transportation	−0.02	−0.64	0.58**	5.04	0.48**	3.45	0.25	0.74
β_6	Prop. Rights × Qual. Transp.	0.00	0.45	−0.08**	−4.37	−0.10**	−3.49	−0.03	−0.35
		n = 37, T = 2–7, N = 247		n = 8, T = 5–7, N = 50		n = 22, T = 2–7, N = 127		n = 9, T = 3–7, N = 55	
	R^2	0.21		0.80		0.29		0.38	
	Adjusted R^2	0.03		0.72		0.10		0.17	
	F-statistic	8.95 on 6 and 201 DF *p*-value: <0.01		23.64 on 6 and 36 DF *p*-value: <0.01		6.84 on 6 and 99 DF *p*-value: <0.01		4.14 on 6 and 40 DF *p*-value: <0.01	

Note: *: 5% significance; . : 10% significance.
Source: Author elaboration.

Discussion

The main finding of our empirical analysis is that the impacts of transportation are higher in countries with less effective property rights systems. This finding is contrary to the theoretical framework. Three possible explanations are presented.

First, in a way, our Land Property Rights Index measures the presence or capacity of the state, which can be a social institution conducive to efficiency, but also creates distortions in markets. Some authors express concerns that protectionist regulations practiced in many developed countries have significant impacts on agricultural productivity (*24*). The Common Agricultural Policy in the European Union and the US Farm Bills are examples. While their motivations are concerns of stability in food supply, fair prices, and security for farmers, they may harm agricultural productivity by limiting competition with foreign farmers (*25*, *26*). This may explain why the impacts of transportation were insignificant for developed countries with little land available and low yield gaps.

Second, the result could be explained by the fact that the benefits of improvements in the property rights system are larger in poor countries, as suggested by Bose et al. (*16*). This effect may be consistent with Alston et al. (*12*). One possibility is that transportation triggers the replacement of customary systems with formal ones. Before and after this institutional change, marginal improvements in the property rights may have little effect on the impacts of transportation. In effect, statistical significance was strongest for countries with suitable land for expansion and low yield gap—middle-income countries mostly in Latin America—where this institutional change may be in progress. In developed countries, this change may have happened long ago, while very poor countries may not yet be experiencing it.

Third, areas with less effective property rights—often land-rich countries—are more likely to house large farms. They may be more productive because they benefit from scale, scope, and perhaps also the presence of sophisticated supply chains, enabling them to tap international markets more easily than family-size farms can. This final explanation may have alarming consequences. If higher agricultural output is easier to achieve in areas with less effective property rights systems, transportation projects may lead to land transactions that are not utility-maximizing for certain agents, violating Pareto optimality. For example, large-scale land acquisitions may originate land conflicts and expropriate or impoverish indigenous peoples, peasants, and other social groups currently living under customary property rights systems (*10*, *18*, *27*). Our findings were strongest for the land-rich countries in the sample, such as Brazil and Argentina. They displayed the highest impacts of transportation and the largest negative marginal effect of property rights. The prevalence of large-scale farms, along with the adoption of capital-intensive production techniques may explain their global competitiveness, but ineffective property rights systems may be the root of land conflicts. Regarding these concerns, Gurara and Birhanu (*4*) state that 51.8% of land globally sought after by foreign investors is located in Sub-Saharan Africa, a region that scores very low in the International Property Rights Index.

Conclusion

This paper empirically analyzed the association between the effectiveness of land property rights and the impacts of transportation on agricultural output using the panel data of 76 countries from 2007 to 2013. An agricultural production function was estimated where both transportation and property rights were treated as constituents of TFP. The results revealed that the impacts of transportation are higher in countries with less effective property rights systems. Three potential reasons are discussed.

Although this study contributed to providing evidence regarding the role of land property rights, future research may seek additional evidence by using different production functions, like the constant elasticity of substitution, or by developing model specifications in which the indicators of property rights and transportation are endogenous. Also, future studies may benefit from longer time series as well as different indicators for property rights and transportation as they become available. More detailed analysis may be possible at the national and local levels. In fact, a myriad of country-specific factors may have been left out of the scope of this research, given its international focus, for instance, specific social and political determinants of investments in transportation. Finally, some countries, especially in the African continent, were not included, given data limitations. Possible biases caused by these omissions were not easy to deduct, so further analysis is required.

Author Contributions

The authors confirm contribution to the paper as follows: study conception and design: Alves & Kato; data collection: Alves & Kato; analysis and interpretation of results: Alves & Kato; draft manuscript preparation: Alves & Kato. All authors reviewed the results and approved the final version of the manuscript.

References

1. U.S. Department of Transportation. *Commodity Flow Survey*. Bureau of Transportation Statistics, Washington, D.C., 2010, p. 254.
2. Fujita, M., P. Krugman, and A. J. Venables. *The Spatial Economy: Cities, Regions and International Trade*. MIT Press, Cambridge, MA, 1999.
3. de Soto, H. *The Mystery of Capital: Why Capitalism Triumphs in the West and Fails Everywhere Else*. Basic Books, New York, 2000.
4. Gurara, D. Z., and D. Birhanu. Large-scale Land Acquisitions in Africa. *African Economic Brief*, Vol. 3, No. 5, 2012, pp. 1–8.
5. Food and Agricultural Organisation. *How to Feed the World in 2050. Insights from an Expert Meeting at FAO*, Food and Agriculture Organization of the United Nations, Rome, 2009, pp. 1–35.
6. Holstein, L. Towards Best Practices from World Bank Experience. Presented at International Conference on Land Tenure and Administration, Orlando, FL, 1996, pp. 1–26.
7. Willy, L. A. *Whose Land Is It? Commons and Conflict States*. Rights and Resources Initiative, EUA, Washington, D.C., 2008.
8. DFID. *Secure Property Rights and Development: Economic Growth and Household Welfare*. 2014, Department for International Development, London, pp. 1–94.
9. Demsetz, H. Toward a Theory of Property Rights. *American Economic Review*, Vol. 57, No. 2, 1967, pp. 347–359.
10. Fenske, J. Land Tenure and Investment Incentives: Evidence from West Africa. *Journal of Development Economics*, Vol. 95, No. 2, 2011, pp. 137–156.
11. Besley, T. J., and M. Ghatak. Property Rights and Economic Development. In *Handbook of Development Economics* (D. Rodrik, and M. Rosenzweig, eds.), Elsevier Ltd, North-Holland, pp. 4525–4595.
12. Alston, L. J., G. D. Libecap, and R. Schneider. *The Determinants and Impact of Property Rights: Land Titles on the Brazilian Frontier*. Working Paper 5405. NBER Research Program in the Development of American Economy, National Bureau of Economic. Research, Cambridge, MA, 1996.
13. Acemoglu, D., S. Johnson, and J. A. Robinson. Institutions as the Fundamental Cause of Long-Run Growth. In *Handbook of Economic Growth* (P. Aghion and S. Durlauf, eds.), Elsevier, Amsterdam, 2005, pp. 385–472.
14. Kerekes, C. B., and C. R. Williamson. Unveiling de Soto's Mystery: Property Rights, Capital Formation, and Development. *Journal of Institutional Economics*, Vol. 4, No. 3, 2008, p. 299.
15. Chong, A., and C. Calderon. Institutional Quality and Income Distribution. *Economic Development and Cultural Change*, Vol. 48, No. 4, 2000, pp. 761–786.
16. Bose, N., A. P. Murshid, and M. A. Wurm. The Growth Effects of Property Rights: The Role of Finance. *World Development*, Vol. 40, No. 9, 2012, pp. 1784–1797.
17. Benin, S., and J. Pender. Impacts of Land Redistribution on Land Management and Productivity in the Ethiopian Highlands. *Land Degradation and Development*, Vol. 12, 2001, pp. 555–568.
18. Deininger, K., and S. Jin. Tenure Security and Land-Related Investment: Evidence from Ethiopia. *European Economic Review*, Vol. 50, No. 5, 2006, pp. 1245–1277.
19. Goldstein, M., and C. Udry. The Profits of Power: Land Rights and Agricultural Investment in Ghana. *Journal of Political Economy*, Vol. 116, No. 6, 2008, pp. 981–1022.
20. Place, F., and S. E. Migot-Adholla. The Economic Effects of Land Registration on Smallholder Farms in Kenya: Evidence from Nyeri and Kakamega Districts. *Land Economics*, Vol. 74, No. 3, 1998, pp. 360–373.
21. Jacoby, H. G., and B. Minten. Is Land Titling in Sub-Saharan Africa Cost-Effective? Evidence from

Madagascar. *World Bank Economic Review*, Vol. 21, No. 3, 2007, pp. 461–485.
22. Deininger, K., D. Byerlee, J. Lindsay, A. Norton, H. Selod, , and M. Stickler. *Rising Global Interest in Farmland.* The World Bank, Washington, D.C., 2010.
23. Arman, S. A., A. S. Manesh, and A. T. Izady. Design of a CGE Model to Evaluate Investment in Transport Infrastructures: An Application for Iran. *Asian Economic and Financial Review*, Vol. 5, No. 3, 2015, pp. 532–545.
24. Bakst, D. *Farms and Free Enterprise: A Blueprint for Agricultural Policy*. Heritage, Washington, D.C., 2016.
25. Rickard, S. *Ploughing the Wrong Furrow: The Costs of Agricultural Exceptionalism and the Precautionary Principle.* Report No. IEA Discussion Paper 75, Institute of Economic Affairs, London, 2016.
26. Cato Institute. *Agricultural Policy*. Cato Institute, Washington D.C., 2009.
27. Brasselle, A. S., F. Gaspart, and J. P. Platteau. Land Tenure Security and Investment Incentives: Puzzling Evidence from Burkina Faso. *Journal of Development Economics*, Vol. 67, No. 2, 2002, pp. 373–418.
28. Baltagi, B. H. *Econometric Analysis of Panel Data*, 5th ed. John Wiley & Sons, Chichester, West Sussex, 2013.
29. Adamopoulos, T. Transportation Costs, Agricultural Productivity and Cross-Country Income Differences. *International Economic Review*, Vol. 52, No. 2, 2011, pp. 489–521.
30. Canning, D., and P. Pedroni. *The Effect of Infrastructure on Long Run Economic Growth.* Harvard University, Mimeo, 2004, pp. 1–30.

The Standing Committee on Transportation and Economic Development (ADD10) peer-reviewed this paper (18-00515).

Article

TRR
JOURNAL OF THE TRANSPORTATION RESEARCH BOARD

Transportation Research Record
2018, Vol. 2672(3) 12–25

DOI: 10.1177/0361198118790373
journals.sagepub.com/home/trr
SAGE

Analysis of Regional Difference and Correlation between Highway Traffic Development and Economic Development in China

Ye Li[1], Jing Fan[1], and Haopeng Deng[1]

Abstract

This study used three-tier architecture to analyze the relationship between the highway traffic development difference and the economic development difference in China from 1997 to 2015. Their time distribution and trends of national and regional differences were quantified via both coefficient of variation and Theil index. Furthermore, the highway traffic connectivity was analyzed for each province. The obtained results showed that the difference of highway traffic development was significantly related to the difference of economic development, and poor connectivity of highway traffic was also strongly related to economic disparity. The investment difference of China's highway infrastructure could be divided into an unstable stage (1997–2010), and a stable stage (2011–2015). The highway traffic development difference could also be divided into two stages (1997–2006 and 2007–2015). Following the construction of central and western regions in around 2006, this difference decreased significantly; however, large differences within region remained. Provinces in the central and western regions have low connectivity and several remote main cities are not connected via high-grade highways. China's national economic difference has been constantly decreasing; however, the three regions follow different trends. Furthermore, with the development of the central and western regions, between-region economic differences decreased rapidly. To improve the level of regional equalization for highway development, the government should reduce differences between developed and undeveloped provinces for each region and connect remote cities, thus ultimately reducing differences in economic development and promoting a balanced regional development.

Transportation construction has been identified as one of the major drivers of the economic growth of China during the past decade (*1*). However, China has extensively pursued a biased development of its transportation infrastructure, concentrating the main portion of public investment on the eastern regions, which directly caused the observed diversity of regional economic growth in China (*2*, *3*). During the "12th Five-Year Plan", which considered the difference between poor and wealthy regions, the government has formulated a different transport development strategy, which generally reflects a fairer approach. In "The plan of poverty alleviation for traffic construction in the contiguous poverty-stricken areas (2011–2020)," the government has formulated the main direction toward promoting the equalization of transportation services, which laid a solid foundation for the achievement of common prosperity. This study investigated the disequilibrium between highway infrastructure development and economic development at three levels (national, regional, and provincial).

The enthusiasm of the economist to discuss the critical function of infrastructure investment on economic development was inspired during the late 1980s (*4*). Since then, considerable research emerged that examined the relationship between highway investment and the resulting economic growth. A significant share of the relevant literature used a production function model to estimate the impact of public capital stock on the resulting output of multiple economic sectors (*5*, *6*). Output elasticities have been reported to vary widely, which is likely due to differences in utilized spatial levels of analysis, definitions of capital stock, estimation techniques, or underlying models (*7*). Various studies have been conducted that specifically examined spillover effects (*5*, *8*) by dividing

[1]School of Transportation Engineering, The Key Laboratory of Road and Traffic Engineering, Ministry of Education, Tongji University, Shanghai, China

Corresponding Author:
Address correspondence to Jing Fan: jing.fan@tongji.edu.cn

the economic impact of the transportation infrastructure into either direct and indirect effects (*9*) or internal and external effects (*10*). Berechman et al. (*11*) used three models and retrieved output elasticities of 0.37, 0.34, and 0.01 for state, county, and municipality levels, respectively, with respect to the obtained public highway capital. However, all these models and studies failed to take reverse causality into account, consequently overestimating the infrastructure variable coefficients in the production function (*12*). Furthermore, by adopting vector auto-regression method and structural equation modeling, many studies tested whether road investments would lead to productivity growth or whether productivity growth entails increased road construction (*13, 14*). In other words, long-term research has investigated the causal relationship between transportation and regional economic growth.

In general, economically underdeveloped areas have been associated with poor infrastructure (*15*). In Europe, differences in the provision of transport infrastructure display a similar geographical pattern with the gross domestic product (GDP) per capita (*16*). Due to the inequality of the regional economic development in China, the exploration of the relationship between this uneven transportation development and the disparity of regional economic development has become a hot topic. It is therefore necessary to conduct an in-depth analysis of the disparity trend of regional economic growth, and to discuss how to effectively utilize the means of infrastructure investment and policies to promote the coordinated development of the regional economy.

Numerous studies examined whether the transportation development of China had differential effects on different parts of the country: Wang verified the uneven distribution of regional railway construction as the main reason for the strong differences in the economic growth rate between eastern, central, and western regions (*2*). Yu et al. calculated the output elasticities of the infrastructures of the eastern, central, and western regions and found that the central region has the highest elasticity, and further suggested that infrastructure investment in the central region should receive more attention by the central government (*17*). Zhang reported that both the transportation infrastructure and the economic growth showed strong spatial concentration characteristics and followed a decreasing trend from eastern to western regions (*18*). Wang et al. utilized exploratory spatial data statistics to analyze the spatial distribution difference of the highway network of northwest China (*19*). Cao et al. adopted the Natural Breaks function of the ArcMap software to divide transportation development levels of 183 cities from 1991 to 2003 and found that the level of urban transportation development was distinctly correlated to the per capita industrial output at the 0.01 level (*20*). Demurger studied the differences in the development of the inter-provincial transport network, using the density of the road network, and analyzed whether underdeveloped infrastructure networks likely led to a growing regional disequilibrium in China (*21*). Fan and Chan-Kang analyzed the relationships between road investment, economic growth, and poverty reduction in China, concluding that areas for road investment should be chosen to balance economic growth and to reduce poverty. The authors further indicated that investing in the coastal areas would boost economic growth, while investing inland (especially in the southwestern region) will reduce poverty (*3*).

Apparently, transport improvement strongly impacts regional development. A previous study reported that the implementation of a national trunk highway system has been divided into two phases. During the first phase (1990–2000), the general trend aggravates the division of Chinese economic regions, although regional disparity would be mitigated to some extent during the second phase (2000–2020) (*22*). The Chinese government committed to a substantial highway investment to benefit a balanced economic growth and to improve equalization, consequently defining poor and remote areas as target areas. However, the trend of regional differentiation in China, and the missing knowledge on how to effectively use highway construction to alleviate this trend, continues to provide major problems for policy makers to achieve a coordinated development of the regional economy. It is therefore important to investigate the relationship between highway traffic development and economic development at different levels and for different regions, as this is of great significance for the formulation of reasonable traffic policies and for the eventual promotion of a sustained economic development.

Regional differences can be described via the coefficient of variation (CV) as well as via Theil's inequality index. As an index of diversity, the CV has been widely applied to the fields of economics and sociology to investigate and compare social inequality across nations, cities, and other social aggregates (*23, 24*). The Theil index is additively decomposable and satisfies several desirable properties as a measure of regional inequality (*25*). This decomposition of the global index in partial factorial contributions is interesting in political terms as well as due to its analytical implications; it furthermore clarifies the global role or contribution of each part (*26*). We chose the CV and the first order decomposition of the Theil index to further study the differences between China's highway investment and the economic development during the past years (1997–2015). The spatial correlation dimension was used to calculate the development of highway networks and the connectivity among main cities, thus testing the correlation between the connectivity of traffic networks and the economic difference within provinces. Via an analysis of the national, regional, and provincial level over time, this study exposed differences between transportation development and economic development, which will help policy makers to make more informed decisions to achieve the equalization of the highway traffic development.

This paper is organized as follows: First, the available data and the data source is described. Next, a framework for analysis and the utilized method is presented. The trends of highway investment growth as well as the regional differences within China are reviewed, and China's highway construction differences and economic differences are analyzed. Finally, the relationship between connectivity and economic differences of the main cities for each province are presented. The paper concludes with a summary of the key findings and a discussion of necessary policy amendments.

Data and Data Source

This study chose 31 provincial administrative units of China (identical to the provinces used in this paper) as study area. According to the level of economic and technological development and its geographical location, the study area can be divided into three economic zones (identical to the economic regions used in this paper). Figure 1 shows the per capita GDP distribution and adjacency relationships of provinces and regions in the research area of China.

The data used for this study are the investments for the completion of highway construction at the provincial level between 1997 and 2016, and at the national level between 1980 and 2016. These data were obtained from the Compilation of National Statistical Data on Traffic and Transportation; the per capita GDP, the permanent resident population, and each grade highway mileage at the provincial level between 1997 and 2015 were supplied from the National Bureau of Statistics, Statistical Yearbook of China; the per capita GDP of prefecture-level cities was obtained from the China City Statistical Yearbooks of 2006, 2012, and 2016; the spatial distance between the main cities in each province was calculated via ArcGIS; and the traffic distance between main cities was obtained from China's Highway and Urban and Rural Road Atlas published in 2012. The atlas provides trip distances on the basis of data measured by 1200 drivers, who prefer to travel higher-grade highways. It should be noted that according to decreasing quality level, highways in China are divided into freeway, Grade I, Grade II, Grade III, and Grade IV; therefore, freeways should be selected first, followed by Grade I highways and so on. The sources of the same category data were identical, and all of these are official publications. Table 1 presents statistical summaries of regions for per capita GDP, permanent resident population, and highway mileage.

Method

This study used a nation–region–province hierarchical structure for the country. Using the province as the underlying regional unit, both the CV and the Theil index were used to quantify national and regional differences, and the spatial correlation dimension was introduced to calculate the development of highway traffic for each province.

Coefficient of Variation

CV indicates the relative dispersion of data, which is calculated via dividing the sample standard deviation by the mean. According to the characteristics of CV, the per capita GDP, the density of the highway network, and the completion investment of the highway construction were selected for each province to measure spatial and temporal differences between China's road traffic construction and its economic development at different scales. Difference indicators were calculated one by one between 1997 and 2015 and the equilibrium degree and the relationship between highway transportation development and economic development of China were dynamically investigated. A larger CV indicates increasing difference between provinces or regions.

CV is defined as a proportion

$$C_v = \frac{1}{\bar{y}}\sqrt{\sum_{i=1}^{n}\frac{(y_i - \bar{y})^2}{n-1}} \quad (1)$$

where n represents the total number of units corresponding to a specific scale (nation scale, region scale, and province scale); y_i represents the per capita GDP, the density of the highway network, or the completion of investment in highway construction of the i unit; and $\bar{y}$ represents the average value of y_i. The variation coefficient of per capita GDP ($C_{v-\text{per capita GDP}}$), the density of the highway network ($C_{v-\text{density of highway}}$), and the completion investment of the highway construction ($C_{v-\text{investment}}$) at national scale and at three economic regional scales were calculated.

Theil Index

First order decomposition of the Theil index can be easily extended to an analysis of the between-region and within-region components of the national inequality. The overall national difference can be measured via the Theil index (T_P) as

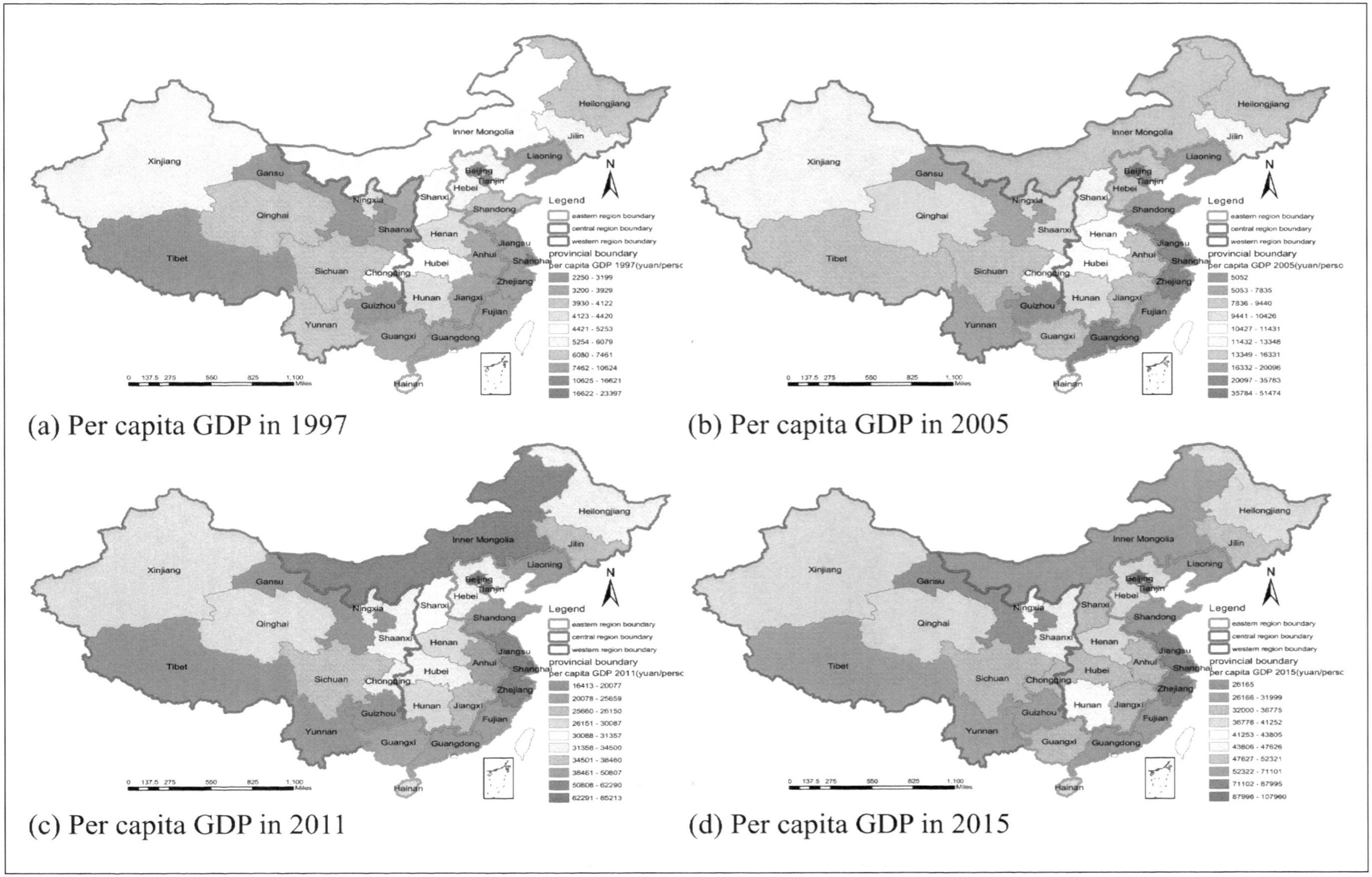

Figure 1. Map of the per capita GDP distribution of provinces in (*a*) 1997, (*b*) 2005, (*c*) 2011, and (*d*) 2015.

Table 1. Summary Statistics of Regions of China

	1997	2005	2006	2011	2015
Provincial level per capita GDP (yuan)					
Average value for China	6704.52	16203.13	18533.65	39441.87	53083.81
Average value for the West	4008	9281.1	10700.9	26209.9	37863.2
Average value for the East	10318.33	25130.42	28542.17	54620.17	72095.5
Average value for the Center	4882.22	11991.22	13892	33906.33	44646.67
Municipal level per capita GDP (yuan)					
Average value for China	–	15412.76	–	38621.51	50961.96
Average value for the West	–	11035.33	–	28534.88	39143.08
Average value for the East	–	20536.40	–	47061.31	62022.64
Average value for the Center	–	12523.51	–	35639.56	46238.91
Permanent resident population of province (million people)					
Average value for China	39.4587	41.48516	41.7816	43.23935	44.2219
Average value for the West	28.228	28.854	28.883	29.095	29.826
Average value for the East	41.9917	46.34083	47.0233	50.0758	51.4142
Average value for the Center	48.56	49.0456	49.1244	49.84	50.6278
Highway mileage of province (km)					
Average value for China	39561.45	62275.55	111516	132464.1	147654.7
Average value for the West	36813.5	63930.9	104188.2	135690	155411.2
Average value for the East	36938.25	52609.33	90317.2	101470.7	113523.2
Average value for the Center	46112.33	73324.56	147923	170204.3	184545.2

$$T_P = \sum_i \sum_j \frac{y_{ij}}{Y} \cdot \log\left[\frac{\left(\frac{y_{ij}}{Y}\right)}{\left(\frac{p_{ij}}{P}\right)}\right] \tag{2}$$

where y_{ij} represents the highway mileage or GDP of province j in region i; Yrepresents the total highway mileage or total GDP of all provinces ($= \sum\sum y_{ij}$); p_{ij} represents the permanent resident population of province j in region i; and P represents the total permanent resident population of all provinces ($= \sum_i \sum_j p_{ij}$).

If we define T_{Pi} to measure between-province difference for region i, T_{Pi} could be expressed as

$$T_{Pi} = \sum_j \frac{y_{ij}}{Y_i} \cdot \log\left[\frac{\left(\frac{y_{ij}}{Y_i}\right)}{\left(\frac{p_{ij}}{P_i}\right)}\right] \tag{3}$$

Then, the national Theil index T_P of Equation 2 can be decomposed into

$$\begin{aligned} T_P &= \sum_i \frac{Y_i}{Y} T_{Pi} + \sum_i \frac{Y_i}{Y} \log\left[\frac{\left(\frac{Y_i}{Y}\right)}{\left(\frac{P_i}{P}\right)}\right] \\ &= \sum_i \frac{Y_i}{Y} T_{Pi} + T_{BR} \\ &= T_{WR} + T_{BR} \end{aligned} \tag{4}$$

where Y_i represents the total highway mileage or total GDP of region i ($= \sum_j y_{ij}$); P_i represents the permanent resident population of region i ($= \sum_j p_{ij}$); T_{BR} represents the difference between western, eastern, and central regions; T_{WR} represents differences within-region. Equation 4 presents the ordinary one-stage Theil inequality decomposition. The total difference T_{WR} can be expressed as the sum of a "within-region" difference term and a "between-region" term. Here, the within-region contribution is itself a weighted sum of the between-province difference of each region (T_{Pi}).

Later, the Theil index of highway mileage was calculated as $T_{P-length\ of\ highway}$ and the GDP as T_{P-GDP}, which provided directions and amplitudes of fluctuation of between-region and within-region differences of highway traffic development and economic development.

Space Relation Fractal Dimension

After analysis of CV and T_P, we obtained both the spatial and temporal differences of highway traffic development and economic development in China as a whole, and as segmented into different economic regions. The geographical location, resource endowment, and the spatial distribution of the inner cities and towns differed for each province. Next, the spatial correlation dimension indicated spatial connectivity of urban systems, and the development and connectivity indices of highway traffic were calculated for provinces; then, correlations with the economic differences of provinces were investigated.

The areal inhomogeneity of cities can be sufficiently characterized via the fractal correlation dimension D of the network (*27*). To obtain D of each province, the spatial correlation function C_r must be considered, expressing the probability that two randomly selected cities of one province are within a distance below r units as

$$C_r = \frac{1}{N^2}\sum_{i,j=1}^{N} \theta(r - d_{ij})(i \neq j) \tag{5}$$

where N represents the number of cities in a province; d_{ij} represents the spatial distance between two cities; θ represents the Heaviside function, which assumes the value of 1, when the distance is lower than r or 0, and when the distance is larger, the value can be calculated via

$$\theta(r - d_{ij}) = \begin{cases} 1 & d_{ij} \leq r \\ 0 & d_{ij} > r \end{cases} \tag{6}$$

The spatial distribution of the city unit has fractal characteristics, and should therefore be scale invariant; then, C_r grows like a power function (*27*), namely: $C_r \propto r^D$, the deformation equation can be obtained as

$$\ln C_r = D \ln r + \text{const} \tag{7}$$

where r represents a given threshold, which belongs to the interval$[r_{min}, r_{max}]$, the value of which depends on the maximum and minimum distance of each province. For each province, C_r was calculated by increasing the value of r. Via linear regression, the correlation dimension D of the spatial distance can be determined, reflecting the inhomogeneity of the main cities in a province. In general, the range of D varies from 0 to 2 and when D approaches 0, the distribution of cities is highly concentrated; when D approaches 1, cities are distributed along a geographical line, and when D approaches 2, the spatial distribution of cities is very uniform. A smaller D value indicates higher concentration of the main cities in a province and a weaker interval resistance; in contrast, lower concentration indicates larger spacing resistance (*28*).

The correlation dimension of the actual traffic distance (D') can be calculated via changing d_{ij} in Equation 5 to the actual traffic distance between both cities, and a closer D' to D indicates better connectivity between cities in the province.

The traffic connectivity index ρ is calculated via

$$\rho = \frac{D'}{D} \tag{8}$$

$\rho = 1$ indicates that the spatial accessibility between cities reached the ideal state. A smaller ρ value indicates decreased accessibility between cities. If ρ is above 1, the traffic network is inadequate.

Analysis and Result

General Analysis of Highway Development and Economic Development of China

The Chinese government invested massively into road construction to accelerate the expansion of the road network throughout counties and towns, to improve the overall quality of roads, and to increase the total mileage of expressways. Since 1984, the Chinese government has further increased its investment in roads infrustructure, particularly for high-quality roads such as highways that connect major industrial centers in coastal areas. During the 1990s, infrastructure investments became a national priority. During the early 2000s, road projects were an important part of China's strategy for the development of the western and the central regions. In 2005, the Western Development Strategy and the Central Rise Policy were launched to improve transportation to the interior of China.

To contribute toward the building of a moderately prosperous society, the highway construction of China has currently entered a vital period. The aim of the "13th Five-Year Plan" was to connect all counties at least via secondary highways as well as to connect counties with freeways where topographical, technical, and economical conditions permit this, thus realizing an inter-village commuter bus in 2020. The completion of this highway construction investment from 1980 to 2016 in China is shown in Figure 2. This reflects that the policy investment effort of highway construction in China increased steadily and that the annual rate of investment growth levelled during recent years. The growth rates averaged 18.5%, 15.8%, and 7.2%, respectively, during 2002–2006, 2007–2011, and 2012–2016.

Both the reform and introduction of the long-term implementation of public investment was concentrated in the coastal regions and in urban areas. Therefore, large regional variations exist within the density and quality of the highway infrastructure in China (Table 2). The western region is poorly connected compared to the central and eastern regions. In 2002, only 19 km of roads existed for every 100 km^2 of land in western China, compared to more than 58 km per 100 km^2 of land in the eastern region. Among all provinces, Tibet and Qinghai are particularly poorly equipped with road infrastructure with a road density of only 6 km and 10 km per 100 km^2 of land in 2015, respectively. Road quality is also worst in the western region and the construction of a freeway in Tibet began in 2014.

In 1997, the investment ratio of the west to the east and to the center was 2:6:3. Completed investments for the center exceeded those of the east in 2003. Furthermore, in 2012, the west exceeded the east. In 2010, completed investments of the east were below the national average. Currently, China is vigorously promoting infrastructure equalization.

In China, most coastal provinces in the eastern region are economically rich, while most of the provincial areas of the center and the west are among the poorer provinces. Furthermore, the state of economic disparity

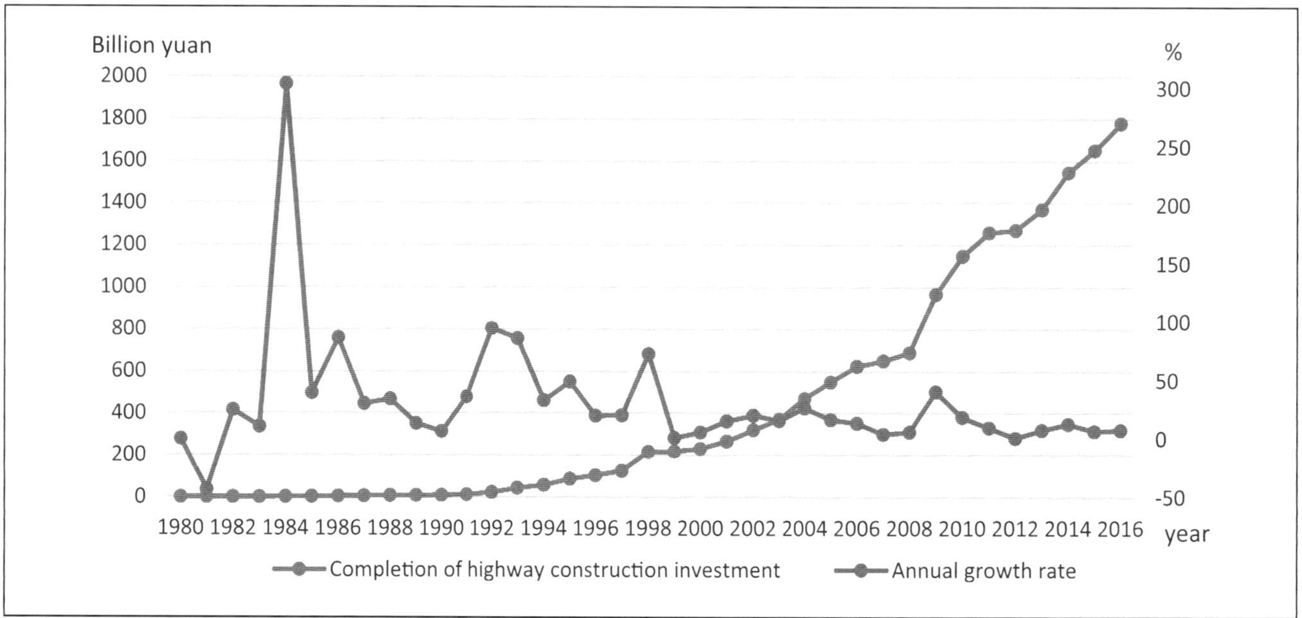

Figure 2. Completion of highway construction in China over time.

Table 2. Comparison of the Highway Development of Different Regions

	Average completion of highway construction investment billion yuan				Density of highway network Km/(100 km^2)				Density of freeway network Km/(100 km^2)			
Year	National	West	East	Center	National	West	East	Center	National	West	East	Center
1997	4.05	2.16	6.16	3.34	27.3	13.4	40.7	21.9	0.19	0.01	0.42	0.08
2000	7.47	5.56	9.49	6.9	31.8	15.5	48.2	24.8	0.53	0.11	1.12	0.22
2002	10.36	7.86	11.95	11.03	38.6	18.9	56.6	32.7	0.82	0.21	1.63	0.43
2003	11.98	8.62	13.38	13.86	39.7	19.4	58.3	33.6	0.97	0.28	1.90	0.51
2004	15.17	10.49	17.06	17.84	41.7	19.8	62.1	34.6	1.18	0.33	2.32	0.61
2009	31.19	25.1	33.06	35.46	80.8	47.1	104.7	78.3	1.36	0.39	2.64	0.76
2010	37.04	34.12	36.15	41.47	83.6	49.3	107.4	81.7	1.50	0.43	2.88	0.85
2011	40.63	38.73	39.27	44.58	85.4	51.0	109.3	83.4	1.68	0.54	3.09	1.05
2012	41.01	39.87	39.52	44.28	88.0	53.0	112.2	85.8	1.88	0.61	3.47	1.18
2013	44.17	45.49	40.74	47.28	90.2	54.7	115.2	87.5	2.06	0.70	3.79	1.27
2014	49.87	55.9	43.8	51.28	92.3	56.7	118.0	88.7	2.26	0.83	4.04	1.49
2015	53.27	59.51	46.6	55.23	94.9	59.5	120.3	90.9	2.46	0.93	4.26	1.74

between regions has continued, which is shown in Figure 1. The lack of transportation infrastructure has caused a development bottleneck, restricting the economic development of the western region as well as that of remote mountainous areas (*3*, *21*).

Difference Analysis Based on the CV

Considering the provincial administrative unit as the basic analytic unit, the national and regional $C_{v-\text{density of highway}}$, $C_{v-\text{per capitaGDP}}$, $C_{v-\text{investment}}$ were calculated and the resulting trends are shown in Figure 3.

Between 1997 and 2015, the development of highways in the western region was quite different, and the national difference was considerable, while the development of the center was relatively balanced. Its development and evolution can be divided into two stages: a strong fluctuation in the difference of highway development existed during 1997–2005, and the difference of highway development decreased and stabilized during 2006–2015. In 2006, the first year of the 11th Five-Year Plan, a highway expansion project and other projects were conducted as a result of which, the national and eastern difference of the highway network was noticeably reduced; however, the difference of western and central regions increased.

The regional economic disparity of China is constantly decreasing, especially among the eastern provinces;

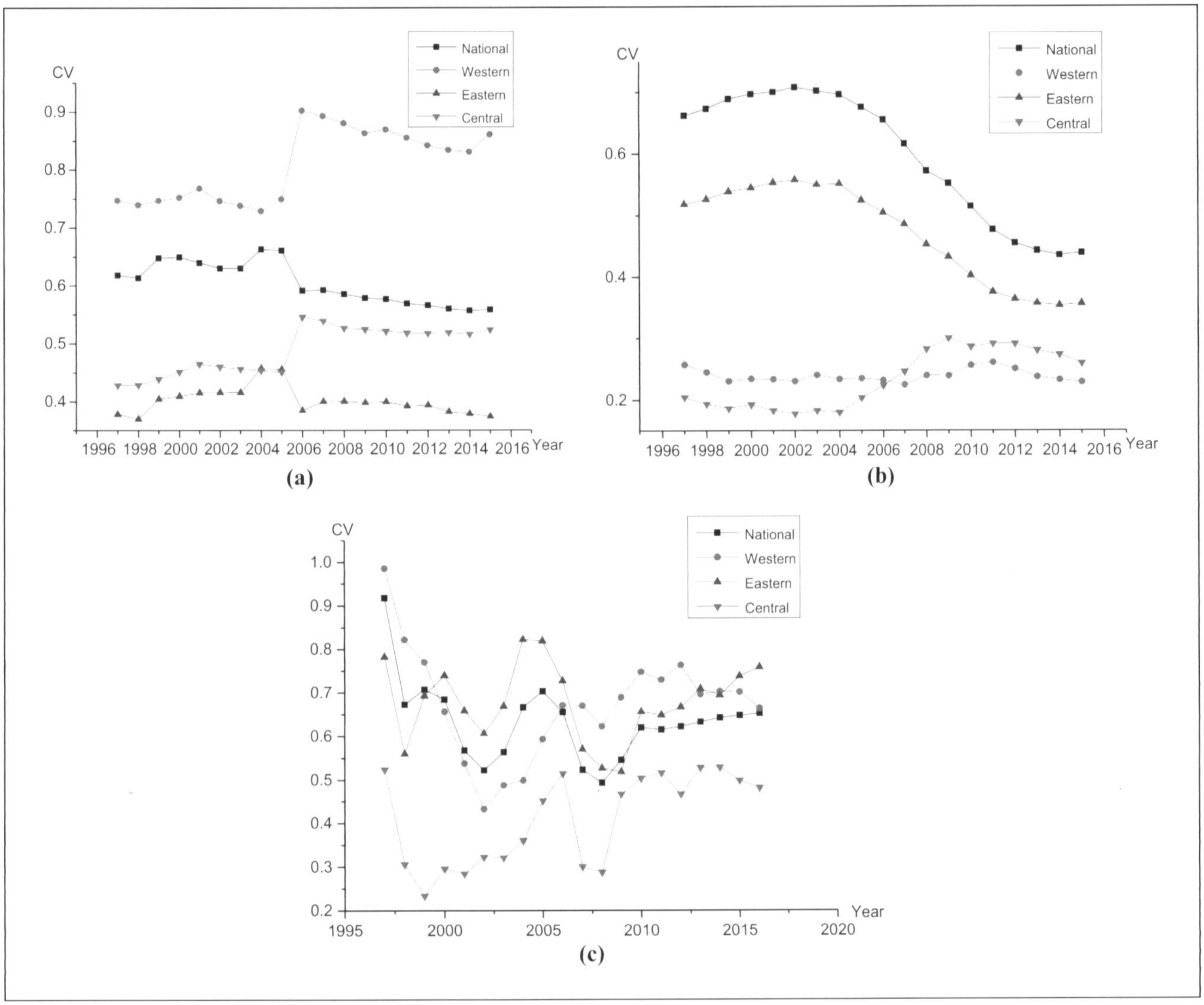

Figure 3. CV of highway traffic development and economic development for different regions for (*a*) density of highway, (*b*) per capita GDP, and (*c*) investment.

however, it still remains at a high level. During 2005–2006, investments were concentrated in the Henan and Hubei provinces of the central region, and as a consequence, the economic difference of the central region first increased, while it recently follows a downward trend; the production levels of western provinces are generally low and economic differences are small and fluctuate slightly.

According to the national $C_{v-\text{investment}}$, the difference of the total social completed investment can be divided into two stages: 1997–2010 formed the great provincial difference in highway investment, as a result of concentrated investment, which was only conducted in some provinces; during 2011–2015, the investment difference remained stable. This trend of regional $C_{v-\text{investment}}$ reflects differences of the regional investment intensity.

The national $C_{v-\text{density of highway}}$ and $C_{v-\text{per capita GDP}}$ are positively correlated at the 0.01 level and the Pearson correlation coefficient reached 0.911. Over the course of highway construction, strengthening of a balanced development of highway transportation can reduce differences of economic development and promote a balanced development of the regional economy.

Difference Analysis Based on the Theil Index

Using the provincial administrative unit as the basic unit of analysis, the national and regional $T_{P-\text{length of highway}}$ and T_{P-GDP}, were calculated between three economic regions, and within-region Thiel index. The obtained trends are shown in Figure 4.

Figure 4*a* provides the same conclusions as the analysis of variance coefficients. The difference of highway development and its evolution can be divided into two stages: 1997–2005 and 2006–2017. The difference of

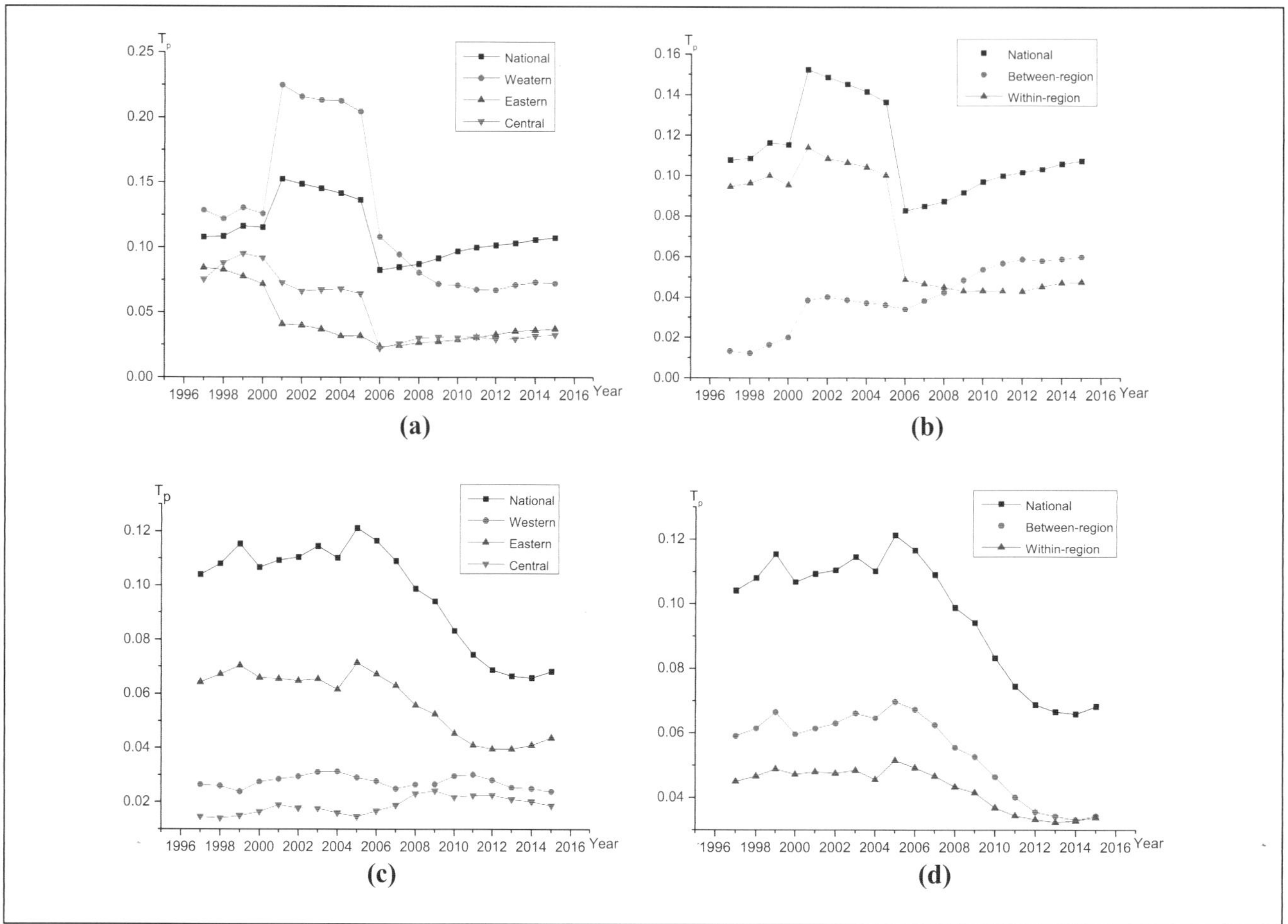

Figure 4. T_p of highway traffic development and economic development in different regions for (*a*) length of highway, (*b*) first order decomposition of national length of highway, (*c*) GDP, and (*d*) first order decomposition of national GDP.

$T_{P-length\,of\,highway}$ changed relative to the population level; however, the sizes of provinces vary widely and with the growth of highway mileage in the western region, the difference of national highway mileage will continue to grow. The density of the highway network can represent the equalization level of highway development.

Since 1997, the difference of highway development in China has mainly manifested as an inter-regional difference, which has been decreasing during recent years, while between-region differences have gradually increased. In 2008, the total length of highways in the western region exceeded that of the eastern region for the first time and between-region differences gradually developed into the main cause, increasing annually until these differences finally steadied. The difference of China's economic development can mainly be ascribed to between-region differences. It decreased annually and is currently equal to within-region. Investments carried out in the western and the central regions have achieved a reduction of this difference.

In the process of equilibrium development of the highway construction, the difference between developed and underdeveloped provinces of regions should be predominantly reduced.

Analysis of Highway Traffic Connectivity and Provincial Economic Difference

The connectivity of highway traffic was calculated via spatial correlation dimension, and the relationship between highway traffic connectivity and economical difference was verified in the provinces. The lower triangular matrixes are known for each province (except for four province-level municipalities) in which nondiagonal elements are equal to the spatial distance and the actual traffic distance between the corresponding vertices provided by main cities, while the diagonal elements all equal zero. This study used a step Δr of 20 km, according to both maximum and minimum of the spatial distance between the main cities in the province. Furthermore, the range of the r value in each province was adjusted, obtaining the series r and $C(r)$; then, the scatter diagrams of $\ln(r)$ and $\ln C(r)$ were drawn, and the spatial correlation dimension D of each province was obtained via least square fit. *D'* was obtained with the same method, simply by changing the space distance into the actual traffic distance. The calculated results are shown in Table 3.

Economic differences within each province are represented by the CV of per capita GDP of prefecture-level cities. The calculated results are also shown in Table 3.

The *D* values indicate the distribution and evenness of the main cities in the province. The distribution characteristics of the main cities in Tibet are most apparent, and the main cities in this province are distributed along a curve. However, only Grade I and II roads are available for communication and exchange between cities in the province or between cities of other provinces that lack a freeway network. The *D* values of Gansu, Ningxia, Xinjiang, Jiangsu, Guangdong, Inner Mongolia, and Heilongjiang are all below 1.35. Due to their long and narrow geographical boundaries, their main cities are distributed along geographical lines. The main cities in the remainder of the province follow a uniform spatial distribution. The ρ values of Jiangsu and Hainan were close to 1, indicating superior highway traffic development. The ρ-value of Tibet was above 1, while for Gansu and Inner Mongolia the ρ-value was below 0.8, indicating poor highway traffic connectivity. Due to the limitation of geographical environment in the provinces of western and central China, the distances between main cities are large, and the connectivity resistance is high; consequently, several main cities are not covered, leading to an incomplete highway network. For example, the freeways of Sichuan are mainly concentrated in the east, near Chongqing, leading to a lack of high-grade road connections in the western region. Figure 5*a* shows that the freeway network of the study area features an apparent spatial agglomeration, which is concentrated in four province-level municipalities (Beijing, Tianjin, Shanghai, and Chongqing) and in Guangzhou City. However, the vast regions of Xinjiang, Gansu, Qinghai, Tibet, and Heilongjiang lack a developed highway network. The connectivity index ρ distribution of study area is shown in Figure 5*b*; here, the ρ-value of Tibet is equal to that of Gansu.

CV-values of GDP per capita decreased in most of the provinces; however, some first increased and then decreased, such as in Hunan, Jiangxi, Shaanxi, and Hebei. The CV of GDP per capita has retained a high level in Yunnan, Gansu, Ningxia, Guangdong, Inner Mongolia, and Heilongjiang, while economic differences were large. The highway traffic network in these 10 provinces shows poor connectivity and several remote main cities were not connected via high-grade highways. Traffic connectivity is the main reason for uneven economic development in the province.

Statistics of GDP per capita are incomplete for the prefecture level cities of Tibet and Qinghai, where the CV has not been calculated yet. However, judging from already existing data, Tibet and Qinghai are economically underdeveloped regions, and economic differences in the province are apparent. We conducted a correlation analysis between ρ-value and CV for 2011 (except for Tibet and Qinghai). The results show that they are significantly negatively correlated at the 0.01 level with a correlation coefficient of -0.581. Thus, the connectivity of highway traffic in the provinces is significantly related to the economic differences of the provinces.

Conclusion and Discussion

During the studied period, the differences between economic development and highway traffic development were strongly related. China is currently improving the disequilibrium of highway traffic development introduced by the previous biased policy. This study calculated and decomposed the difference of highway traffic development and economic development of China by using the CV and the Theil index. For future highway construction policies, the obtained findings offer clues regarding the suitability of implementing rebalancing policies in terms of national or regional design.

The investment difference can be divided into two stages: during 1997–2010, the government focused on investing into central or western provinces and consequently the variation of investment was large and unstable; during 2011–2016, the investment difference remained between 0.62–0.65. The difference of highway traffic development was also divided into two stages: 1997–2006 and 2007–2015. In approximately 2006, the national and eastern difference was significantly reduced, while the difference between western and central regions firstly increased and then decreased.

With the implementation of highway infrastructure construction projects, the national economic difference decreased and the resulting reaction of the three economic regions differed. The economic difference of the eastern region continued to decrease, while that of the central region first increased and then decreased (following an inverse U-shape), and that of the western region generally remained at a low level of productivity, where economic differences only fluctuated slightly.

For a long time, within-region differences contributed predominantly to the differences of highway traffic development. With the increasing highway mileage in the western region, the difference between-region has been increasing as well, which is an inevitable phenomenon and will eventually reach a steady state. The implementation of a novel investment policy has played a significant role in reducing the once dominant between-region economic differences.

Table 3. Highway Traffic Connectivity and Economic Differences of Provinces

	Spatial correlation dimension			CV		
	D (R^2)	D' (R^2)	ρ	2005	2011	2015
West				West		
Sichuan	1.4579 (0.93)	1.3055 (0.92)	0.8955	0.5463	0.4203	0.4274
Guizhou	1.5707 (0.88)	1.449 (0.92)	0.9225	0.6222	0.4470	0.4271
Yunnan	1.4649 (0.95)	1.3765 (0.96)	0.9397	0.7360	0.6399	0.5838
Tibet	0.9426 (0.97)	1.0404 (0.97)	1.1038	—	—	—
Shaanxi	1.6728 (0.90)	1.5097 (0.93)	0.9025	0.4945	0.5454	0.3957
Gansu	1.2062 (0.88)	0.8939 (0.89)	0.7411	1.0026	0.9260	0.5906
Qinghai	1.5035 (0.91)	1.3304 (0.91)	0.8849	—	—	—
Ningxia	1.2605 (0.97)	1.0858 (0.98)	0.8614	0.6704	0.6045	0.5527
Xinjiang	1.2218 (0.95)	1.0601 (0.94)	0.8677	0.7817	0.5949	0.3903
East				East		
Hebei	1.3996 (0.94)	1.2611 (0.95)	0.9010	0.3659	0.4251	0.3812
Liaoning	1.6425 (0.94)	1.5316 (0.92)	0.9325	0.5471	0.4313	0.4549
Jiangsu	1.2591 (0.84)	1.2581 (0.85)	0.9992	0.6853	0.4337	0.3694
Zhejiang	1.7356 (0.95)	1.6481 (0.96)	0.9496	0.3941	0.3083	0.2914
Fujian	1.6007 (0.92)	1.4298 (0.95)	0.8932	0.5697	0.2647	0.1963
Shandong	1.7395 (0.93)	1.5312 (0.92)	0.8803	0.6190	0.5027	0.4843
Guangdong	1.2406 (0.91)	1.1656 (0.93)	0.9395	0.8416	0.6792	0.6762
Guangxi	1.5321 (0.97)	1.4375 (0.98)	0.9383	0.3822	0.3753	0.4299
Hainan	1.4188 (0.94)	1.4098 (0.93)	0.9937	0.1290	0.1701	0.07582
Center				Center		
Shanxi	1.5295 (0.91)	1.3604 (0.92)	0.8894	0.4110	0.3304	0.3701
Inner Mongolia	1.346 (0.94)	1.0102 (0.92)	0.7505	0.6409	0.6250	0.6013
Jilin	1.393 (0.94)	1.354 (0.93)	0.9720	0.3678	0.2365	0.2390
Heilongjiang	1.3368 (0.91)	1.1735 (0.91)	0.8778	0.9273	0.8904	0.6725
Anhui	1.5565 (0.89)	1.5014 (0.92)	0.9646	0.6441	0.5945	0.4543
Jiangxi	1.629 (0.98)	1.5483 (0.96)	0.9505	0.4817	0.5411	0.4676
Henan	1.766 (0.91)	1.6295 (0.91)	0.9227	0.4292	0.3761	0.3462
Hubei	1.4029 (0.97)	1.2725 (0.95)	0.9070	0.5175	0.4606	0.4707
Hunan	1.7404 (0.93)	1.4231 (0.94)	0.8177	0.4645	0.5714	0.5637

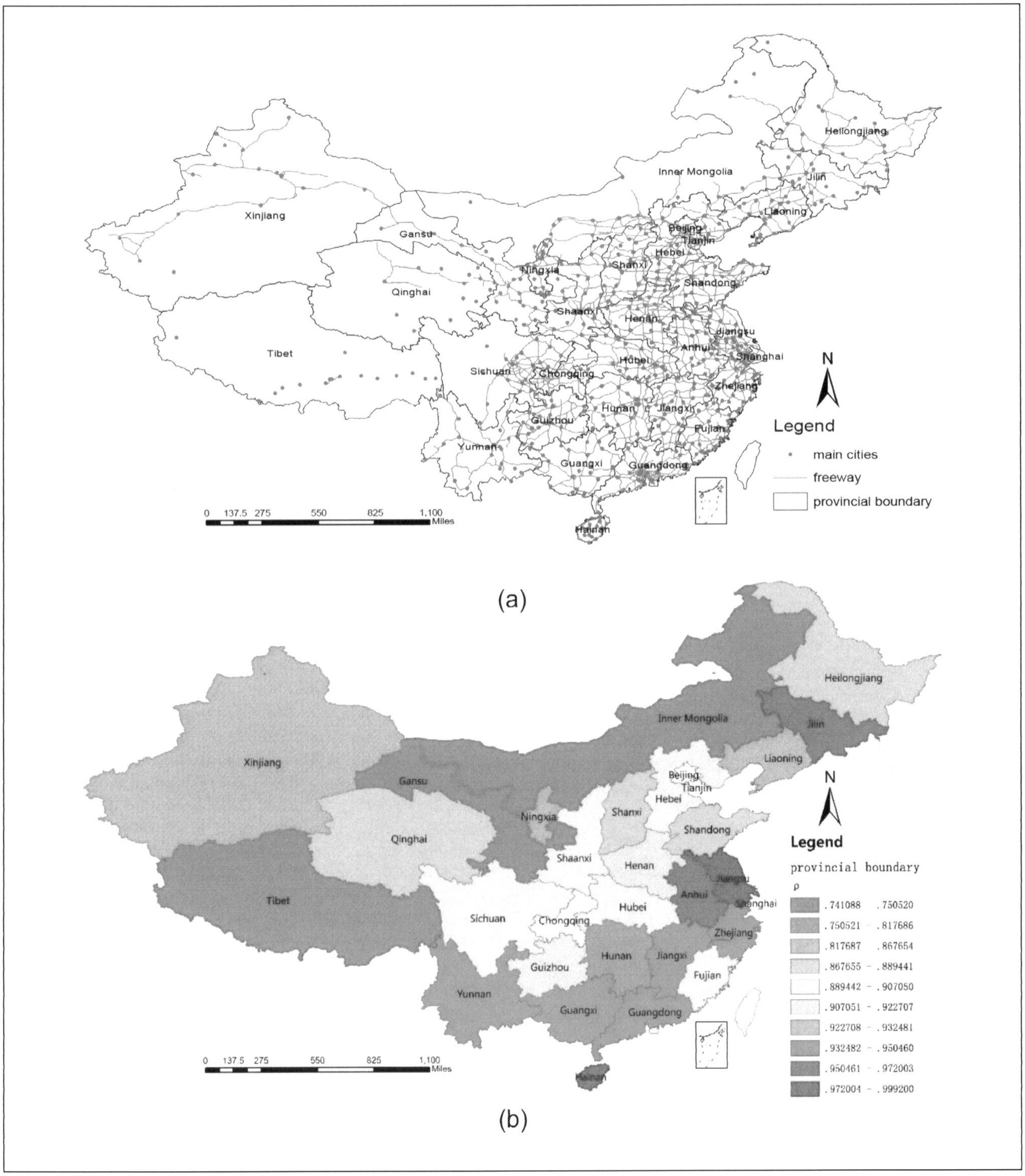

Figure 5. Development of highway traffic in provinces showing (*a*) freeway network and main cities, and (*b*) connectivity of provinces.

The connectivity of highway traffic strongly correlates with the economic difference of a province, and improved connectivity leads to smaller economic difference in that province. However, traffic connectivity of provinces in the central and western regions remains poor. Due to many restrictions of the geographical environment, the distances between cities are large, and some main cities are not connected via high-grade highways.

Intercity railways are currently rapidly developing; however, highway transportation can provide a door-to-door service, which enables a more vigorous economic development for the region and has a significant potential to promote a balanced economic development. In future, the equalized development of highway traffic in China should first reduce the existing differences between developed and underdeveloped areas of each region. Then, high-grade highways should be built deep into remote areas.

The change of traffic mileage was less sensitive to the development of a highway network, which makes it more advantageous to study intercity connectivity or accessibility in response to travel time. With the recent development of map services (e.g. Baidu map and API interface), data support has been provided and as a consequence, studying the relationship between multimodal transportation development and economic or industrial development at different scales became convenient.

Acknowledgments

This study was supported by the projects of the National Natural Science Foundation of China (71774118, 91546115).

Author Contributions

The authors confirm contribution to the paper as follows: study conception and design: Ye Li, Jing Fan, Haopeng Deng; data collection: Jing Fan; analysis and interpretation of results: Jing Fan; draft manuscript preparation: Ye Li, Jing Fan, Haopeng Deng. All authors reviewed the results and approved the final version of the manuscript.

References

1. Fan, S., L. Zhang, and X. Zhang. *Growth, Inequality, and Poverty in Rural China: The Role of Public Investments*, Vol. 125. International Food Policy Research Institute, Washington, D.C., 2002.
2. Wang, H. Z. Transportation and Regional Economic Discrepancy-Empirical Analysis of Railways in China. *Journal of Shanxi Finance and Economics University*, Vol. 33, No. 2, 2011, pp. 61–68.
3. Fan, S., and C. Chan-Kang. *Road Development, Economic Growth, and Poverty Reduction in China*, Vol. 12. International Food Policy Research Institute, Washington, D.C., 2005.
4. Aschauer, D. A. Is Public Expenditure Productive? *Journal of Monetary Economics*, Vol. 23, No. 2, 1989, pp. 177–200.
5. Arbués, P., F. B. José, and M. Mayor. The Spatial Productivity of Transportation Infrastructure. *Transportation Research Part A: Policy and Practice*, Vol. 75, 2015, pp. 166–177.
6. Lakshmanan, T. R. The Broader Economic Consequences of Transport Infrastructure Investments. *Journal of Transport Geography*, Vol. 19, No. 1, 2011, pp. 1–12.
7. Ozbay, K., D. Ozmen-Ertekin, and B. Joseph. Contribution of Transportation Investments to County Output. *Transport Policy*, Vol. 14, No. 4, 2007, pp. 317–329.
8. Cantos, P., M. Gumbau-Albert, and J. Maudos. Transport Infrastructures, Spillover Effects and Regional Growth: Evidence of the Spanish Case. *Transport Reviews*, Vol. 25, No. 1, 2005, pp. 25–50.
9. Boarnet, M. G. *The Direct and Indirect Economic Effects of Transportation Infrastructure*. University of California Transportation Center, 1996.
10. Mikesell, J. L., J. Q. Wang, Z. J. Zhao, and Y. He. Impact of Transportation Investment on Economic Growth in China. *Transportation Research Record Journal of the Transportation Research Board*, 2016. 2531: 9–16.
11. Berechman, J., D. Ozmen, and K. Ozbay. Empirical Analysis of Transportation Investment and Economic Development at State, County and Municipality Levels. *Transportation*, Vol. 33, No. 6, 2006, pp. 537–551.
12. Chen, Z., and K. E. Haynes. Transportation Infrastructure and Economic Growth in China: A Meta-Analysis. In *Socioeconomic Environmental Policies and Evaluations in Regional Science* (Shibusawa, H., K. Sakurai, T. Mizunoya, and S. Uchida, eds.), Springer, Singapore, 2017.
13. Pereira, A. M., and J. M. Andraz. Public Investment in Transportation Infrastructure and Industry Performance in Portugal. *Journal of Economic Development*, Vol. 32, No. 1, 2007, p. 1.
14. Jiang, X., X. He, L. Zhang, H. Qin, and F. Shao. Multimodal Transportation Infrastructure Investment and Regional Economic Development: A Structural Equation Modeling Empirical Analysis in China from 1986 to 2011. *Transport Policy*, Vol. 54, 2017, pp. 43–52.
15. Straszheim, M. R. Researching the Role of Transportation in Regional Development. *Land Economics*, Vol. 48, No. 3, 1972, pp. 212–219.
16. Vickerman, R., K. Spiekermann, and M. Wegener. Accessibility and Economic Development in Europe. *Regional Studies*, Vol. 33, No. 1, 1999, pp. 1–15.
17. Yu, N., M. De Jong, S. Storm, and J. Mi. The Growth Impact of Transport Infrastructure Investment: A Regional Analysis for China (1978–2008). *Policy and Society*, Vol. 31, No. 1, 2012, pp. 25–38.
18. Zhang, X. L. Regional Comparative Analysis on the Relationship between Transport Infrastructure and Economic Growth in China. *Journal of Finance and Economics*, Vol. 33, No. 8, 2007, pp. 51–63.
19. Wang, J. W., T. Mao, and X. Fu. Spatial Disparities of the Distribution of Highway Network in Northwest China based on Exploratory Spatial Data Analysis. *Arid Land Geography*, Vol. 36, No. 2, 2013, pp. 329–336.
20. Cao, X., L. Zhang, D. Xue, and D. Wang. The Changes in Disparity of Urban Transportation Development Level Rank in China. *Acta Geographica Sinica-Chinese Edition*, Vol. 62, No. 10, 2007, pp. 1034–1040.
21. Demurger, S. Infrastructure Development and Economic Growth: An Explanation for Regional Disparities in China? *Journal of Comparative Economics*, Vol. 29, No. 1, 2001, pp. 95–117.

22. Li, S. M., and Y. M. Shum. Impacts of the National Trunk Highway System on Accessibility in China. *Journal of Transport Geography*, Vol. 9, No. 1, 2001, pp. 39–48.
23. Chen, P. Y., and X. G. Zhu. Regional Inequalities in China at Different Scales. *Acta Geographica Sinica*, Vol. 67, No. 8, 2012, pp. 1085–1097.
24. Bedeian, A. G., and K. W. Mossholder. On the Use of the Coefficient of Variation as a Measure of Diversity. *Organizational Research Methods*, Vol. 3, No. 3, 2000, pp. 285–297.
25. Bourguignon, F. Decomposable Income Inequality Measures. *Econometrica*, Vol. 47, No. 4, 1979, pp. 901–920.
26. Duro, J. A., and J. Teixidó-Figueras. Ecological Footprint Inequality across Countries: The Role of Environment Intensity, Income and Interaction Effects. *Ecological Economics*, Vol. 93, No. 3, 2013, pp. 34–41.
27. Capecchi, V., A. Crisci, S. Melani, M. Morabito, and P. Politi. Fractal Characterization of Rain-gauge Networks and Precipitations: An Application in Central Italy. *Theoretical & Applied Climatology*, Vol. 107, No. 3–4, 2012, pp. 541–546.
28. Liu, J. S., and Y. G. Chen. A Study on Fractal Dimensions of Spatial Structure of Networks and the Methods of Their Determination. *Acta Geographica Sinica-Chinese Edition*, Vol. 54, 1999, pp. 471–478.

The Standing Committee on Transportation and Economic Development (ADD10) peer-reviewed this paper (18-02800).

Article

TRR
JOURNAL OF THE TRANSPORTATION RESEARCH BOARD

Transportation Research Record
2018, Vol. 2672(3) 26–36
© National Academy of Sciences:
Transportation Research Board 2018
Article reuse guidelines:
sagepub.com/journals-permissions
DOI: 10.1177/0361198118783863
journals.sagepub.com/home/trr
SAGE

Changes in Individual Economic Activities and Regional Market Structures Caused by Rural Road Improvements in Cambodia

Rika Idei[1] and Hironori Kato[1]

Abstract

In rural areas of developing countries, most individuals engage in farming of staple foods and trading them in their local markets for daily earnings. If a road is improved, what would happen to those markets and the local populations' economic activities? This study aimed to examine the effects of rural road improvements on local economic activities and market structures in rural areas of Cambodia using quantitative and qualitative data collected from face-to-face interviews. The data included both individual economic status and households' travel behavior around their markets. This study showed that improving road conditions helps low-income individuals visit a market frequently and releases rice farmers from the time and cost burdens of transporting their rice to markets or enables customers to now approach them directly for trades. The enhancement of economic activities was also confirmed among local populations on the roadsides, particularly groups with economic capacities, and this effect was distributed to markets even beyond the roadsides. Our findings suggest that rural road improvements enhance the economic status of individuals who engage in dominant economic activities along the improved roads and that these benefits extend beyond roadside markets themselves. However, such positive impacts were not realized by households that did not own any land for rice farming and whose economic status remained poor. Instruments should be integrated into future rural road improvement projects to include households of small economic capacities in the local economic development process.

Many people in rural areas of developing countries are engaged in subsistence farming as their primary source of income, and markets are their economic centers for trading their products and engaging in social–economic-purposed interactions (*1*). Assuring people's accessibility to markets is critical in these areas to improve their individual economic statuses (*2*, *3*).

Various studies have analyzed the impact of rural road improvements on people's economic activities around markets, and have highlighted in particular farm produce trading. When roads are of low quality, farmers are reluctant to transport their produce to markets, and instead sell them to traders at their fields or residences. However, prices are often favorable to traders who are skilled in negotiating and familiar with price information, which disadvantages farmers (*4*). Rural road improvements are expected to reduce such inequitable trading practices by enabling farmers to transport their products efficiently and to integrate themselves into market economies (*5*–*7*). Prior studies have not identified how the improvements cause such positive impacts on individual socio-economic conditions and regional economic development status and whether those impacts are distributed equally.

This paper aims to empirically examine the impacts of rural road improvements in Cambodia on the behavior of the local population around their markets along the roads. In those areas, single-season rice farming is the dominant economic activity through which markets are recognized to function as economic centers for the local population (*7*, *8*). This study uses two types of survey data—quantitative data obtained from a socio-economic survey conducted about one year after the completion of civil works, and qualitative data obtained from extensive follow-up interviews by the authors nine months after the first study.

The remainder of this paper is organized as follows. The next section reviews previous findings and discussions on the effects of rural road improvements on local populations' economic activities and market environments. The dataset used in the empirical analysis is then described, followed by model estimations of individual travel behavior around markets. Observations

[1]Department of Civil Engineering, The University of Tokyo, Tokyo, Japan

Corresponding Author:
Address correspondence to Rika Idei: idei-r@ip.civil.t.u-tokyo.ac.jp

of the changes in the local populations' livelihoods are detailed using data from the follow-up interview survey. Finally, discussions and suggestions for future research are provided.

Literature Review

In rural areas of developing countries, strategies for farmers to sell their farm produce and for traders to purchase or sell that produce are affected by road conditions because market prices are influenced by costs and time to transport the produce (*9*). Olsson (*10*) observed in the rural Philippines that a major direct effect of rural road improvements was the cost and time saved in transporting goods or products, and their secondary effects were lowering trade costs, easier information flows, and consequently reductions in consumer prices. These findings were supported by Wondemu and Weiss (*11*), who found in rural Ethiopia that road improvements could positively influence households' (HHs) farm activities by: (i) increasing farm-gate prices which are negotiated between farmers and traders on the farmers' property; (ii) improving productivity; and (iii) influencing subsequent marketing strategy changes. Casaburi et al. (*12*) examined in Sierra Leone the impacts of rural road improvements on transportation costs and staple crop prices in local markets, and found that traders tended to react more sensitively than farmers to such transportation cost changes.

Rural road improvements often stimulate individuals' mobility and economic activities, thus transforming the market environment. Regarding changes in markets adjacent to improved roads, Khandker et al. (*9*) indicated the possibility that markets facilitate interactions among the local people. Using a time-series statistical dataset, Mu and van de Walle (*6*) illustrated changes in markets after a rural road improvement project in Vietnam. Their findings indicated: (i) an increase in the number of individuals engaged in the service and trade sectors; (ii) diversification of service sectors; and (iii) changes in the ethnic minority's economic transaction practices from barter to cash-based trades. Using Nicaragua as a case study, Rand (*13*) presented another type of change evident in local economies around markets: job opportunity augmentation in the agricultural sector and increases in individual working hours.

Previous studies have shown that improving rural roads can enhance individual economic activities and fuel regional economic growth by letting markets act as a catalyst. However, some of these studies noted that such positive impacts could be generated only when specific conditions were satisfied, such as HH composition and socio-economic conditions, and markets' institutional or policy conditions, particularly related to price determination systems (*5*, *11*). For instance, Gachassin et al. (*14*) found in Cameroon that the frequency of individuals' visits to markets was inversely proportional to distances from their homes because people in remote areas tend to be self-sufficient instead of participating in trade economies in the markets, thus losing opportunities to receive cash.

Surveys

Post-completion Questionnaire Survey

Survey Organization. Some of the data used for the empirical analysis were collected through a post-completion survey implemented during a rural road improvement project in Cambodia. The project was executed by the Government of Cambodia from 2010 to mid-2016, aiming to improve 25 rural roads for all-weather use. The roads totaled approximately 500 km around the Tonle Sap region—the poorest area in the country, mainly because of its large population and overreliance on rice farming as the primary source of income (*15*). The project rehabilitated those roads using a double bitumen surface treatment. Three survey areas were selected on the basis of the provincial-level poverty rates in 2012: the lowest poverty rate (area along road TK, hereafter called "TK"), the mid-point (area along road KC, "KC"), and the highest (area along road BB, "BB") among the project areas (*16*). Figure 1 illustrates the locations of the selected three roads and markets that respondents visited.

The post-completion survey was conducted for approximately two weeks through face-to-face interviews from February to March 2016. One of the survey's goals was to examine the impacts of the roads' improvements on the local people's livelihoods during the post-completion stage and focused on: (i) socio-economic profiles, including HH composition, income in wet and dry seasons, and assets or properties acquired; (ii) economic activities including market access and associated travel behaviors; and (iii) access to social service facilities. To compare the conditions before and after the project, respondents were asked to answer items (ii) and (iii) for both 2011 (before the civil works commenced) and 2015 (after the civil works were completed).

Before identifying the potential HHs, the surveyors visited commune chiefs and obtained permission to conduct interviews with their people. Next, they selected potential HHs, considering residents' distances from the study road (hereafter called "the road"), locations, house exteriors, and economic capacities. Then, they visited and sought one HH member to participate in the survey. Finally, individuals who agreed to participate were interviewed by the surveyors who used a paper-based structured questionnaire. If the HH head was available, they were asked to provide comprehensive ideas of their HH's economic conditions. The survey team calculated the sample sizes allocated to each area using the population statistics of the government's Commune Database as of 2014 (*17*).

In parallel to this survey, from February 26 to March 3, 2016, one of the authors interviewed 25 leaders and council members of the 13 communes on the three roads. These interviews aimed at collecting the communes' demographic

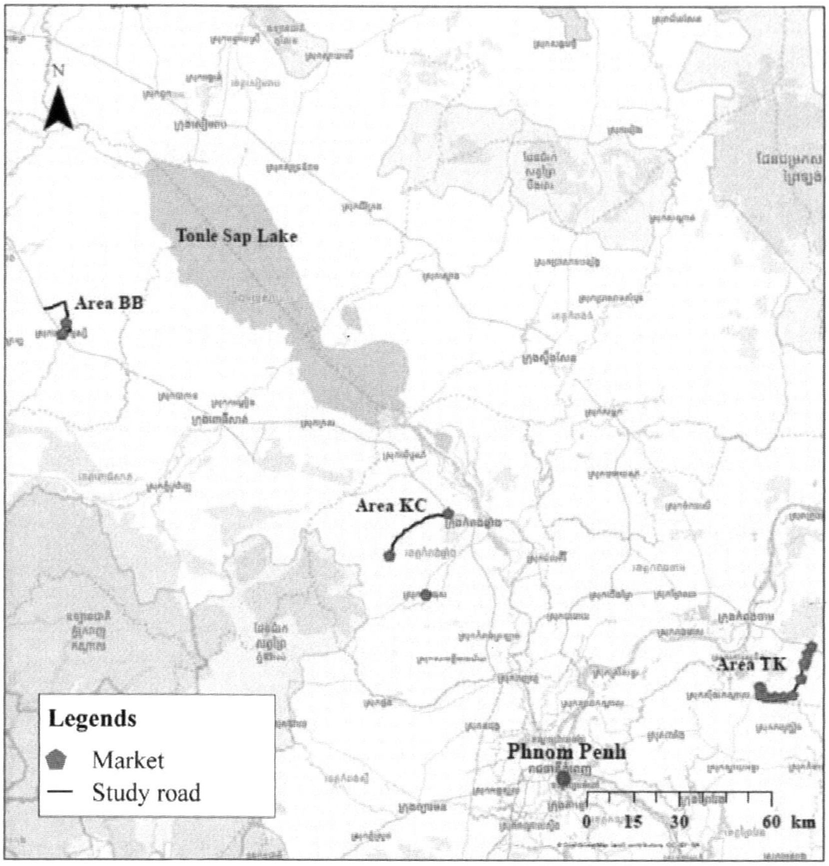

Figure 1. Market locations in three study areas.

data and understanding changes observed in their local populations' travel and economic activities after the road improvements.

Dataset Constructed for Empirical Analysis. Tables 1 and 2 present the consolidated dataset of local people's behavior around their markets, and contain 265 subsamples with valid responses regarding behavior around markets out of the originally collected 400 samples. The average commune-level poverty rate for each area was 15.5% in Area TK, 20.3% in Area KC, and 24.6% in Area BB, respectively, all of which were improved from the 2011 values of 22.2%, 28.0%, and 27.8%.

As shown in Table 1, most respondents (74.2%) were engaged in either self-employed or employed farming, whereas the second largest job category (10.3%) were vendors of goods or farm produce. Regarding respondents' educational backgrounds, 52.5% had never attended school or had failed to complete a primary education, followed by those who had completed only a primary education (27.2%). Of the respondents, 99.6% were Khmer and 0.4% were Cham (Khmer-Muslim). The majority (98.5%) were Buddhist, with the remainder being Muslim (1.1%) or Christian (0.4%). Area BB had the highest average income and largest standard deviation (S.D.) despite having the highest poverty rate among the three areas, while Area TK had the lowest average income and the smallest S.D. despite having the lowest poverty rate. The average distance from respondents' houses to their most frequently visited market was shortest in Area TK and longest in Area KC. This finding corresponds to the geographical distributions of the markets visited by the respondents, which were located along the roads BB and TK, whereas three markets were scattered and one was located 17.9 km from the endpoint of the road KC. A correlation was found between distances from individuals' homes to their most frequently visited markets and the commune-level poverty rate of 0.40, implying that market accessibilities are poorer for people residing in areas with higher poverty rates.

Table 2 compares respondents' behavior related to visiting a market before and after the road improvements. First, the table shows that mobile phone ownership significantly increased across all areas. It should be noted that all respondents had one or more mobile phones in their HHs, which reflects the recent nationwide penetration following the enhancement of competitiveness in the telecommunication markets with the active participations of private firms in 2014 (*18*). Time-series surveys by Phong et al. (*19*, *20*) also

Table 1. Descriptive Statistics of Respondents' Characteristics

Survey areas	Total (N = 265)		TK (N = 89)		KC (N = 104)		BB (N = 72)	
Total HHs in the villages alongside	10,465		4,766		1,875		3,824	
Families in the villages alongside	12,138		5,543		2,201		4,394	
Population in the villages alongside	53,086		25,161		8,896		19,029	
Average population density of the communes alongside (m^2)	215.70		327.42		135.64		108.16	
Average poverty rate among the communes alongside	19.41%		15.54%		20.29%		24.57%	
Number of markets answered	18		13		3		2	
Gender								
Female	56.98%		55.06%		59.62%		55.56%	
Male	43.02%		44.94%		40.38%		44.44%	
Age								
10–19	0.38%		1.12%		0.00%		0.00%	
20–29	7.92%		3.37%		15.38%		2.78%	
30–39	21.51%		23.60%		19.23%		22.22%	
40–49	23.40%		20.22%		24.04%		26.39%	
50–59	23.40%		22.47%		23.08%		25.00%	
60–69	17.74%		20.22%		14.42%		19.44%	
70–79	5.65%		9.00%		3.85%		4.17%	
Employment type								
Farmer	74.20%		74.70%		69.23%		80.65%	
Daily labor	3.97%		6.90%		2.88%		1.61%	
Vendors of goods or farm products	10.32%		3.45%		16.35%		11.29%	
Employed in manufacturing/service	2.78%		3.45%		3.85%		0.00%	
Government officer	1.98%		0.00%		4.81%		0.00%	
Teacher	0.40%		1.15%		0.00%		0.00%	
Student	1.19%		3.45%		0.00%		0.00%	
No job	1.59%		0.00%		0.96%		4.84%	
Others	3.57%		6.90%		1.92%		1.61%	
Ethnicity								
Khmer	99.62%		100.00%		99.04%		100.00%	
Cham	0.38%		0.00%		0.96%		0.00%	
Education								
No education	52.45%		61.80%		36.54%		63.89%	
Primary	27.17%		28.09%		27.88%		25.00%	
Lower secondary	15.86%		7.87%		28.85%		6.94%	
Higher secondary	3.77%		1.12%		5.77%		4.17%	
College or above	0.75%		1.12%		0.96%		0.00%	
Religion								
Buddhist	98.49%		100.00%		96.16%		100.00%	
Muslim	1.13%		0.00%		2.88%		0.00%	
Christian	0.38%		0.00%		0.96%		0.00%	
HH income	Average	S.D.	Average	S.D.	Average	S.D.	Average	S.D.
('000 Cambodian Riel)	18.06	22.98	14.52	15.08	14.60	22.30	27.45	28.68
Distance to market	Average	S.D.	Average	S.D.	Average	S.D.	Average	S.D.
('000 meters)	6.60	6.67	4.09	5.39	9.49	7.97	5.52	3.48

Note: S.D. = standard deviation.

revealed the recent technical evolution of mobile phones to enable people to send and read short message services in Khmer—giving the population better and more efficient access to information. Second, the results revealed that respondents' daily market visits increased in all areas. All respondents traveled to markets after the road improvements, although a few had not previously visited. Third, in all areas in both 2011 and 2015, the highest percentages of respondents' reasons for visiting the markets was buying food and commodities, followed by selling their own produce. Fourth,

Table 2. Changes in Travel Behavior from 2011 to 2015 in the Survey Areas.

	Total (N = 265)		TK (N = 89)		KC (N = 104)		BB (N = 72)	
Survey areas	2011	2015	2011	2015	2011	2015	2011	2015
Ownership of mobile phones	42.54%	100.00%	51.11%	100.00%	33.02%	100.00%	45.83%	100.00%
Ownership of transportation modes								
Vehicle and truck	1.94%	2.64%	5.62%	5.62%	0.00%	1.92%	0.00%	0.00%
Motorcycle	40.31%	74.72%	47.19%	78.65%	41.35%	85.58%	26.39%	55.56%
Bicycle	40.31%	71.32%	40.45%	49.44%	57.69%	84.62%	58.33%	79.17%
Tuk-tuk	1.94%	2.26%	3.37%	4.49%	0.96%	1.92%	0.00%	0.00%
Frequency of using the road								
Daily	60.38%	76.23%	71.91%	91.01%	62.50%	73.08%	43.06%	62.50%
3–4 times a week	33.21%	20.37%	24.72%	7.87%	29.81%	21.15%	48.61%	34.72%
Less	5.66%	3.40%	3.37%	1.12%	5.77%	5.77%	8.33%	2.78%
Never	0.75%	0.00%	0.00%	0.00%	1.92%	0.00%	0.00%	0.00%
p-value	0.886		0.971		1.000		0.971	
Purpose of visiting markets								
To sell your products	13.58%	12.45%	13.49%	12.36%	14.42%	11.54%	12.50%	13.89%
To buy food and commodities	77.74%	79.62%	82.02%	83.15%	76.93%	79.81%	73.61%	75.00%
To purchase commodities for business	7.17%	6.42%	4.49%	4.49%	7.69%	7.69%	9.72%	6.94%
To meet with others	1.13%	1.13%	0.00%	0.00%	0.96%	0.96%	2.78%	2.78%
Other	0.38%	0.38%	0.00%	0.00%	0.00%	0.00%	1.39%	1.39%
p-value	0.984		1.000		1.000		1.000	
Primary travel mode								
Vehicle (truck and van)	3.02%	3.77%	1.12%	2.25%	5.77%	6.73%	1.39%	1.39%
Motorcycle	62.26%	66.04%	67.42%	77.53%	75.00%	73.08%	37.50%	41.66%
Tuk-tuk	15.85%	14.72%	2.25%	1.12%	0.00%	0.96%	55.55%	51.39%
Walk	9.81%	7.92%	21.34%	14.61%	5.77%	6.73%	1.39%	1.39%
Bicycle	9.06%	7.55%	7.87%	4.49%	13.46%	12.50%	4.17%	4.17%
p-value	0.841		0.984		0.818		1.000	

Note: p-values were computed using the Wilcoxon-Mann-Whitney test, representing the probability of rejecting the null hypothesis that travel behavior before road improvements are the same as after.

more than 75% of the respondents used motorcycles in Areas KC and TK before and after the road improvements, and more than half of the respondents used a tuk-tuk—a three-wheeled motorized rickshaw (*21*)—to visit their markets in Area BB. Note that no respondent in the area privately owned a tuk-tuk, indicating that the tuk-tuk in Area BB was generally used as shared transportation. Following the motorcycle, the most common travel mode used by respondents was walking in Area TK, bicycle in Area KC, and motorcycle in Area BB, following the tuk-tuk. Walking was preferred in Area TK because it has the highest market density.

Follow-up Interview Survey

Following the abovementioned socio-economic survey, another survey was carried out from December 12 to 27, 2016 by the authors' study group, through which face-to-face semi-structured interviews were held with 45 individuals along or around the roads and adjacent markets. This survey aimed at understanding how project beneficiaries and local stakeholders perceived the changes, mainly on the economic and travel behavior of the respondents and their HH members, modal ownership, and accessibility to education and health facilities. This enabled a deeper understanding of the mechanisms regarding the impact of the rural road improvements from qualitative viewpoints.

Out of 45 interviewees, 34 were from HHs residing along the three roads and the remaining 11 interviewees were shop owners—three rice mill firm staff, one operator of a family-run restaurant, and seven shop owners or staff. Those shop owners were interviewed mainly on the changes in their business environments after the road improvements, including sales and types of customers. Five interviewees ran their businesses at markets but not on the roadsides. To select the interviewees, one of the authors first drove the roads to understand the environment, and next, selected potential interviewees, considering their houses' exteriors, residential places, HH compositions, economic activities, and ownership of travel modes, as well as the geographical balance of the number of interviewees. The authors then asked the potential interviewees to participate voluntarily and, if they agreed, the interview was carried out. The number of female interviewees was larger than the number of males, mainly because the survey was conducted during the harvest season for rice farming, when most male HH members work in the fields during the daytime.

Empirical Analyses

Model Estimations with Questionnaire Survey Data

This study employs two models for analyzing the impacts of rural road improvements on travel frequency to markets and on travel mode choices for their market visits. The first analysis is intended to examine whether HH income influences travel frequency when controlling for road improvements. The second analysis is motivated by the fact that more than 20% of the respondents used a shared mode despite increases in their HHs' ownership of a private mode of travel. Several reasons could be behind this fact: (i) privately owned modes were often used for their HH members' commuting to a workplace or school and were therefore not always available for market visits; and (ii) shared modes tend to be more spacious and able to accommodate bulk shopping. Note that the potential available shared modes of travel in the study areas are motorcycle taxis or tuk-tuks, both of which are privately operated. Both analyses were carried out using the dataset compiled for 2015 (after the civil works completion).

First, an individual's frequency of market visits after the road improvements is analyzed by formulating the two simple ordered logit models by differentiating the set of the explanatory variables, namely Models 1 and 2. In both models, the choice set is assumed to be: "3: traveling every day"; "2: traveling three or four days a week"; and "1: traveling less than three days a week." The common potential variables are: "Distance to market most frequently visited ('000 meters)"; "Visit using a private mode (dummy)," which takes the value of 1 if an individual visits using a privately owned wheeled mode—bicycle, motorcycles, or automobile—and 0 otherwise; "Visit using a shared mode (dummy)," which takes the value of 1 if an individual visits using a shared wheeled mode—motorcycle, automobile, or tuk-tuk—and 0 otherwise; "Percentage of paved road access to market most frequently visited"; and "Low income (dummy)," which takes the value of 1 if the daily per capita income is below the average income in Cambodia of 2015 (US$ 1,093.46 per capita per annum, equivalent to Cambodian Riels 12,168.86 per capita per day (*22*)) and 0 otherwise. Model 1 employs a combination variable of "Percentage of paved road access to market most frequently visited" and "Low income (dummy)."

Table 3 shows the models' estimation results, whose coefficients and statistical significances are almost the same. First, the estimated coefficient of "Distance to market most frequently visited" is significantly negative, indicating that people residing far from market(s) visit less frequently. Second, "Visit using a private mode (dummy)" and "Visit using a shared mode (dummy)" are not statistically significant on the frequency of market visits. Third, the pavement rate improvement significantly impacted the frequency, indicating that the smoother road conditions promoted people's visits to markets. Fourth, people whose income was below the national average visited markets more frequently than the others. This explanation is in line with previous findings that lower-income people's purchasing power is not sufficient to buy groceries beyond their immediate daily needs and they are therefore forced to visit markets frequently (*23*). Finally, the combined pavement rate with low-income variable, which is unique to Model 1, was estimated to be insignificant, indicating that the road improvement would not necessarily enhance the frequency only for low-income people to visit markets. The authors carried out the likelihood ratio of the two models to check their performance and the null hypothesis that all the coefficients are zero is statistically rejected in both models.

Next, a multinomial logit model is used to formulate the travel model choice. The choice set is assumed to consist of three options: privately owned mode, shared mode, and walk. The shared mode includes motorcycle, taxi, and tuk-tuk, whereas the privately owned mode includes motorcycle and bicycle. Potential explanatory variables are "Distance to market most frequently visited ('000 meters)"; "Daily income per capita in 2015 (KR '00000)"; "Number of members in a household"; "Ownership of rice fields by HH (dummy)," which takes the value of 1 if a HH owns rice fields and 0 otherwise; "Ownership of motorcycle(s) by HH (dummy)," which takes the value of 1 if a HH owns motorcycle(s) and 0 otherwise; and "Percentage of paved road from improvement in access to market most frequently visited."

Table 4 shows the model estimation results. First, "Distances to a market most frequently visited" and "Daily income per capita" are estimated to be significantly positive, and both are mode-specific variables for the privately owned mode and the shared mode. They indicate that those who reside far from markets or who earn higher incomes tend to choose a wheeled mode rather than walking. Such a choice is reasonable because a motorcycle or bicycle enables individuals to travel more efficiently when accessing a faraway market, whereas wealthier individuals can afford to purchase private vehicles and pay for using a shared mode. Second, the statistical significance of "Number of members in a HH" was confirmed, indicating that people from large HHs are likely to use wheeled modes probably for economic efficiencies or to allow them to be accompanied on their visits. Third, "Ownership of rice fields by HH" positively affected the use of both the privately owned and shared wheeled modes to visit a market, which could be interpreted as follows: using these modes enabled people to bring their rice to a market to sell or to transport purchased fertilizers or chemicals such as pesticides to maintain product quality (*7*).

Table 3. Estimation Results of Ordered Logit Model for Frequency of Market Visits after Road Improvements (n = 265).

	Model 1		Model 2	
Variable	Coefficient	*t*-statistics	Coefficient	*t*-statistics
Distance to market most frequently visited ('000 meters)	−0.071	−3.134**	−0.070	−3.091**
Visit using a private wheeled mode (dummy)	−0.736	−0.912	−0.785	0.973
Visit using a shared wheeled mode (dummy)	−1.090	−1.324	−1.158	−1.405
Percentage of paved road access to market most frequently visited	1.171	2.178*	0.911	2.243*
Low income (dummy)	1.596	2.389**	1.163	3.680***
(Percentage of paved road access to market most frequently visited) × (Low income (dummy))	−0.589	−0.741		
Intercepts				
1|2	−3.431	−3.644***	−3.648	−4.037***
2|3	−1.014	−1.127	−1.249	−1.471
Initial log-likelihood (L0)	−171.178		−171.178	
Final log-likelihood (LL)	−152.543		−152.820	
−2(L0–LL)	37.269		36.715	
Akaike Information Criterion	321.087		319.641	
Adjusted ρ^2	0.109		0.107	

***: <0.001; **: <0.01; *: <0.05.

Fourth, "Ownership of motorcycle(s) by HH" also positively and significantly influenced the choice of a motorcycle as a private mode to travel to markets, which is a very reasonable choice. Lastly, the estimated coefficient of "Percentage of pavement improvement," a mode-specific variable for the privately owned mode and the shared mode, was significantly positive. Therefore, the local population is more likely to choose the wheeled mode for visiting markets on an improved road because the smooth surface enables greater comfort and higher speeds. Note that the authors attempted to employ other variables related to bicycle and automobile ownership in the utility functions of each mode, but statistical significance was not found. The insignificance of bicycle ownership is probably because of its limited loading capacity, whereas that of automobiles may be due to congestion on roads near the market from newly induced traffic demand after the completion of the civil works. The model's adjusted ρ^2 indicates that it explains people's modal use for visiting their market. This is verified by the likelihood ratio test as the null hypothesis that all the coefficients are zero is statistically rejected.

Findings from Follow-up Interview Survey

Changes in Behavior of Local Population around Markets. First, this survey revealed that seven individuals commenced new business activities in their residence during or after the civil works, out of which five changed their primary income source from rice farming to shop operation. This is because they found growth in the number of road users who could be potential customers for a profitable business. All of them owned rice fields and continued producing rice for sale or their own consumption to secure their livelihoods. Furthermore, their business variety was noted to be limited to selling vegetables, snacks, bottled drinks or water, meat, fish, and cooked foods, probably because starting such a business might require only small initial investments rather than sophisticated technical skills and education. In fact, four shop owner respondents had completed only a primary education and one had never attended school, but they generated satisfactory profits. Our interviews found that successful entrepreneurship with a limited educational background might have discouraged their children from studying at higher education institutes. The children had interests in helping their HHs' business or starting new businesses on their own. This contrasts with the belief of interviewees who had grandchildren or children that higher education attainment could help their grandchildren or children to get employed in better-paid jobs than rice farming or contracted labor work. In this study, rural road improvements may have unexpected side effects on the capabilities of the next generation.

The interviews further found that 15 HHs obtained bicycles and 19 HHs obtained motorcycles during or after the road improvements, regardless of their income. 33 HHs purchased these modes using their personal savings, and one HH received financial support from relatives residing outside the area. It was found that the motivation to obtain private wheeled modes was not always related to a HH's economic activities, but for enhancing their HH members' travel efficiencies and easing their grandchildren and children's schooling. Therefore, these modes were purchased as

Table 4. Estimated Results of Travel Mode Choice Model when Visiting Markets after Road Improvements (n = 265)

Variable	Option	Coefficient	*t*-statistics
Distance to market most frequently visited ('000 meters)	S, P	0.135	2.144*
Daily income per capita in 2015 (KR '00000)	S, P	4.666	2.397*
Number of members in a HH	S, P	0.253	1.761.
Ownership of rice fields by HH (dummy)	S, P	1.362	2.516*
Ownership of motorcycle(s) by HH (dummy)	P	3.395	8.371***
Percentage of paved road access to market most frequently visited	S, P	1.409	1.761.
Constant	S	−3.305	−2.924**
Constant	P	−4.926	−4.176***
Initial log-likelihood (L0)	−291.132		
Final log-likelihood (LL)	−157.725		
−2(L0–LL)	266.814		
Akaike Information Criterion	331.450		
Adjusted ρ^2	0.431		

***: < 0.001; **: <0.01; * <0.05; .<0.1.
Note: S = shared mode; P = privately owned mode.

HH assets. Four HHs purchased automobiles by borrowing money from a nearby microfinance institute (one HH) and by using their own savings (three HHs). Consequently, those who own the new motorized mode(s) traveled to large markets and purchased a larger variety of items.

Finally, the interviews found that three HHs reported no change in their economic status. None of them owned land for rice farming or had completed primary education, and they earned income by engaging in daily labor or selling fish or animals. This finding reveals that the road improvements could lead to increases in income, particularly for owners of rice fields, the local dominant economic activity.

Changes in Economic Environments. First, the field interviews showed that shop owners in small markets regularly purchase items in bulk at larger markets on behalf of their customers, generally once a week, using their privately owned or shared modes, such as tuk-tuks and motorcycle taxis.

Second, 13 HHs remained engaged in rice farming as their primary income source, and all of them saw a growth in their 2015 incomes from selling rice. Ten HHs recognized increases in unit prices, two HHs increased the amount of rice they sold, and one HH found positive changes in both. Out of those HHs, five used to bring their produce to markets nearby and one HH used to bring their produce directly to customers' houses before the road improvements. However, they stopped this practice after the road improvements because their customers came to them, releasing them from financial and time burdens related to transporting their rice. These may be signs that the road improvements transformed the local market structure. Additionally, they were able to determine the prices of their rice by referring to the market price information rather than through individual negotiations, because owning mobile phones enabled them to obtain such information more easily from their business partners and relevant government institutions.

Third, after the road improvements, new rice mill operators emerged in all study areas, particularly mobile operators. One reason was that the improved road surfaces enabled people to physically drive trucks or vans loaded with rice mill machines. Another reason was the increase in the needs of rice farmers for polishing rice themselves before selling it. Doing the polishing themselves might allow them to sell their rice at a better price. However, one of the HHs was concerned about inflation in the prices of other products. In fact, the government's public market statistics on rice prices indicated no significant increases during 2014 and 2015 (*24*, *25*). This implies that the positive changes reported in the survey might be limited to their areas and their communities.

Fourth, the interviewees selling vegetables, fruit, and rice at large markets commonly confirmed significant increases in customers residing along the roads and the gradual emergence of competitive business environments after the road improvement. Further, the authors found that some traders in markets at road junctions with other national roads started direct transactions with traders from larger cities, such as Phnom Penh and provincial or district capitals. Therefore, the road improvements affected economic structures not only along the improved roads themselves but also beyond their immediate areas.

Fifth, increases in motorcycle taxis and tuk-tuk operators were observed along the roads, which was supported by another finding from two male interviewees who started motorcycle taxi businesses during and after the civil works and offered their services at more reasonable prices and with shortened travel times.

Sixth, the interviews also found a clear contrast in the significance of impacts on local socio-economic environments between Area TK, the wealthiest area with 13 markets, and the other two areas with fewer markets. This implies that the positive economic effects that emerged in individual markets are bolstered and extended over networks among markets along the roads, and these interactive effects could be observed in areas of high market densities.

Lastly, it was further found that many similar-style grocery shops had emerged along all the study areas, and they expected to face tough competition among themselves in the near future. To survive such eventualities, they should make immediate efforts to diversify their income sources, as reported by Yamauchi et al. (*26*).

Discussion

The follow-up survey revealed that the HHs' reasons for visiting markets after the rural road improvements were varied and mixed. For instance, some individuals who had recently opened shops visited markets mainly to purchase items to sell in their own businesses, whereas others visited for networking. Another possible reason is that mobile shops selling daily food appeared along all the roads, reducing people's need to visit markets to satisfy their daily needs, although the shops sometimes offered higher prices than those at the markets. Because the rural road improvements involved changes in people's decision-making, regarding their daily economic activities and use of time, local market functions could be continuously transformed beyond those of a trading activity platform.

Second, the survey suggested that the use of wheeled modes for visiting markets was not significantly changed, as shown in Tables 3 and 4, despite the increased ownership of private modes during and after the civil works. The road improvements led to an increase in the number of shared mode operators, since such services were found to be easy to start and flexible in responding to local people's needs for more multi-purposed trips around markets, as previously discussed (*27*). Considering the services' demand-responsive nature, the recent growth is likely to lead to reductions in user fees and enhance the local people's mobility and participation in interactions in markets.

Third, as for imbalances between farmers and traders in rural markets that we were originally aware of, findings from this study suggest that the rural road improvements made farmers more powerful as price setters than before, because those farmers could be approached by more traders and, using a newly owned mode or shared modes, could participate in larger markets and expand their business networks with potential partners within or outside their localities.

Lastly, the follow-up survey showed that the economic situation of low-income households or those with limited economic resources might worsen because of recent increases in product prices in their areas, resulting in increases in their daily expenditure despite their incomes remaining unchanged after rural roads improvements. Table 3 indicates that low-income people frequently visit markets and that they have the potential to become integrated into market-based economies by establishing personal networks. However, its sustainability is questionable if they continue to participate in the markets only as consumers. We now appreciate that the rural road improvements could have negative consequences for poorer individuals, making them more vulnerable than others.

Conclusions

This paper examined how rural road improvements impacted local people's economic activities around local markets using a case study on rural areas of Cambodia. The model estimation results indicated that the improvements more significantly enhanced the people's frequency of visits to markets and their use of bi-wheeled modes of transport, like bicycles or motorcycles. The follow-up interview survey revealed that the improvements enhanced individuals' incomes, particularly owners of rice fields, promoted economic activities along the roads, and expanded such activities around markets even beyond the roadside areas by fostering business relationships and stimulating entrepreneurship in people and potentially their next generations. These findings imply that rural road improvements are a contributive factor in regional development—one of the strategic aims of international aid communities in providing developing countries with continuous help to improve rural roads.

For the future, the structures of markets adjacent to improved rural roads and the profiles of traders from the outside need to be further studied because changes were acknowledged in local people's trading patterns and the prices of their farm produce after the road improvements. These studies would assist in predicting how local economies are transformed through future rural road improvement projects and would allow local beneficiaries to prepare themselves for their future economies. To better understand the impact of rural road improvement on regional economies, a comparative study across countries or regions of various geographical and socio-economic conditions could be proposed. Additionally, because this study found that people without rice farming assets failed to benefit significantly and remained poor, special instruments need to be implemented to integrate them into the local economy's development process. Possible instruments include: (i) access to local financial associations for starting new economic activities and maintaining a basic minimum income, and (ii) institutional support for their children to attend school at least until the end of their basic education to improve employment opportunities in non-farming jobs. Doing so may help to narrow economic disparities among the local population. Lastly, the models employed on people's behavioral patterns when visiting markets could be elaborated by performing diagnostic tests, which could enable us to understand the impact the improvements have on regional trade and traffic patterns on and beyond the roads. Quasi-experimental studies are also suggested, employing similar datasets to other areas where the improvements had not been implemented to examine more explicitly the impact of rural road improvements by controlling any possible exogenous factors, like inflation or policy instruments.

Acknowledgments

The survey in February and March 2016 was implemented by the Ministry of Rural Development, Kingdom of Cambodia, with the support of the Asian Development Bank. The authors sincerely thank them for allowing us to access their data. Furthermore, the authors appreciated the financial support from the Fuji Xerox Co., Ltd. Kobayashi Fund to conduct the follow-up study, as well as the survey members who offered extensive support.

Author Contributions

The authors confirm contribution to the paper as follows: study conception and design: Idei & Kato; data collection: Idei & Kato; analysis and interpretation of results: Idei & Kato; draft manuscript preparation: Idei & Kato. All authors reviewed the results and approved the final version of the manuscript.

References

1. Smith, D. R., A. Gordon, K. Meadows, and K. Zwick. Livelihood Diversification in Uganda: Patterns and Determinants of Change across Two Rural Districts. *Food Policy*, Vol. 26, No. 4, 2001, pp. 421–435.
2. Stifel, D., B. Minten, and B. Koro. *Economic Benefits and Returns to Rural Feeder Roads: Evidence from a Quasi-Experimental Setting in Ethiopia*. Addis Ababa: Ethiopia Strategy Support Program, International Food Policy Research Institute, 2012.
3. van de Walle, D. Impact Evaluation of Rural Road Projects. *Journal of Development Effectiveness*, Vol. 1, No. 1, 2009, pp. 15–36.
4. Abebe, G. K., J. Bijman, and A. Royer. Are Middlemen Facilitators or Barriers to Improve Smallholders' Welfare in Rural Economies? Empirical Evidence from Ethiopia. *Journal of Rural Studies*, Vol. 43, 2016, pp. 203–213.
5. Gibson, J., and S. Rozelle. Poverty and Access to Roads in Papua New Guinea. *Economic Development and Cultural Change*, Vol. 52, No. 1, 2003, pp. 159–185.
6. Mu, R., and D. van de Walle. Rural Roads and Local Market Development in Vietnam. *Journal of Development Studies, Vol.* 47, No. 5, 2011, pp. 709–734.
7. Cai, J., L. Ung, S. Setboonsarng, and P. Leung. *Rice Contract Farming in Cambodia: Empowering Farmers to Move Beyond the Contract Toward Independence*, 2008. https://www.adb.org/publications/rice-contract-farming-cambodia-empowering-farmers-move-beyond-contract-toward. Accessed July 25, 2017.
8. Thath, R. *Factors Affecting Cost Efficiency of Cambodian Rice Farming Households*, 2014. http://ir.nul.nagoya-u.ac.jp/jspui/handle/2237/20952. Accessed July 25, 2017.
9. Khandker, S. R., Z. Bakht, and G. B. Koolwal. The Poverty Impact of Rural Roads: Evidence from Bangladesh. *Economic Development and Cultural Change*, Vol. 57, No. 4, 2009, pp. 685–722.
10. Olsson, J. Improved Road Accessibility and Indirect Development Effects: Evidence from Rural Philippines. *Journal of Transport Geography*, Vol. 17, No. 6, 2009, pp. 476–483.
11. Wondemu, K. A., and J. Weiss. Rural Roads and Development: Evidence from Ethiopia. *European Journal of Transport and Infrastructure Research*, Vol. 12, No. 4, 2012, pp. 417–439.
12. Casaburi, L., R. Glennerster, and T. Suri. *Rural Roads and Intermediated Trade: Regression Discontinuity Evidence from Sierra Leone*, 2013. https://papers.ssrn.com/sol3/papers.cfm?abstract_id=2161643. Accessed July 25, 2017.
13. Rand, J. Evaluating the Employment-Generating Impact of Rural Roads in Nicaragua. *Journal of Development Effectiveness*, Vol. 3, No. 1, 2011, pp. 28–43.
14. Gachassin, M., B. Najman, and G. Raballand. *The Impact of Roads on Poverty Reduction: A Case Study of Cameroon*. Working Paper Series 5209. World Bank Policy Research, Washington, D.C., 2010. http://papers.ssrn.com/sol3/papers.cfm?abstract_id=1559726. Accessed July 25, 2017.
15. Asian Development Bank. *Report and Recommendation of the President to the Board of Directors: Proposed Loan for Kingdom of Cambodia for Rural Roads Improvement Project*. Asian Development Bank, Manila, 2010.
16. Ministry of Planning, Kingdom of Cambodia. *Based on Commune Data Base (CDB), Year 2004–2012 Selected to Present on Maps and Graphics*, Phnom Penh, 2012. http://www.mop.gov.kh/DocumentEN/Subnational_Poverty_Rate_2004-2012_Eng.pdf. Accessed July 25, 2017.
17. Ministry of Rural Development, Kingdom of Cambodia. *Completion Socio-economic Survey Report for Project Performance Monitoring and Evaluation*. Unpublished, 2016.
18. Carter, A. Mobile Communications in Cambodia, 2016. https://www.linkedin.com/pulse/mobile-communications-cambodia-anh-truong/. Accessed November 3, 2017.
19. Phong, K., L. Srou, and S. Javier. *Mobile Phones and Internet in Cambodia*, 2015. Asian Foundation, Phnom Penh. https://asiafoundation.org/resources/pdfs/MobilePhonesinCB2015.pdf. Accessed July 25, 2017.
20. Phong, K., L. Srou, and S. Javier. *Mobile Phones and Internet Use in Cambodia*, 2016. Asian Foundation, Phnom Penh. http://asiafoundation.org/publication/mobile-phones-internet-use-cambodia-2016/. Accessed July 25, 2017.
21. Kolnberger, T. Between Mobility and Immobility: Traffic and Public Space in Phnom Penh, Cambodia. *Pacific Geographies*, Vol. 37, No. 6, 2012, pp. 4–9.
22. CEIC Data. *Cambodia Household Income per Capita*. https://www.ceicdata.com/en/indicator/cambodia/annual-household-income-per-capita. Accessed November 7, 2017.
23. Singh, R., R. Ang, and J. A. Sy-Changco. Buying Less, More Often: An Evaluation of Sachet Marketing Strategy in an Emerging Market. *The Marketing Review*, Vol. 9, No. 1, 2009, pp. 3–17(15).
24. Food and Agriculture Organization. *Global Information and Early Warning System: Country Briefs. Cambodia*, 2016. http://www.fao.org/giews/countrybrief/country.jsp?code=KHM. Accessed July 9, 2017.
25. World Food Programme. *Cambodian Food Price and Wage Bulletin*, 2015. http://documents.wfp.org/stellent/groups/public/

documents/ena/wfp275279.pdf?_ga=2.265148870.899342172.1510120113-454732459.1509699495. Accessed November 8, 2017.

26Yamauchi, F., M. Muto, S. Chowdhury, R. Dewina, and S. Sumaryanto. Are Schooling and Roads Complementary? Evidence from Income Dynamics in Rural Indonesia. *World Development,* Vol. 39, No. 12, 2011, pp. 2232–2244.

27Lee, Y., M. Hickman, and S. Washington. Household Type and Structure, Time-Use Pattern, and Trip-Chaining. *Transportation Research Part A: Policy and Practice*, Vol. 41, No. 10, 2007, pp. 1004–1020.

The Standing Committee on Transportation and Economic Development (ADD10) peer-reviewed this paper (18-01155).

Article

TRR
JOURNAL OF THE TRANSPORTATION RESEARCH BOARD

Location Affordability and Regional Economic Resilience: Evidence from the U.S. Foreclosure Crisis

Transportation Research Record
2018, Vol. 2672(3) 37–45

DOI: 10.1177/0361198118777082
journals.sagepub.com/home/trr
SAGE

Ann M. Hartell[1]

Abstract

The form and function of cities is in large part defined by the interrelationship between land development and transportation systems. This relationship has important implications for households, including how it shapes economic costs for housing and transportation. Understanding these cost burdens is relevant for transportation agencies that are developing policies, plans, and programs with the intent of improving affordability. This paper presents an analysis of the relationship between high housing cost burdens, high levels of vehicle ownership, demographic characteristics, and mortgage foreclosure. Using data from before the recent U.S. economic crisis, the analysis estimates the effects of pre-crisis characteristics of census tracts and metropolitan regions on foreclosure rates. Multi-dimensional measures of urban form are included to evaluate the effect of metro-level characteristics. The models indicate high shares of Black homeowners, high shares of Hispanic homeowners, and lower income levels are associated with increased foreclosure. Also important is the share of households with high housing cost burdens. High shares of households with high vehicle ownership levels are significantly and positively associated with default and foreclosure. This suggests the costs associated with the ownership of multiple vehicles contributed to households' economic vulnerability, a vulnerability revealed by mortgage foreclosures. The results for the urban form factors are mixed. Although increased compactness of jobs and housing has a negative effect on foreclosure, increased development intensity across the entire metro region has a positive effect. These differential effects challenge broad-brush assertions of negative impacts of urban sprawl.

Policy makers, planners, and advocates have made an increasingly strong case for the need to link housing costs and transportation costs to define, understand, and address issues of social welfare and equity. This linked approach takes the position that joint consideration of these two, interdependent household cost burdens offers a more complete understanding of the economic implications of household location choices, spatial planning decisions, housing policy, and transportation investments.

The appeal of considering the two cost categories together stems from classic ideas of the trade-off between the costs of distance and the value of proximity, which are shaped by the characteristics of urban systems (*1*). The value of proximity to attractive destinations, such as employment centers, is capitalized into land (and housing) prices, while locations more distant have lower land costs to compensate for the increased costs incurred by more travel. Theoretically, this means housing close the central business district will be more costly; offset, however, by shorter trips and the potential for using less costly modes of transportation (transit, walking or cycling). Conversely, while housing in the suburbs or exurbs will be less costly, longer trip distances and few alternatives to the private car will increase transportation costs. Whether a given location is affordable depends on how great a share of household income is required to pay for the combined costs.

Location affordability is emerging as an issue of concern in current transportation planning practice; it appears in the goals and indicator frameworks of several regional comprehensive and transportation plans that describe goals and objectives to improve affordability by linking transportation interventions with land use development and affordable housing, by expanding transit systems, and by infrastructure investments for low-cost modes of walking and cycling. A common thread is providing alternatives to the high costs of owning and operating a private car (see e.g., *2–4*). A number of these plans cite the highly influential Housing + Transportation (H + T) Index, first published in 2006 by the Center for Neighborhood

[1]Institute for Multi-Level Governance and Development, Wirtschaftsuniversität Wien/Vienna University of Economics and Business, Vienna, Austria

Corresponding Author:
Address correspondence to Ann M. Hartell: ahartell@nas.edu

Technology, which brought location affordability into sharp focus (*5*).

This analysis probes the relationship between household cost burdens for transportation and housing and patterns of urban form using data from before and after the recent U.S. housing crisis. Previous work in this topic included a limited number of metros and categorical data on urban form (*6*); this study builds on that work by using a dataset of over 250 metros and more detailed, quantitative metrics of four dimensions of urban form. The current paper uses multi-level regression models to simultaneously estimate tract-level and metro-level effects. From this perspective, the results provide insights into household, neighborhood, and regional characteristics that contributed to or eroded the ability of households to weather the economic downturn without experiencing default or foreclosure. Specifically, the study takes up three research questions: What patterns in household cost burdens are associated with foreclosure? Do the effects of cost burdens persist when demographic characteristics are controlled for? Are characteristics of urban form significant factors for foreclosure? The answers to these questions offer insights for transportation policies, plans, and programs designed to address location affordability and strengthen economic resilience.

Background and Literature Review

In the United States, housing is generally considered affordable when housing costs account for 30% or less of household income. While there are some important critiques of the ratio approach (see e.g., *6*, *7*), the 30% ratio is widely used in planning and policy as a threshold for affordability. Thresholds for transportation affordability of 15%–20% of income have been proposed (*9*, *10*). The H + T Index defines location affordability as housing and transportation costs that, combined, do not exceed 45% of household income (*5*).

In the research literature, several studies have noted the differences in affordability patterns when location affordability is compared with housing affordability. For example, Mattingly and Morrissey (*11*) found when estimates of commuting costs were added to housing costs for neighborhoods in Auckland, NZ, the differences between total cost burdens for central locations compared to the urban fringe were greatly reduced. Acolin and Green (*12*) developed a location affordability measure for São Paulo and argue for its use in developing housing policy to better address social welfare concerns.

An active area of research is the effect of urban form—the joint configuration of land use and the transportation network—on location affordability. A substantial body of work has developed around the idea that compact, mixed-use development, with a multimodal transportation network will reduce household transportation cost burdens by shifting households away from auto dependence and toward public transit, walking, and cycling (see e.g., *13–15*). Such shifts in transportation mode are expected to reduce overall household transportation costs (*9*). Yet the empirical evidence is somewhat mixed on the effect of urban form on travel behavior (see e.g., *16–18*). A recent meta-regression consolidates estimates across studies of the effect of urban form on vehicle miles traveled (VMT); results suggest urban form does reduce driving although the effect is small (*19*).

Similarly, the effects of urban form on housing-related outcomes are also unclear. A policy evaluation notes that anti-sprawl policies such as constraining urban expansion, promoting high-density infill development, requiring higher design standards, and remediating brownfields can be expected to increase housing costs, especially for single-family dwelling units (*20*). This is aligned with Glaeser and Kahn's (*21*) work on the role of sprawl in making housing consumption more affordable and suburban locations feasible because of the use of private automobiles as the predominant mode of transport. In contrast, Gilderbloom et al. (*22*) found higher values of Walk Score™ (a proprietary indicator of proximity to amenities, density, and street grid density) reduced the number of foreclosure sales in Louisville, KY. Using data from investor mortgage defaults for multifamily housing, Pivo (*23*) found lower risk of default for properties located in census block groups with higher numbers of retail establishments, where residents had shorter commutes, or where high shares of workers commuted using a subway or elevated train. These effects are attributed to stronger cash flows and higher valuation for such properties. Explicitly taking up the question of different measures and the variability of effects in 29 cities, Kelly-Schwartz et al. (*24*) found no significant relationship between composite measures of sprawl and health outcomes. However, higher street network connectivity was associated with better health, while density alone had a negative effect on health.

The diversity of approaches to measuring urban form and sprawl no doubt contribute to the mixed results. Hamidi et al. (*25*) note that the definitions of sprawl vary to such an extent that they can result in completely contradictory rankings of cities by the degree of sprawl. Galster et al. (*26*) critique the anti-sprawl literature for a lack of clarity, with sprawl conceptualized as both a cause of and an effect on urban development. Cutsinger et al. (*27*) also make the case for greater rigor in conceptualizing and measuring sprawl. This has important implications beyond the academic literature because inconsistent and ad hoc approaches produce inconsistent results which cloud interpretations for practitioners and policy makers.

One proposed improvement is the adoption of multi-dimensional indices. Multi-dimensional indices are increasingly feasible given the vast amount of spatial data available enabling the assembly of detailed national-level datasets for quantitative studies. One example is the 2010 Sprawl Index, published by the advocacy organization Smart Growth America (*28*), which draws from the literature to select measures of development density, land use mix, centrality, and road network characteristics in a multi-dimensional index. Taking a more data-driven approach, Sarzynski, Galster, and Stack (*29*, *30*) use principal component analysis to develop measures of urban form. Other multi-dimensional indices are developed by Galster et al. (*26*), Cutsinger et al. (*27*), and Ewing et al. (*31*). While methods vary, there is a trend among scholars to take advantage of increasing data availability and a growing range of methods to construct measures that are promising for policy-relevant research applications.

Data and Method

In this study, the dependent variable is the estimated rate of mortgage foreclosures from the first round of the U.S. Department of Housing and Urban Development (HUD) Neighborhood Stabilization Program, (NSP1; *32*). NSP1 data were developed as part of HUD's efforts to support state and local agencies in targeting foreclosure relief programs. Because the intent was to evaluate risk and proactively identify troubled neighborhoods, HUD used data on mortgage defaults, actual foreclosure filings, plus additional factors known to contribute to foreclosure to estimate small-area foreclosure rates for 2007 and the first six months of 2008. These factors include the number of vacant addresses, the share of mortgage loans that are high-cost loans, housing price declines from the area price peak between 2000 and 2008, and the county unemployment rate in June 2008, plus a control for investor loans versus owner-occupied housing (for full documentation, see *33*). Thus the model casts a somewhat wider net than formal default filings by accounting for some important economic drivers of foreclosure.

Still, as estimates, these data may underestimate numbers of distressed households: they do not include information about households that were able to avoid full default through a short sale, renegotiation of a mortgage, changes in living arrangements (taking in family members or others, i.e., "doubling up"), or households that faced high cost burdens without approaching the extreme situation of mortgage default. Alternatively, these data may underestimate the number of homeowners who chose to default on a mortgage strategically, not because they could no longer afford their mortgage, but because it made more financial sense to abandon a property that had dramatically dropped in value. More generally, the use of estimated data introduces a degree of error, the nature and magnitude of which is unknown. Despite these caveats, the NSP1 data are an interesting resource for analyses of conditions associated with location unaffordability because of their comprehensive geographic coverage at fine resolution (see, e.g., *34* for a study of rent control policies using the NSP1 data).

For this study, predictor variables for demographic characteristics are tract-level data from the 2000 U.S. Census (*35*). In the analysis, variables are selected to describe basic neighborhood characteristics: homeowner median income, share of Black homeowners, and share of Hispanic/Latino homeowners. Cost variables are percent of homeowners with mortgages reporting housing costs of 30% or more of household income, and percent of homeowners with high levels of automobility (three or more vehicles available to a household). An interaction term, multiplying median income by the automobility variable is also included. Finally, two control variables are included: the number of housing units as a control for tract size, and the number of housing units occupied by homeowners as a control for the number of mortgages. The time lag between the predictor variables and the dependent variable (2000–2007/08) allows for the analysis of pre-crisis conditions, avoiding problems of reverse causality. The list of variables and descriptive statistics is provided in Table 1.

The emphasis on automobility (the use of and reliance on private autos) as an indicator of high transportation costs follows from the positive relationship between household transportation expenditures and auto ownership (*9*). In terms of absolute costs, the private automobile represents the highest-cost mode, although it does provide convenient mobility on a ubiquitous network. According to the Bureau of Labor Statistics Consumer Expenditure Survey (CES; *36*), the national average annual expenditure to own and operate a motor vehicle was $3,679 in 2000 and $4,045 in 2008, with minor variability by census region. Multiplying these values by three gives an estimate for national average cost for high automobility of $11,037 in 2000 and $12,137 in 2008, equivalent to nearly 25% of average income in 2000 and just over 19% in 2008. It is noteworthy that the CES reports average national income increased dramatically from 2000 to 2008, by just over 40%; yet the share of income expended on vehicles dropped by only about 6%, in large part because expenditures on gas and oil doubled. This highlights the vulnerability of households with high levels of automobility for whom volatility in the variable costs of driving erode income gains, with implications for economic resilience.

Although census data report the number of vehicles "available" to a household, which may not equate to vehicle ownership, it is likely that households with routine access to a vehicle do in fact own it, or at minimum make significant contribution to the costs of having access to and using it. The census does not collect information on transportation costs or impute transportation costs as a share of household income (as is done for housing cost burdens), yet high vehicle ownership levels represent sizeable expenditures for transportation that disproportionately increased during the study period compared with income, increasing this cost burden to households.

The characteristics of urban form are represented in the model by a set of variables measuring urban form characteristics developed by Sarzynski et al. (*29*). Sarzynski and colleagues compiled a set of 14 land use metrics for 257 U.S. metropolitan regions (U.S. Census Urbanized Areas (UAs)

Table 1. Variable Definitions and Descriptive Statistics (N = 39,274)

Tract-level variables	Min	Max	Mean	Median	SD
Foreclosure rate (estimated foreclosure rate as share of mortgages, logistic transformed; dependent variable)	−7.6623	−1.0986	−3.1265	−3.0243	0.8205
Untransformed foreclosure rate (for reference; not used in models)	0.0005	0.2500	0.0534	0.0463	0.0349
Owner occupied (% of occupied units that are owner occupied)	0.0017	1.0000	0.6352	0.6820	0.2398
Housing units (number of occupied housing units divided by 100)	0.0200	111.6800	16.5179	15.3700	8.0990
Income (median income for homeowners divided by 10,000)	0.0200	20.0000	5.7605	5.3750	2.2820
Hispanic (% owner-occupied housing units occupied by Hispanic or Latino owner-occupant household)	0.0000	1.0000	0.1015	0.0273	0.1808
Black (% owner-occupied housing units occupied by Black owner-occupant household)	0.0000	1.0000	0.1230	0.0169	0.2455
High housing cost (% of specified, owner-occupied housing units, with mortgages and where 30% or more of income spent on housing in 2000)	0.0000	1.0000	0.2151	0.2000	0.1082
High vehicle ownership (% of owner-occupied housing units with three or more vehicles in 2000)	0.0000	1.0000	0.2015	0.1998	0.1013
Income * vehicle (interaction variable: income * high vehicle ownership)	−4.5838	5.4935	0.0632	0.0188	0.2651
Metro-level variables					
Intensity (value of intensity dimension in 2000)	−2.1999	4.0765	–	–	–
Compact (value of compactness dimension in 2000)	−1.8526	4.7724	–	–	–
Mixed use (value of land use mix dimension in 2000)	−2.7681	5.2120	–	–	–
Core dominance (value of core-dominance dimension in 2000)	−2.4150	2.2713	–	–	–

Note: SD = standard deviation. Urban form variables are standardized (mean = 0, SD = 1).

plus any territories that are functionally dependent on an urban center; dependency defined as areas in which 30% or more of residents commute to the UA) for the year 2000. Using these scholars' data and replicating their method in Stata, a factor analysis reduces the 14 metrics to a solution retaining four factors (see *29* for full details), which provide an index of four common conceptualizations of urban sprawl:

- Intensity: housing density, job density, continuity of development within the metropolitan area, and share of jobs that need to move to create an even distribution of jobs across the metro (latter metric loads negatively).
- Compactness: the degree to which housing and jobs are concentrated toward the central core of the city and the density of development at the center compared to the periphery.
- Mixed use: the degree to which housing and jobs are integrated at a fine scale.
- Core-dominance: the ratio of jobs located in the central business district compared to jobs located in all other job concentrations.

These four factors explain 84% of the variation in their original data. Additionally there is little correlation among the factors, suggesting they measure distinct phenomena.

The scoring coefficients for each factor are then used to calculate a score for each city for each dimension. This analysis includes the 257 metros for which the urban form dimensions were calculated. The data include very large and relatively small Metropolitan Statistical Areas (MSAs) in all parts of the country. (A complete list of MSAs is available on request from the author.)

The unit of analysis is the 2000 census tract, with tracts selected by census boundaries for MSAs in 2000. This varies somewhat from the geography used in defining the urban form factors. Here, data limitations require the assumption that the amount and degree of change across a UA plus its functionally dependent area provides valid information about urban form characteristics and change for the territory of the census-defined MSA, which may be geographically larger or smaller. In cases where the MSA is larger, this means that areas are included in the model that may have different urban form characteristics than the territory used to calculate the four factors. For those metros where the MSA is smaller than the territory used to calculate the four factors, characteristics of the larger geography are considered to also apply to the smaller area covered by the MSA. In the analysis, the values of the dimensions of urban form serve as metro-level variables. This does not account for localized differences in development patterns or the transportation system, but rather focuses the analysis on the metro as an important context for household cost burdens and, therefore, on foreclosure rates.

Method

This study uses a series of multi-level regressions to investigate the relationship between demographic, household cost, and urban form variables and estimated foreclosure rates. The models are estimated using R (version 3.3.1), an open source statistical computing language, with the lme4 package (*37*) on the RStudio platform (version 0.99.903; *38*).

A multi-level modeling approach addresses both conceptual and technical considerations. An important motivation is the interest in the effects of metro-wide characteristics on a tract-level outcome. From a technical standpoint, because tracts in the same metro are subject to many of the same contextual effects, they are expected to have similar values, violating a basic assumption of linear models: independence of observations. One way to address this is to aggregate census tracts by metro, using metro-level averages for the demographic and cost-burden variables; this approach results in considerable loss of information about variation in conditions within metros. The alternative is multi-level (or hierarchical) models that retain both tract-level and metro-level variation while estimating the effects of both levels on the dependent variable. Multi-level models also partition the variance, allowing for an assessment of how well variables capture the effects on the outcome variables. See Raudenbush and Bryk (*39*) for full discussion of these models.

Because the dependent variable is expressed as a percentage with its range bounded at zero, it is logistically transformed to reduce distortions that can result from modeling bounded data in regression models. As a further adjustment, 87 tracts with outlier conditions were removed from the dataset; 13 tracts with fewer than 10 mortgages, and thus estimated foreclosure rates of zero, preliminary analysis revealed these tracts produced large errors. Additionally, 74 sparsely populated tracts had negative or zero values for median income. This is because the census allows reporting of self-employment income minus certain business expenses, and the reporting of income from estates or trusts, both of which can be negative or zero. Thus reported income for these tracts represents the value of a tax subsidy or an accounting loss rather than a meaningful value for household income.

Results

As a first step, the variation of foreclosure rates between metros is estimated (see left-hand columns, Table 2). This null model estimates a unique (random) intercept for each metro to describe the difference between the overall average foreclosure rate and how each metro differs from that average, without including any other predictors. The variation in foreclosure rates by metro accounts for 41% of the total variance in the data [0.2928/(0.2928 + 0.4201)], strongly indicating that important explanatory information about foreclosure is related to metro-level characteristics. These differences are undoubtedly more than characteristics of urban form, but also include economic factors, mortgage lending patterns, state foreclosure laws, and differences in the timing and severity of the downturn.

A second model adds the tract-level variables as fixed effects (a single coefficient is estimated for effects across all metros; see center columns, Table 2); metro-level estimates of random intercepts account for the unobserved metro characteristics, including the unique internal spatial patterns of the demographic and cost burden variables. Overall, the results for the tract-level fixed effects are as expected, and in line with the existing literature. Tract median income has a negative and significant effect on foreclosure rates. Shares of Black and of Hispanic homeowners have a positive effect on foreclosure. The control variable for the number of housing units has a small but negative effect, although the interpretation of this is unclear. The share of households with housing cost burdens greater than 30% of income is positive and significant, even when accounting for metro-level housing market characteristics (e.g., median housing prices) in the random intercepts. The high vehicle ownership variable is also positive and significant. Further, when this effect is combined with the additional positive effect of the interaction between vehicle ownership and income, the overall effect of high car ownership increases at an increasing rate.

Notably, adding the tract-level variables halves the residual variance compared to the null model. Although this second model accounts for differences in foreclosure rates among metros in the random intercept, this specification offers no substantive insights into these differences. Thus, the next step is to add the metro-level variables for urban form to explore the ability of these variables to further reduce the unexplained variation among metros by moderating the random intercepts for each metro.

The columns on the right-hand side of Table 2 present the results for a model with a random intercept for each metro, and fixed effects for tract-level characteristics as well as for metro-level urban form variables. In this specification, the urban form variables estimate the effect of these metro-level characteristics on the metro-level intercept, moderating it either up or down from the overall mean (reported as the fixed effect intercept). The variation in the random intercepts for the metros is reduced by 16% compared to the previous model [(0.2486 − 0.2078)/0.2486]. Altogether, the urban form variables explain 29% of the variation in foreclosure rates among metros [compare with the null model: (0.2928 − 0.2078)/0.2928]. As individual predictors, the urban form variables have differential effects. The largest effect is from the compactness dimension, which has a negative effect on foreclosure rates. In contrast, intensity has a positive effect, while the dimensions of core dominance and mixed use have lower or no significance.

Model selection criteria point to the improved fit from including the urban form variables: values for AIC, BIC, and log-likelihood are all reduced. The error term displays a somewhat wider dispersion for low foreclosure rates. Durbin Watson tests for serial dependence of the error indicate a degree of positive correlation in the residuals. The source of this remaining heteroskedasticity is uncertain, but may stem from the fact that the foreclosure data used are estimated. This may introduce bias into the models if there is some systematic bias between the estimated foreclosure rate and the

Table 2. Multi-Level Models (Dependent Variable: Foreclosure Rate; N = 39,274; Groups = 257)

	Null model		Tract-level fixed effects, random intercept		Tract- and metro-level fixed effects, random intercept	
Random effects	Variance (SD)	Share of variance	Variance (SD)	Change in variance	Variance (SD)	Change in variance
Metro	0.2928 (0.5411)	41%	0.2486 (0.4986)	−15% (vs null model)	0.2078 (0.4559)	−16% (vs tract-level fixed effects only)
Residual	0.4201 (0.6481)	59%	0.1870 (0.4325)	−56% (vs null model)	0.1870 (0.4325)	–
Fixed effects	Coefficient (SE)		Coefficient (SE)		Coefficient (SE)	
Intercept	−3.17260*** (0.03278)		−2.6250*** (0.0335)		−2.6240*** (0.0313)	
Tract-level variables						
Owner occupied	–		0.0025 (0.0120)		0.0024 (0.0120)	
Housing units	–		−0.0029*** (0.0003)		−0.0029*** (0.0003)	
Income	–		−0.1652*** (0.0013)		−0.1654*** (0.0013)	
Hispanic	–		0.7105*** (0.0181)		0.7088*** (0.0181)	
Black	–		0.9462*** (0.0111)		0.9455*** (0.0111)	
High housing cost	–		0.2226*** (0.0248)		0.2211*** (0.0248)	
High vehicle ownership	–		0.7269*** (0.0294)		0.7272*** (0.0293)	
Income vehicle	–		0.0707*** (0.0087)		0.0706*** (0.0087)	
Metro-level variables						
Intensity	–		–		0.0919** (0.0287)	
Compact	–		–		−0.1637*** (0.0291)	
Mixed use	–		–		−0.0477 (0.0287)	
Core dominance	–		–		−0.0573* (0.0287)	
Model comparison						
AIC	78421		46822		46784	
BIC	78446		46916		46913	
Log likelihood			−23400		−23377 (chi-sq. = 45.2 ***)	

Note: Although there is some disagreement on the appropriateness of p-values in connection with coefficient estimates in multi-level models, they are included here for their familiarity to readers.
SD = standard deviation; SE = standard error.
*** = 0.001. ** = 0.01. * = 0.05.

other variables used in this analysis, or if there is some relevant omitted variable for tract-level characteristics.

For the second and third models, estimates for the tract-level variables are largely unchanged and continue to model internal variation. Of particular interest are the results for the variable for the rate of high vehicle ownership, which remains positive and significant. Even when controlling for metro-level conditions that may influence auto dependence, variation in levels of high automobility are still significant for foreclosure. In interpreting this result, it is important to note that the high vehicle variable is not an aggregate or generalized rate of vehicle ownership. Rather, it is a measure of rates of high transportation cost levels, calculated as the share of owner-occupied households with three or more vehicles. Such households exist in all types of metros, even those with very low levels of sprawl. In addition, visual inspection of bivariate plots finds no evidence of correlation between the rate of high vehicle ownership and the four urban form variables.

As a further check for confounding effects, the model is repeated without the high vehicle ownership variable and without its interaction with income. This explores whether effect of high levels of vehicle ownership are simply strong proxies for an actual effect of sprawl, overpowering the metro-level effects in the models. The results indicate this is not the case: the estimates, signs, and significance for the four urban form variables change very little, while overall model fit is slightly reduced. Comparing the variances, the high vehicle variable explains 1.8% of the residual variance of foreclosure rates after accounting for all metro-level characteristics in the intercepts and the urban form variables. For comparison, a similar model is estimated without the high housing cost burden variable, which is found to similarly reduce the residual variance by 1.7%. Thus a tract-level increase in high levels of high vehicle ownership has a positive effect across a wide range of metro types, independent of the contextual effects of urban form, and the explanatory power of high vehicle ownership levels is comparable to that of high housing cost burdens. (Results of these additional models are available on request.)

Discussion and Conclusion

The recent foreclosure crisis presents an opportunity to investigate economic resilience through the lens of location affordability by examining conditions and characteristics associated with mortgage default and foreclosure. The results for the effects of demographic characteristics are as expected: higher median income reduces foreclosure, while increasing shares of Black and Hispanic homeowners are associated with higher foreclosure rates. The share of households with high housing cost burdens is also associated with increased foreclosure rates. Here, housing cost burdens are defined according to the common policy threshold of 30% of household income. Repeating the analysis using a higher threshold for affordability, at say 40% or even 50% of income, may reveal further information about the relative influences of demographics, cost burdens, and urban form.

Perhaps the most intriguing result is the significant and positive relationship between high levels of automobility and estimated foreclosure rates. This quantified link between high transportation cost and a negative housing outcome holds across a range of metro types. Notably, tract-level variation in rates of high vehicle ownership is not collinear with the urban form variables. The robust relationship between levels of high vehicle ownership and a housing outcome lends support for the concept of location affordability. More specifically, the results suggest value in policies, planning, and programs to promote reduced rates of high vehicle ownership.

While transit service is frequently suggested as the remedy to auto dependence, emerging alternatives such as car sharing or ride sharing could be even more promising, allowing households to replace one or more owned vehicles with access to a vehicle on a flexible, as-needed basis thus reducing fixed costs. Changes in communication technology and widespread wireless communications can increase the potential and adoption rate for these on-demand mobility strategies, deploying them in a targeted way within a metro. Such alternatives may be feasible even in places where metro-level urban form limits the feasibility of transit or where pockets of high auto dependence exist in otherwise transit-supportive metros. Additional disincentives for high automobility could include reduced parking requirements for housing developments, on-street parking restrictions, or higher road/registration taxes for high-car households. From the mortgage financing side, the results lend support for mortgage lending tools such as the Location Efficient Mortgage, first developed by the Center for Neighborhood Technology, which increased the amount a household could borrow for homes purchased in locations that were pedestrian friendly and transit accessible, where households were expected to have reduced transportation costs (*38*). Note, however, that high levels of automobility may simply serve as an indicator for other patterns of household expenditures, such as high levels of consumer spending (and perhaps debt) in general, which increase the risk of foreclosure. If this is the case, this variable can still serve as an important proxy to help identify neighborhoods at risk, especially in an economic downturn.

In addition to these tract-level factors, this analysis included four measures of urban form as contextual variables expected to shape household cost burdens. Although the factors involved in the recent foreclosure crisis are complex and multifaceted, the variables provide insights into the relationship between a range of urban form characteristics and their differential effects on foreclosure. The models reveal compactness is associated with reduced foreclosure, while overall intensity is associated with increased foreclosure. Further, while increasing concentration of jobs and housing towards the central core (part of the compactness dimension) may support location affordability, increasing density across the entire urban system is expected to have the opposite effect. Two other dimensions of urban form, the degree of core-dominance in job distribution and the level of mixed use are, respectively, less significant and not significant for foreclosure. Certainly, increases along all dimensions may be desirable for various economic, social, or environmental reasons. The results of this study, however, indicate trajectories of increasing intensity should be coupled with interventions to reduce negative implications for affordability. Overall, these results do not support arguments of a general link between sprawl and location unaffordability.

Alternatively, the mixed results for the urban form variables may be viewed as a consequence of their not fully capturing and quantifying the characteristics of urban form that are relevant for modeling location affordability, despite their high level of detail and comprehensive geographic coverage. Or perhaps the effects of urban form on household cost burdens are so attenuated that they are not linked. Indeed, the complex and dynamic nature of metros, shaped by their unique historical, political, and economic pathways, poses a challenge to developing measures of urban form suitable for quantitative analyses. There are, however, several ways the urban form variables could be augmented to extend the current work. The focus on land use patterns leaves characteristics of the transportation system in the metro-level random intercepts with the rest of the measured differences among metros. Therefore, adding information on the characteristics of the transportation system at the metro level (such as transit accessibility) or tract-level patterns of travel behavior (such as transit use) might capture more information about factors that shape transportation cost burdens, especially for households below the three-vehicle threshold. More detailed information on the absolute numbers and types of jobs and housing units in addition to the pattern of their distribution is another potential improvement. Models that use the 14 individual metrics rather than the four composite dimensions might yield additional insights, although at the expense of model tractability.

Despite these limitations, this study finds evidence of relationships between high transportation costs, urban form, and mortgage default and foreclosure, suggesting a need for coordination across the land use and transportation sectors. The results also caution against broad-brush, anti-sprawl generalizations. Instead, effective strategies for reducing the joint cost burdens of housing and transportation will require consideration of metro-specific conditions, including cost burden patterns and land development characteristics. Finally, the relationship between high vehicle ownership and foreclosure supports the concept of location affordability and indicates the transportation sector has an important role in this emerging area of policy and practice.

Acknowledgments

The author thanks Dr. Andrea Sarzynski, University of Delaware, for generously providing the urban form data for this study. Dr. Gunther Maier and Dr. Thomas Rusch, of WU/Vienna University of Economics and Business, offered important insights on methods.

References

1. Alonso, W. *Location and Land Use: Toward a General Theory of Land Rent*. Harvard University Press, Cambridge, MA, 1964.
2. Chicago Metropolitan Agency for Planning. *Go To 2040. Comprehensive Regional Plan*. Chicago Metropolitan Agency for Planning, Chicago, IL, 2010.
3. Metropolitan Transportation Commission. *Plan Bay Area. Final Performance Assessment Report*. San Francisco, CA, 2013.
4. San Diego Association of Governments. *San Diego Forward*. San Diego, CA, 2015.
5. Center for Neighborhood Technology. *Housing + Transportation Index*. http://htaindex.cnt.org. Accessed July 18, 2016.
6. Hartell, A. M. Evaluating the Concept of Location Affordability: Recent Data on the Relationship Between Transportation, Housing, and Urban Form. *Housing Policy Debate*, Vol. 27, No. 3, 2016, pp. 356–371.
7. Stone, M. E. What Is Housing Affordability? The Case for the Residual Income Approach. *Housing Policy Debate*, Vol. 17, No. 1, 2006, pp. 151–184.
8. Chaplin, R., and A. Freeman. Towards an Accurate Description of Affordability. *Urban Studies*, Vol. 36, 1999, pp. 1949–1957.
9. Litman, T. *Transportation Affordability*. Victoria Transport Policy Institute, Victoria, BC, Canada, 2016.
10. Lucas, K., G. Marsden, M. Brooks, and M. Kimble. An Assessment and Critique of Capabilities for Examining the Long-Term Social Sustainability of Transport and Land-Use Strategies. Presented at 86th Annual Meeting of the Transportation Research Board, Washington, D.C., 2007.
11. Mattingly, K., and J. Morrissey. Housing and Transport Expenditure: Socio-Spatial Indicators of Affordability in Auckland. *Cities*, Vol. 38, 2014, pp. 69–83.
12. Acolin, A., and R. K. Green. Measuring Housing Affordability in Sao Paulo Metropolitan Region: Incorporating Location. *Cities*, Vol. 62, 2017, pp. 41–49.
13. Smith, K. R., B. B. Brown, I. Yamada, L. Kowaleski-Jones, C. D. Zick, and J. X. Fan. Walkability and Body Mass Index: Density, Design, and New Diversity Measures. *American Journal of Preventive Medicine*, Vol. 35, No. 3, 2008, pp. 237–244.
14. Ewing, R., and R. Cervero. "Does Compact Development Make People Drive Less ?" The Answer Is Yes. *Journal of the American Planning Association*, Vol. 4363, 2017, pp. 19–25.
15. Khan, M., D. M. Kockelman, and X. Xiong. Models for Anticipating Non-Motorized Travel Choices, and the Role of the Built Environment. *Transport Policy*, Vol. 35, 2014, pp. 117–126.
16. Khattak, A. J., and D. Rodriguez. Travel Behavior in Neo-Traditional Neighborhood Developments: A Case Study in USA. *Transportation Research Part A: Policy and Practice*, Vol. 39, No. 6, 2005, pp. 481–500.
17. Sarzynski, A., H. L. Wolman, G. Galster, and R. Hanson. Testing the Conventional Wisdom About Land Use and Traffic Congestion: The More We Sprawl, the Less We Move? *Urban Studies*, Vol. 43, No. 3, 2006, pp. 601–626.

18. Eid, F., H. G. Overman, D. Puga, and M. A. Turner. Fat City: Questioning the Relationship Between Urban Sprawl and Obesity. *Journal of Urban Economics*, Vol. 63, No. 2, 2008, pp. 385–404.
19.Stevens, M. R. Does Compact Development Make People Drive Less? *Journal of the American Planning Association*, Vol. 4363, 2017, pp. 6–18.
20. Litman, T. *Analysis of Public Policies That Unintentionally Encourage and Subsidize Urban Sprawl*. Report for the New Climate Economy Cities Research Programme. http://static.newclimateeconomy.report/wp-content/uploads/2015/03/public-policies-encourage-sprawl-nce-report.pdf. Accessed July 20, 2017.
21. Glaeser, E. L., and M. E. Kahn. Sprawl and Urban Growth. In *Handbook of Regional and Urban Economics*, Vol. 4, 2004, pp. 2481–2527.
22.Gilderbloom, J. I., W. W. Riggs, and W. L. Meares. Does Walkability Matter? An Examination of Walkability's Impact on Housing Values, Foreclosures and Crime. *Cities*, Vol. 42, 2015, pp. 13–24.
23. Pivo, G. The Effect of Transportation, Location, and Affordability Related Sustainability Features on Mortgage Default Prediction and Risk in Multifamily Rental Housing. *Journal of Sustainable Real Estate*, Vol. 5, No. 1, 2013, pp. 50–67.
24Kelly-Schwartz, A. C., J. Stockard, S. Doyle, and M. Schlossberg. Is Sprawl Unhealthy? A Multi-Level Analysis of the Relationship of Metropolitan Sprawl to the Health of Individuals. *Journal of Planning Education and Research*, Vol. 24, No. 2, 2004, pp. 184–196.
25.Hamidi, S., R. Ewing, I. Preuss, and A. Dodds. Measuring Sprawl and Its Impacts: An Update. *Journal of Planning Education and Research*, Vol. 35, No. 1, 2015, pp. 35–50.
26.Galster, G., R. Hanson, M. R. Ratcliffe, H. Wolman, S. Coleman, and J. Freihage. Wrestling Sprawl to the Ground: Defining and Measuring an Elusive Concept. *Housing Policy Debate*, Vol. 12, No. 4, 2001, pp. 681–717.
27. Cutsinger, J., G. Galster, H. Wolman, R. Hanson, and D. Towns. Verifying the Multi-Dimensional Nature of Metropolitan Land Use: Advancing the Understanding and Measurement of Sprawl. *Journal of Urban Affairs*, Vol. 27, No. 3, 2005, pp. 235–259.
28. Smart Growth America. *Measuring Sprawl 2014*. http://www.smartgrowthamerica.org/measuring-sprawl. Accessed July 27, 2015.
29. Sarzynski, A., G. Galster, and L. Stack. Typologies of Sprawl: Investigating United States Metropolitan Land Use Patterns. *Urban Geography*, Vol. 35, No. 1, 2014, pp. 37–41.
30. Sarzynski, A., G. Galster, and L. Stack. Evolving United States Metropolitan Land Use Patterns. *Urban Geography*, Vol. 35, No. 1, 2014, pp. 25–47.
31. Ewing, R., R. Pendall, and D. Chen. Measuring Sprawl and Its Transportation Impacts. *Transportation Research Record: Journal of the Transportation Research Board*, 2003. 1831: 175–183.
32. U.S. Department of Housing and Urban Development. *HUD Provided Local Level Data*. http://www.huduser.org/portal/datasets/nsp_foreclosure_data.html. Accessed April 7, 2015.
33. U.S. Department of Housing and Urban Development. *Neighborhood Stabilization Program: Methodology and Data Dictionary for HUD Provided Data*. http://www.huduser.org/portal/datasets/nsp_foreclosure_data.html. Accessed April 28, 2013.
34. Ambrosius, J. D., J. I. Gilderbloom, W. J. Steele, W. L. Meares, and D. Keating. Forty Years of Rent Control: Reexamining New Jersey's Moderate Local Policies After the Great Recession. *Cities*, Vol. 49, 2015, pp. 121–133.
35. U.S. Census Bureau. *Census 2000 Gateway*. http://www.census.gov/main/www/cen2000.html. Accessed June 10, 2013.
36.Bureau of Labor Statistics. *Table 8. Region of Residence: Average Annual Expenditures and Characteristics, Consumer Expenditure Survey, 2000*. http://www.bls.gov/cex/csxreport.htm#annual. Accessed July 11, 2013.
37.Bates, D., M. Maechler, B. Bolker, and S. Walker. Fitting Linear Mixed-Effects Models Using lme4. *Journal of Statistical Software*, Vol. 67, No. 1, 2015, pp. 1–48.
38Center for Neighborhood Technology. *Location Efficient Mortgages*. http://www.cnt.org/projects/rethinking-mortgages. Accessed November 5, 2017.
39.Raudenbush, S. W., and A. S. Bryk. *Hierarchical Linear Models. Applications and Data Analysis Methods*, 2nd ed. Sage, Thousand Oaks, p. 512.

The Standing Committee on Transportation and Economic Development (ADD10) peer-reviewed this paper (18-02714).

The views in this paper are those of the author and do not reflect those of the National Cooperative Highway Research Program or the Transportation Research Board.

Article

TRR
JOURNAL OF THE TRANSPORTATION RESEARCH BOARD

Transportation Research Record
2018, Vol. 2672(3) 46–55

DOI: 10.1177/0361198118788191
journals.sagepub.com/home/trr
SAGE

Facilitating Smart Growth with Context-Sensitive Transportation Review of Site Development

Daniel Hardy[1], Eric Graye[2], and Caleb Van Nostrand[1]

Abstract

In 2016, Montgomery County, Maryland revamped the approach with which the planning department reviews transportation impacts of site development. The development of the 2017 *Local Area Transportation Review Guidelines* included a comprehensive and collaborative review of alternative approaches to growth management and resulted in the development of more robust and multimodal analytic approaches, a set of context-sensitive and multimodal trip generation rates, and a commitment to continue the development of pro-rata share districts in lieu of traditional traffic impact studies. The 2017 guidelines were developed with the benefit of a review of contemporary best practices nationwide and a 30-person working group representing various transportation impact review stakeholder interests that convened 21 times over the course of 3 years. The new approach was developed to balance three potentially competing objectives: streamline the process in smart growth areas of the county, increase the degree of multimodalism, and increase the degree of robustness in the analytic approach and data. The key additions to the process included establishing policy area groupings based on transportation demand characteristics; developing a set of context-sensitive, multimodal trip generation rates; and introducing quantitative tests for non-auto impacts. This paper highlights both the process and outcomes of the updated approach.

The Montgomery County Planning Board maintains a set of local area transportation review (LATR) guidelines (*1*) to be used for preparation and review of transportation impact studies for development. The document is used by transportation engineers, planners, public agency reviewers, and community members participating in the development review process.

In late 2014, the Montgomery County Planning Board, at the direction of the Montgomery County Council, initiated a comprehensive review of their processes for reviewing the transportation impacts of site development. The review had three primary objectives:

- Streamline the development review process to increase predictability (for both applicants and community and business stakeholders) in "smart growth" areas of the county where economic development is most efficiently supported by prior investments in the transportation system, including but not limited to the regional METRORail system.
- Increase the degree by which multimodal impacts are assessed quantitatively.
- Make analytic approaches more robust, reflecting the capabilities of improved analytic tools and data resources.

There is a tension between the three objectives; the first seeks to simplify and improve legibility for the most urban locations in the county where transportation systems are the most complicated. The other two objectives seek to reflect the urban complexity through more intensive quantitative processes and tools. This report describes how the changes to the County's development review approach met these objectives, through examination of the project context, approach, and key outcomes.

Project Context

Montgomery County is an urbanizing jurisdiction adjacent to Washington, DC with a long history of applying progressive growth-management strategies. Both the definition and achievement of transportation system adequacy are topics of almost continuous debate among county residents and policy makers. The definition of transportation system adequacy for new development discussion has become increasingly important as the county matures from decades of active greenfield development to an increased reliance on infill development and redevelopment prompted by a scarcity of

[1]Renaissance Planning, Arlington, VA
[2]Montgomery County Planning Department, Functional Planning and Policy Division, Silver Spring, MD

Corresponding Author:
Address correspondence to Daniel Hardy: dhardy@citiesthatwork.com

buildable raw land. This article describes the growth-management changes enacted by the county council in November 2016 after a 2-year study of different options.

Geographic Context

During the past four decades, as Montgomery County repositioned itself from a bedroom community of commuters to a regional job center, change in the county was marked by its population growth. Now, this diverse, populous jurisdiction has settled into a mature growth stage. The county has an annual population growth rate of around 1%, which is expected to slow over the next 30 years. It is important to note, however, that 1% growth still equates to almost 200,000 new residents by 2045 (*2*).

Policy Context

The Subdivision Staging Policy (SSP) is a set of policy tools that guide the timely delivery of public facilities (schools, transportation, water, sewer, and other infrastructure) to serve existing and future development. These policy tools are the guidelines for the administration of the Adequate Public Facility Ordinance (APFO). These tools are reviewed and updates considered every 4 years.

Although commonly referred to as a separate ordinance, the APFO is actually part of Montgomery County's subdivision regulations: Section 50-35 (k) of the county code. The introductory sentence of the APFO states, "A preliminary plan of subdivision must not be approved unless the Planning Board determines that public facilities will be adequate to support and service the area of the proposed subdivision" (*3*).

At the start of the 2016–2020 SSP review, there were four means by which the development approval process affects the provision of transportation capacity, described below from the broadest to the narrowest focus:

- The transportation impact tax assesses the degree to which all development contributes to funding the provision of significant master-planned transportation projects that the County is responsible for constructing. The impact tax, governed by Section 52 of the county code, is not technically part of the SSP, but it is integral to the consideration of transportation impacts. The Planning Board Draft recommends changes to the impact tax that would be implemented concurrently with the SSP revisions.
- The policy area review process, called Transportation Policy Area Review (TPAR), assesses the degree to which travel conditions in the development site's policy area are adequate from an aggregate perspective.
- The LATR process assesses the degree to which travel conditions in the immediate vicinity of the development site are adequate, in which the vicinity of the site is determined by the size of the project.
- Finally, many site development approval conditions related to transportation are derived from other elements of the regulatory process, notably site design, access, and circulation, based on design standards that are independent of the SSP.

Study Working Group and Literature Review

The project benefited from the participation of a 30-person working group that met 21 times between fall of 2014 and spring of 2017. The group consisted of representatives of state, regional, and local agencies involved in the review of transportation impact studies and delivery of public transportation projects and services; representatives of the development community; and representatives of civic groups.

The project team conducted an extensive literature review of the state of the practice in growth management and transportation impact study approaches from a variety of jurisdictions to identify transferable concepts applicable to Montgomery County. The literature review examined concepts either adopted or under consideration in some 30 jurisdictions nationwide.

The key findings from the literature review were that many jurisdictions are, like Montgomery County, seeking more multimodal approaches to site impact mitigation, yet most retain an autocentric analytic approach to defining adequacy or concurrency.

The most promising ideas from the literature review were assembled into an evolving list of concepts that were examined by the working group. Many concepts were incorporated into the *2017 LATR Guidelines* (*4*) and the references for these concepts are incorporated in subsequent sections of this paper.

Several other concepts were considered by the working group, the Planning Board, or the county council and not incorporated for several reasons as noted below:

- The use of vehicle miles traveled (VMT) as a transportation system adequacy metric was discussed at length, particularly given the focus on VMT in California's Senate Bill (SB) 743. Ultimately, while the working group members were nearly unanimous on the value of VMT as a planning measure of effectiveness, they were almost nearly as opposed to its use as an adequacy format, due to the fact that there are no accepted measures of VMT impact in absolute terms. The application of VMT in relative terms, as promoted by the California Office of Planning and Policy (*5*), was also found to create concerns in relation to artificial boundary conditions in defining average rates.
- In response to interest in SB 743, the project team developed a "very low VMT" approach based on an analysis of a 200-unit hypothetical residential development in the Bethesda Central Business District (CBD), where the jobs/housing ratio is roughly 4:1,

limiting the ability of office workers to live near their workplace; the typical Bethesda CBD employee generates about 20 VMT per weekday (including those who do not drive). The hypothetical new development might be expected to generate 3,400 VMT, but if the prevailing 16% local-employment rate applied, then 32 of those units would generate short trips to work and replace employees making longer trips, reducing regional VMT by 650. Therefore, if the parking spaces were limited to less than 20% of the dwelling unit (DU) count, the building might be considered a zero-VMT development for policy purposes. This concept was not implemented in part due to considered duplication with the proposed pro-rata share approach in red policy areas to be developed over time. One aspect of the concept was retained; if a proposed residential development contains ancillary retail (less than 10% of the building square footage and also less than 15,000 gross square feet [GSF] total) and provides no on-site parking for that retail, then vehicle trip generation does not need to be considered for the retail use.

- The concept of "protected intersections," for which vehicular capacity cannot be increased by policy due to the adverse effect on the pedestrian environment, was proposed for certain CBD areas based on the policy in effect in San Jose, CA (*6*). As with the very low VMT approach, this concept was felt to be duplicative of the proposed approach to implement pro-rata share districts in the most urban areas of the county over time.

Key Project Outcomes

The *2017 LATR Guidelines* implement the more context-sensitive and multimodal procedures and analysis methods reflected in the 2016–2020 Subdivision Staging Policy as they relate to the determination of adequacy of local intersection performance in the context of the development review process. The guidelines contain many new ideas that essentially rethink how the County approaches the evaluation of local transportation system performance. The following key changes are reflected in this document.

- Recognizing that there is not a "one size fits all" set of rules that applies countywide but rather that the expectations for transportation system adequacy and the types of appropriate mitigation need to be applied in a context-sensitive manner.
- Organizing policy areas into four groups (i.e., red, orange, green, and yellow) that recognize current land use patterns, the prevalence of modes of travel other than the single occupant vehicle, and the planning vision for different parts of Montgomery County.
- Updating vehicle trip generation rates and developing person-trip generation rates that reflect the diversity of land use patterns and travel behavior across the county.
- Creating the ability to adjust trip generation rates based on reduced parking where such reductions are supported by the zoning code.
- Streamlining the approach by eliminating the separate TPAR process and adjusting impact taxes commensurately to replace the expected revenue loss.
- Establishing a new system for evaluating local area transportation conditions that emphasizes the application of delay-based measures reflecting the experience of travelers, rather than focusing on critical lane volume.
- Expanding LATR to include a set of multimodal (i.e., bicycle, pedestrian, and transit) transportation adequacy tests beyond those that focus on motor vehicle travel.

The following paragraphs describe the approach to implementing these key changes.

Establishing Policy Area Contexts

Montgomery County established a series of transportation policy areas in the 1980s to recognize that transportation needs for the more urban, transit-oriented districts differed substantially from the suburban and rural areas of the county. These policy area definitions have evolved over time; there are currently 38 separate policy areas (including two for the independent cities of Rockville and Gaithersburg for which the Montgomery County Planning Board does not have planning and zoning authority, but for which collaborative planning and zoning coordination occurs). One primary use for the policy area definition has been to establish different congestion standards for different areas of the county. The *2017 LATR Guidelines* refined the policy area definitions to reflect four different contexts as indicated by travel behavior metrics.

The County has long used the terminology "metro station policy areas" for the most densely developed business districts served by the regional METRORail system. For many constituents, metro station policy areas represented the areas of the county where development should be focused and multimodal transportation system measures applied. That perception often complicated discussions of development and transportation system objectives throughout the county. Communication with both stakeholders and decision makers about the continuum of place types was therefore a critical element of study success. Figure 1 summarizes the development densities (in discrete vertical bars) and journey-to-work (JTW) non-auto driver mode share (NADMS) percentages (represented by the continuously decreasing line) for the County's policy areas. The changes in development definitions reinforce the common perception that the metro station policy areas have far greater densities than most other areas in the county. But the NADMS curve demonstrated that residents of even lower density areas benefit significantly from

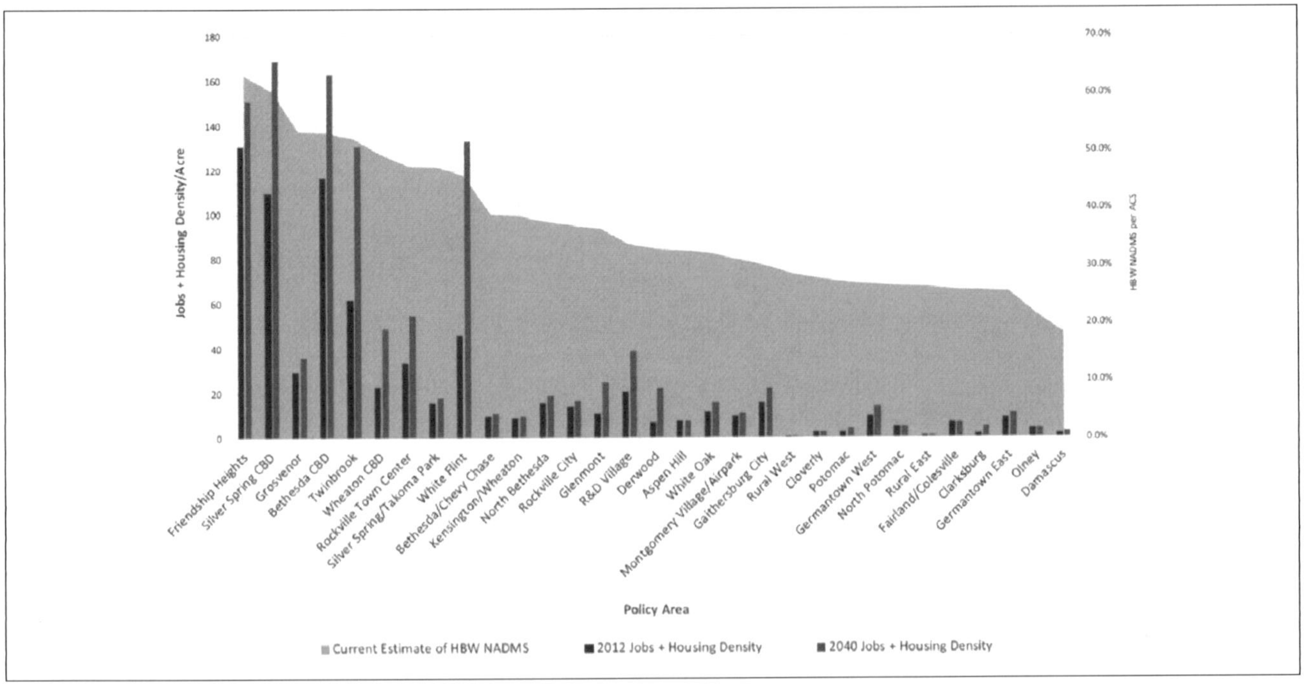

Figure 1. Continuum of places: development density and mode share by policy area.

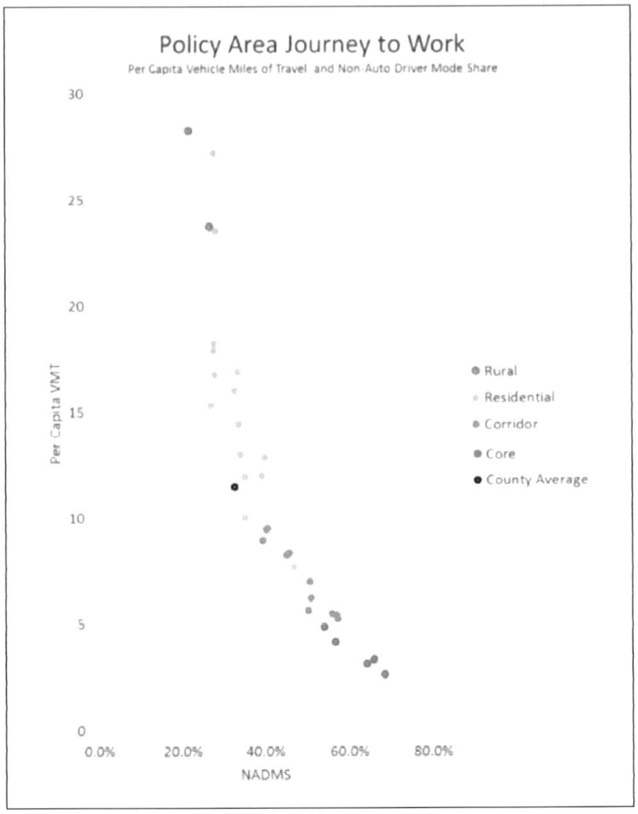

Figure 2. Relationship between journey-to-work mode share and VMT.

multimodal transportation services and that there is a far smoother continuum of placetype from a transportation behavior perspective than from a land use perspective. This graphic became an iconic reference point during community and stakeholder discussions.

Using results derived from the application of the planning department's regional travel demand model utilizing Metropolitan Washington Council of Governments year 2040 Round 8.3 Cooperative Forecast information, Montgomery County's policy areas were categorized by a combination of home-based work VMT per household (VMT/HH) and home-based work NADMS.

Figure 2 shows the clear inverse relationship between VMT/HH and NADMS (i.e., NADMS increases as VMT/HH decreases). Using the combination of a visual interpretation of this relationship to find logical cut-off points for each SSP and some local and professional judgment, each policy area was categorized into four SSP areas that are represented by colors ranging from red (most dense/urban) to green (least dense/urban). Figure 2 shows an example in which colored dots were associated with descriptive labels (i.e., "core," "corridor," "residential," and "rural").

Many constituents were concerned that any policy area labels such as shown in Figure 2 reflected judgment calls about the place they lived or worked. After several public discussions about alternative labels, the Planning Board decided to simply retain the color-coded descriptions. Figure 3 shows the final organization of the county's 38 policy areas into four

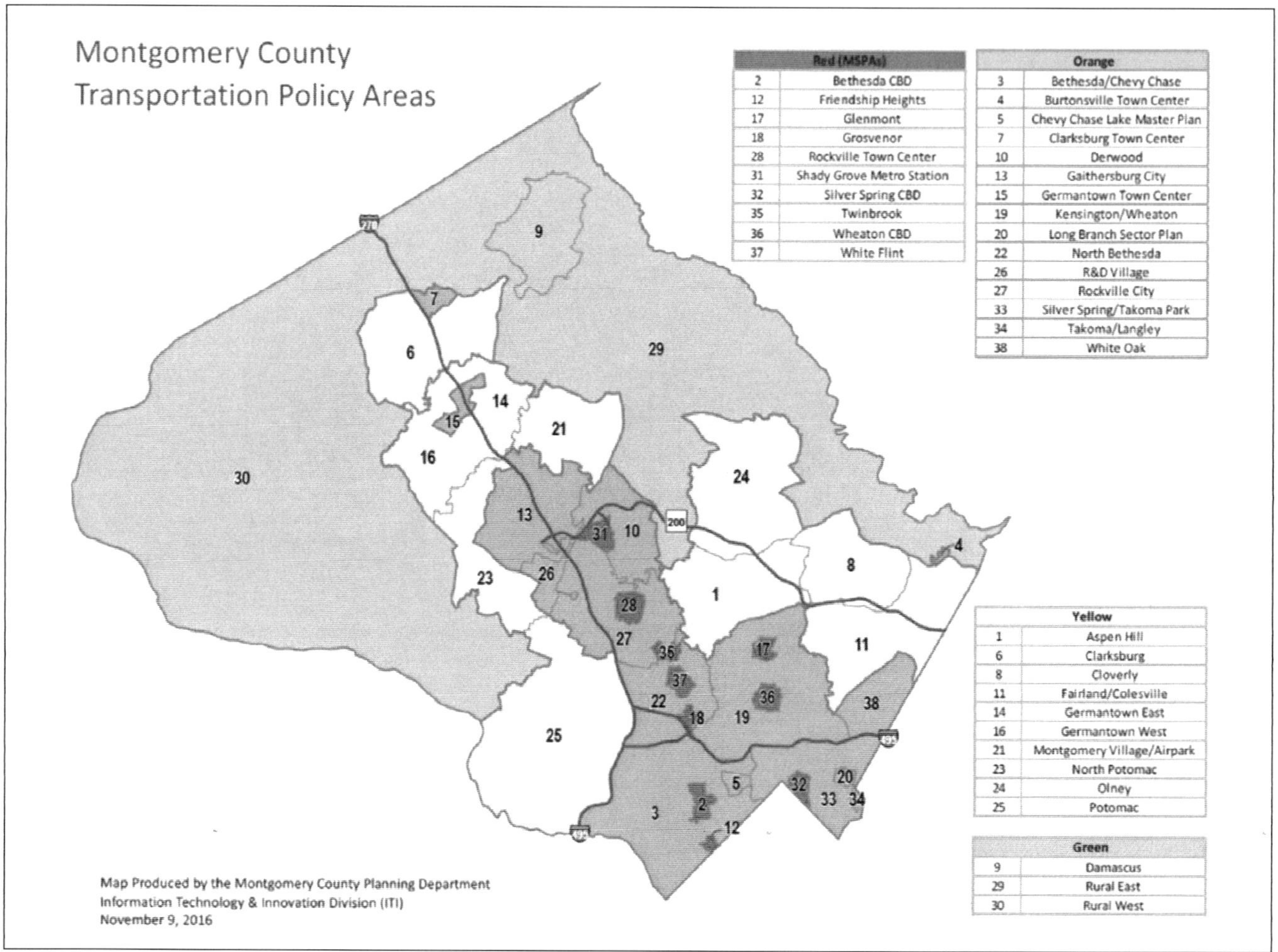

Figure 3. Subdivision staging policy areas.

categories with their color-coded descriptions. Red policy areas are the down county CBDs and metro station policy areas, which are characterized by high-density development and the availability of premium transit service (i.e., METRORail or MARC). Orange policy areas are the corridor cities, town centers, and emerging transit-oriented development (TOD) areas in which premium transit service (i.e., Corridor Cities Transitway, Purple Line/Bus Rapid Transit) is planned. Yellow policy areas are the lower density areas of the county characterized by mainly residential neighborhoods with community-serving commercial areas. Green policy areas are the county's agricultural reserve and rural areas.

Advancing Pro-Rata Share Districts

A pro-rata share district leverages the private sector role toward implementing a comprehensive plan holistically, rather than incrementally. In a pro-rata share district, each development contributes resources (whether facility construction or funding) toward a well-defined set of projects with the contribution defined in proportion to the relative level of demand contributed by that development. The pro-rata share concept can be expressed as a fraction in which the numerator is the private sector funding for total system supply and the denominator is the unit of development demand. Beyond this basic concept, the details of defining the numerator and denominator vary from place to place; they are dependent upon the physical, environmental, and political context.

The primary advantages of a pro-rata share district are to:

- Limit the "free rider" or "last in" problem associated with typical threshold-based transportation impact study approaches in which exactions are based more on timing than impact (i.e., 95% of available or remaining system supply might be used by the free-riders with the full burden of improvement imposed upon the applicant using the last 5%).
- Focus exaction efforts on planned system improvements rather than identifying ad hoc improvements (even though based logically on transportation impact analyses) that may not contribute to the desired end state.

- Measure success more through implementation of both the planned private infrastructure (planned land development per zoning) and public infrastructure (a multimodal transportation network in a comprehensive plan) than by mitigation based on site-specific level of service (LOS) or quality of service (QOS) objectives.

Montgomery County established its first pro-rata share district for the White Flint Policy Area in 2011 and the second pro-rata share district for the White Oak Policy Area in 2016. Other jurisdictions with pro-rata share districts include Kissimmee, FL (*7*); Baltimore, MD (*8*); and Portland, OR (*9*). The 2016–2020 Subdivision Staging Policy recommends that Montgomery County further evolve over time through the incremental implementation of additional pro-rata share districts.

Multimodal and Context-Sensitive Trip Generation

The methodology used by the project team to develop the context-sensitive factors to apply to Institute of Transportation Engineers (ITE) trip generation rates is similar to that described in Section 4.6 of NCHRP Report 758 (*10*) and is divided into seven steps:

1. The number of person trips produced and attracted by transportation analysis zone (TAZ) was computed for motorized and non-motorized trips using trip generation production and attraction files from the regional travel demand model. These person trips were separated by trip purpose, including home-based work, home-based shopping, and home-based other, and by motorized and non-motorized trips.
2. The mode split for motorized trips was further summarized from the output person-trip tables from the mode choice module of the travel demand model using a custom Cube Voyager application. These files were summarized by zone, purpose, and mode including drive alone, shared ride, shared ride with three or more persons, commuter rail, METRORail, a combination of bus and METRORail, and bus only.
3. Average mode share for all trip purposes was calculated as a sumproduct of the computed mode split by purpose (calculated from the mode choice model) and the trip purpose split by zone (calculated from the trip generation model), like Table 4.3 from NCHRP Report 758. These calculations were performed for both productions and attractions.
4. The percent split of trip purposes by land use and area type are taken from the calibration report for the Transportation Planning Board's travel demand forecasting model v2.3 for productions and attractions separately. The land uses include population, office employment, retail employment, industrial employment, and other employment. Using this, the percent split of attractions by land use, area type, and purpose can be calculated by zone.
5. Using the mode share calculated in the Step 3 and the percent split of trips in Step 4, mode split by land use and policy area is computed.
6. The mode split computed in Step 5 is summarized into peak hour, bi-directional mode split. For household/residential, it is assumed that 85% of a.m. trips are productions and 15% are attractions (based on typical a.m. peak directional mode split for residential uses) to separately adjust productions/attractions in the model dataset.
7. The vehicle trip factor is computed by using the Rural West policy area as a pivot point for suburban area types found in the ITE *Trip Generation* manual (*11*).

The outcome of this process is a series of trip generation rates by mode and land use in the *2017 LATR Guidelines* appendix for each of the policy areas. Figure 4 provides an excerpt of the types of tabulation included in the guidelines' appendix identifying the mode split assumptions to be used for each of the 38 policy areas in the county.

It is valuable to restate the fact that the NCHRP 758 process synthesizes the results of multiple trip purposes so that the NADMS applicable to site driveway counts is notably different from the mode shares for JTW trips, a metric more immediately familiar to many transportation planners. In particular, the inclusion of all trip purposes results in a notably higher proportion of auto passenger trips.

The *2017 LATR Guidelines* provide an example of the application of this approach by development applicants using a hypothetical 100,000 GSF office building in the Germantown East policy area.

First, from pages 1,260 and 1,261 of the 9th edition of *Trip Generation* (*11*) and the guidelines' vehicle trip factor of 0.95 for Germantown East, the site is estimated to generate 156*0.95 = 148 a.m. peak hour vehicle trips and 149*0.95 = 142 p.m. peak hour vehicle trips. The a.m. peak hour is the critical peak hour for person-trip generation analysis as the ITE vehicle trip rate is higher for the a.m. peak hour than for the p.m. peak hour.

The number of trips generated by any mode is the total number of total person trips multiplied by the mode share for that mode. The first step is to convert ITE vehicle trips to policy-area-specific total person trips. For the a.m. peak hour, the number of person trips is the number of vehicle trips divided by the auto driver mode share (148/72.1% = 205). The number of person trips exceeds the threshold of 50 meaning a quantitative auto analysis is required.

The next steps are to convert person trips to transit trips and non-motorized trips for considering transit system adequacy, pedestrian system adequacy, and bicycle system adequacy. The number of transit trips (205 * 1.8% = 4) is fewer

Appendix Table 1b: Mode Split Assumptions by Policy Area							
Policy Area #	**Development Type**	**Auto Driver**	**Auto Passenger**	**Transit**	**Non-Motorized**	**Total**
1 Aspen Hill	Residential	62.5%	25.8%	5.3%	6.4%	100%
	Office	74.2%	18.2%	2.9%	4.7%	100%
	Retail	72.1%	23.4%	1.3%	3.2%	100%
	Other	74.0%	18.2%	2.5%	5.2%	100%
2 Bethesda CBD	Residential	50.9%	20.8%	11.7%	16.6%	100%
	Office	47.9%	12.6%	23.8%	15.7%	100%
	Retail	44.2%	16.9%	10.9%	27.9%	100%
	Other	47.3%	13.2%	23.0%	16.5%	100%
3 Bethesda/Chevy Chase	Residential	56.1%	23.6%	7.6%	12.6%	100%
	Office	61.8%	17.4%	11.5%	9.3%	100%
	Retail	61.6%	24.7%	3.2%	10.5%	100%
	Other	60.5%	17.1%	12.6%	9.9%	100%
4 Burtonsville Town Center	Residential	62.3%	25.9%	4.9%	6.9%	100%
	Office	73.0%	19.8%	2.8%	4.3%	100%
	Retail	71.6%	24.3%	1.0%	3.1%	100%
	Other	73.9%	19.4%	2.5%	4.2%	100%
5 Chevy Chase Town Center	Residential	56.1%	23.6%	7.6%	12.6%	100%
	Office	61.8%	17.4%	11.5%	9.3%	100%
	Retail	61.6%	24.7%	3.2%	10.5%	100%
	Other	60.5%	17.1%	12.6%	9.9%	100%

Figure 4. Examples of mode split guidelines by policy area.

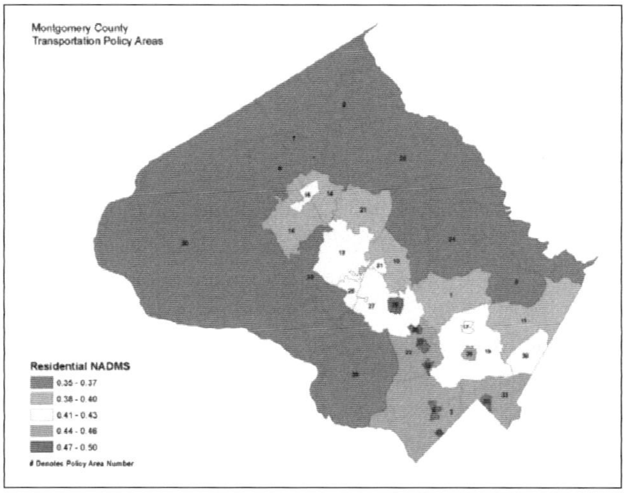

Figure 5. Residential non-auto driver mode share (NADMS) by policy area.

than the threshold of 50 so that a quantitative transit analysis is not required. The number of non-motorized trips (205 * 5.0% = 10) plus the number of transit trips (4, from above) totals 14, or fewer than the threshold of 50 so a quantitative pedestrian analysis is not required.

Figure 5 summarizes the NADMS for each policy area for residential development. The same general geographic patterns that describe policy area types in Figure 2 are evident in Figure 5; the NADMS shares are highest in the most transit served areas along the METRORail Red Line and lowest in the rural policy areas along the county's northern boundaries.

Many development applications will be small enough, or located in a sufficiently suburban area of the county, that they will not need to conduct any quantitative transit or non-motorized impact analyses. These applicants will still benefit from the context-sensitive trip generation rates that will reduce estimated vehicle trips, and therefore vehicle trip impacts. Figure 6 presents this information for the office land uses in a graphic format. The southernmost policy areas shown in red or tan have official vehicle trip generation rate assumptions that are 63% to 90% of the ITE vehicle trip generation rate.

There are two ways by which applicants can further reduce their vehicular trip generation. First, research suggests that there is a correlation between parking supply and vehicle trip generation, particularly when applied in a supportive parking-pricing environment with alternative transportation options.

For residential uses, each 2% reduction in parking below the minimum number of spaces yields a 1% reduction in vehicle trip generation rates for that use. This relationship is based on the equation in Table 2-9 of Transportation Research Board's Transit Cooperative Research Program, Report 128, "Effects of TOD on Housing, Parking, and Travel" (*12*). Applying this equation to a prototypical TOD site with 10 DU/acre, a ratio of one parking space per DU would yield 0.24 peak hour vehicle trips, and a ratio of 0.5 parking spaces per DU would yield 0.18 peak hour vehicle trips (in other words, a 50% reduction in parking yields a 25% reduction in vehicle trips).

For office uses, each 3% reduction in parking below the minimum number of spaces yields a one percent reduction in vehicle trip generation rates for that use. This relationship is based on the relationships shown in Figure 6-9 of a 2004 report by Lund, Cervero, and Willson for Caltrans "Travel Characteristics of Transit-Oriented Development in

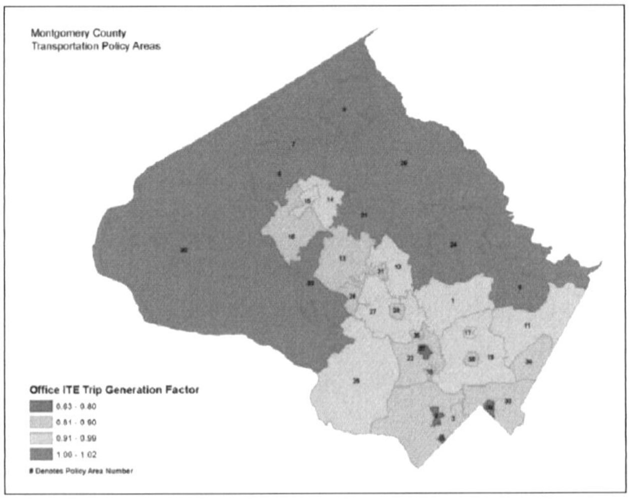

Figure 6. Vehicle trip generation factor to adjust ITE office land use rates.

California" (*13*), which shows that in a transit-rich environment supported by travel demand management strategies a similar reduction from 1.0 to 0.5 parking spaces at an office site could be expected to increase transit mode share from 41% to 50% (which for simplicity's sake is assumed to equal a reduction in auto mode share from 59% to 50%). In other words, in this case a reduction of 50% of parking spaces reduces auto trips by about 15%, or roughly a 3:1 ratio.

Secondly and alternatively, an applicant may develop a customized TDM program and enter into a Traffic Mitigation Agreement to monitor program success.

Developing Multimodal Adequacy Approaches

To achieve an approximately equivalent transportation LOS in all areas of the county, greater vehicular traffic congestion is permitted in policy areas with greater transit accessibility and usage. Non-motorized QOS is prioritized in areas where higher pedestrian and bicyclist volumes are expected.

Areas with pro-rata share districts satisfy adequacy by participating in their district requirements. Elsewhere, the definition of adequacy, and the associated required mitigation actions, seeks to balance the three project objectives of streamlining smart growth development, increasing multimodalism, and increasing the level of robustness of the technical analysis in the following manners:

- Motor vehicle adequacy requires an assessment of a more robust operational analysis in congested areas, but the operational analysis can be conducted with the goal to maintain background conditions across an analysis network rather than attaining a specific level of delay for each intersection in that network. Vehicle impacts can also be reduced by reducing trip generation by parking reduction or TDM program described above, or by implementing non-auto facilities at a rate of US$16,000 per peak hour vehicle trip.
- Transit system adequacy is limited to examining bus passenger/capacity ratios at nearby bus stops and mitigation is negotiated with the transit system provider, with the value of the improvement based on the capital cost of providing additional buses that would reduce the peak load to acceptable LOS D.
- Pedestrian system adequacy is based on crosswalk LOS, plus a requirement to address existing Americans with Disabilities Act (ADA) non-compliance issues based on the logic that there is a nexus between new development and existing inadequate conditions when the new development generates sufficient pedestrian activity to substantially affect exposure to the inadequacy.
- The thresholds for transit system and pedestrian system adequacy are set at levels that will not trigger analytic requirements for smaller infill developments. The use of thresholds for quantitative analysis for transit and non-motorized system impacts was influenced by guidance from policies in New York City (*14*) and Washington DC (*15*).

The specific requirements are described below:

Motor vehicle adequacy is defined by a series of intersection LOS standards by policy area and includes volume to capacity ratios, average vehicle delay, and critical lane volume (CLV). For intersections located within red or orange policy areas, the *Highway Capacity Manual* (HCM) (*16*), operational (delay-based) LOS standard applies to all study intersections. For intersections located within yellow or green policy areas, the CLV-based LOS standard applies to study intersections with a CLV of 1,350 or less and the HCM delay-based LOS standard applies to study intersections with a CLV of more than 1,350. The concept of travel time reliability was introduced to the 2017 guidelines through a reference to the regional congestion management report: intersection on roadway segments that have an observed travel time index of 2.0 must perform a network analysis that considers queueing and spillback.

Pedestrian system adequacy is defined as providing LOS D service or better for any signalized crosswalk. The methodology for evaluating pedestrian LOS is described in the 2010 HCM in Chapter 18. Any site that generates more than 50 pedestrian peak hour trips (including trips to transit) must:

- Fix (or fund) ADA non-compliance issues within a 500-ft radius of site boundaries.
- Ensure LOS D for crosswalk pedestrian delay (or no more delay than existing) at LATR study intersections within 500 ft of site boundaries or within areas where the County's road construction code specifies use of urban design standards.

Regardless of the development size and location, if an intersection operational analysis is triggered for any intersections within a RCUA/BPPA, mitigation must not increase average pedestrian crossing time at the intersection.

Bicycle system adequacy is defined as providing a low level of traffic stress (LTS) for bicyclists. For any proposed development generating at least 50 peak hour non-motorized trips and located within a quarter mile of an educational institution or existing/planned bikeshare station, the applicant must make improvements needed to provide low LTS-2 conditions (a level of traffic stress that most adults will tolerate) that link the site to or otherwise extend an LTS-2 facility within 750 ft of a development site boundary or implement a master-planned improvement that provides an equivalent improvement in LTS.

Transit system adequacy for LATR is defined as providing a peak load of LOS D for bus transit service routes (1.25 transit riders per seat) during the peak period (in the peak direction). For any development generating at least 50 peak hour transit riders the applicant must inventory bus routes at stations/stops within 1,000 ft of the site and identify the peak load for each route at that station. The applicant must coordinate with the transit service provider to identify and implement (or fund) improvements that would be needed to address conditions worse than LOS D due to additional patrons generated by the development.

Synthesizing Travel Demand and Impact Taxes

The current transportation impact tax rate varies by land use and a simplified district approach using countywide vehicle trip generation rates as a basis, but with an estimate that red policy areas generated about half the vehicle trips as those elsewhere in the county, and separating the Clarksburg policy area based on specific development needs. The project team analyzed both residentially generated VMT and person-miles of travel (PMT) in motor vehicles (autos and transit) using model data for the 2010 validation year. The analysis focused on residentially generated VMT and PMT for JTW, given the complications associated with dividing VMT between productions and attractions, particularly considering inter-county versus intra-county travel. To examine the attraction end of the JTW the team also examined the JTW NADMS.

The findings were that for household JTW trips, the countywide average trip length is 11.45 mi. The red areas generate only 37% of countywide VMT, orange areas generate 79% of countywide VMT, yellow areas generate 134% of countywide VMT, and green areas generate 226% of countywide VMT. The countywide JTW NADMS for the attraction end of the trip is 32.6%. The red areas have an auto mode share that is 81% of the countywide average (in other words, the NADMS changes from 32.6% to 45.2%, which means the auto mode share increases from 67.4% to 54.8%, and 54.8% is 81% of 67.4%). The orange areas have an auto mode share 106% of the countywide average, the yellow areas are at 124% of the countywide average and the green areas are at 133% of the countywide average.

The Planning Board recommended establishing transportation impact tax rates by policy area group for residential development based on the VMT sensitivity test and for commercial development based on the NADMS sensitivity test. The county council ultimately voted to retain the current structure, in part due to other equity concerns and public testimony regarding impact tax changes and the complexity of integrating the elimination of TPAR. The county council also found that the analysis of VMT and NADMS demonstrated that the previous rates were not inappropriate in scale.

Next Steps

Locally, the adoption of the *2017 LATR Guidelines* introduces a new phase of development review that will benefit from continued dialogue among all stakeholders in the economic development process. Several initiatives are underway to facilitate these conversations. The Montgomery County Department of Transportation is initiating studies for the next round of pro-rata share districts to be implemented. The Montgomery County Planning Department has a biennial reporting mechanism to the county council on growth management. The planning department will also be documenting decisions on cases under the *2017 LATR Guidelines* to help identify what clarifications may be needed and to ensure consistency in the application of the guidelines. The county code requires that the APFO be revisited at least every 4 years so that any key findings through the implementation and monitoring approach can be assembled and the process adjusted in 2020.

From an international perspective, several concepts in the *2017 LATR Guidelines* are highly transferable to other jurisdictions, notably:

- Pro-rata share districts for small, smart growth, areas where streamlined applications are most desired
- Multimodal, context-sensitive trip generation rates
- Area-specific and mode-specific adequacy definitions and thresholds triggering study
- Monetization of impacts to facilitate cross-modal mitigation strategies

Although the current project benefited from the wide range of resources in the literature review, some of the resources are now more than a decade out of date. Future research on trip behavior (including trip purpose information not readily captured by "big data" sources) is needed to update key behavioral relationships with post-recession data that also reflect the shifting of the millennial generation into adulthood.

Author Contributions

The authors confirm contribution to the paper as follows: study conception and design: EG and DH; data collection: EG, DH, and CVN; analysis and interpretation of results: EG, DH, and CVN; draft manuscript preparation: DH. All authors reviewed the results and approved the final version of the manuscript.

References

1. Montgomery County Planning Board, Maryland-National Capital Park and Planning Commission. *Local Area Transportation Review and Transportation Policy Area Review Guidelines*. Maryland-National Capital Park and Planning Commission, Silver Spring, 2013.
2. Montgomery County Planning Board, Maryland-National Capital Park and Planning Commission. *Subdivision Staging Policy*. Maryland-National Capital Park and Planning Commission, Silver Spring, 2016. http://montgomeryplanning.org/wp-content/uploads/2016/12/SSP-PBD-Master-Correction.pdf. Accessed July 20, 2017.
3. Montgomery County Council. *Montgomery County Code*. Montgomery County Council, Rockville, Md. http://www.montgomerycountymd.gov/mcg/countycode.html. Accessed July 20, 2017.
4. Graye, E., and P. Dunn. *Draft 2017 Local Area Transportation Review Guidelines*. Montgomery County Planning Board, Maryland-National Capital Park and Planning Commission, 2017. http://www.montgomeryplanningboard.org/agenda/2017/documents/REVISED_item10_2017LATRGuidelinesContinuatonStaffReport5-25-2017.pdf. Accessed July 23, 2017.
5. California Governor's Office of Planning and Research. *Updating Transportation Impacts Analysis in the CEQA Guidelines*. California Governor's Office of Planning and Research. Sacramento, 2014.
6. City of San Jose. *Traffic Impact Analysis Handbook*. City of San Jose, 2009.
7. State of Florida Department of Community Affairs. *Transportation Concurrence Requirements and Best Practices*. State of Florida Department of Community Affairs, Tallahassee, 2006.
8. City of Baltimore Department of Transportation. *Rules and Regulations for Traffic Mitigation in the City of Baltimore Pursuant to Ordinance #06-345 As Amended by Ordinance #11-529*. City of Baltimore Department of Transportation, 2012.
9. City of Portland. *Portland Code, Chapter 17.15 Transportation System Development Charge*. Portland Oregon. https://www.portlandoregon.gov/citycode/?c=28848#cid_180236. Accessed July 27, 2017.
10. Daisa, J. M., M. Schmitt, P. Reinhofer, K. Hooper, B. Bochner, and L. Schwartz. *NCHRP Report 758: Trip Generation Rates for Transportation Impact Analyses of Infill Developments*. Transportation Research Board of the National Academies, Washington D.C., 2013.
11. Institute of Transportation Engineers. *Trip Generation*, 9th ed. Institute of Transportation Engineers, Washington D.C., 2012.
12. Arrington, G. B., and R. Cervero. *TCRP Report 128: Effects of TOD on Housing, Parking, and Travel*. Transportation Research Board of the National Academies, Washington D.C., 2008.
13. Lund, H., and R. Cervero. *Travel Characteristics of Transit-Oriented Development in California*. Final Report. University of California, Berkeley, CA and California State Polytechnic University, Pomona, 2004.
14. Mayor's Office of Environmental Coordination. *City Environmental Quality Review Technical Manual*. New York City Mayor's Office of Environmental Coordination, 2014.
15. District of Columbia Department of Transportation. *DDOT Guidelines for Comprehensive Transportation Review*. District of Columbia Department of Transportation, Washington D.C., 2012.
16. *Highway Capacity Manual 2010*. Transportation Research Board of the National Academies, Washington, D.C., 2010.

The Standing Committee on Transportation and Land Development (ADD30) peer-reviewed this paper (18-01755).

Article

TRR
JOURNAL OF THE TRANSPORTATION RESEARCH BOARD

Transportation Research Record
2018, Vol. 2672(3) 56–68

DOI: 10.1177/0361198118791866
journals.sagepub.com/home/trr
SAGE

Impact of Major Road Supply on Individual Travel Time Expenditure: An Exploration with a 30-Year Variation of Infrastructure and Travel

Ryosuke Abe[1] and Kay W. Axhausen[2]

Abstract

This study estimates the impact of major road supply on individual travel time expenditures (TTEs) using data that cover 30-year variations in transportation infrastructure and travel behavior. The impacts of the supply of road and rail infrastructure are estimated with a data set that combines records of large-scale household travel surveys in the Tokyo metropolitan area conducted in 1978, 1988, 1998, and 2008. Linear and Tobit models of individual TTEs are estimated by following the behavior of birth cohorts over the 30-year period. The models incorporate the changes in transportation infrastructure, measured as lane kilometers of two levels of major road stock and vehicle kilometers of urban rail service. The results show significant negative effects of lane kilometers for higher-level and lower-level major roads on the TTEs for all travel purposes and for commuting, after controlling for socioeconomic backgrounds and generations of individuals. This study discusses that, in Tokyo, the estimated effect is more likely to reflect the effect of a major road network per se on individual TTEs than the (indirect) effect of major road supply on individual TTEs working through land development activities (i.e., induced car travel demand). For example, the caveat is that actual road investment decisions still need to consider the induced component of road traffic in addition to the (direct) effect that is estimated in this study.

Travel time expenditure (TTE) is a measure of the amount of time actually spent traveling (*1*). TTE is an essential indicator of the benefit of an investment in transportation infrastructure and, as such, has been analyzed intensively in transportation research. Mokhtarian and Chen's (*2*) review of travel time budget studies over 40 years shows that individual TTEs vary by individual and household characteristics, attributes of activities at the destination, and the characteristics of residential areas.

This study revisits the question of how the supply of transportation infrastructure influences individual TTEs over the long run. In many cases, today's transportation investment decisions assume a negative impact of the supply of transportation infrastructure on TTEs of the trips between specific places. The question that remains is whether greater transportation infrastructure ultimately means fewer minutes spent in daily travel for individuals (*3*). Few studies have investigated this impact with data covering long-run variations in transportation infrastructure and individual travel behavior. The long-run variation in infrastructure covers many large-scale projects that take a long period and it adds critical components to the transportation network after their completion (Figure 1 shows how the length of Tokyo's road and rail networks has changed over time). Using the long-run variation may contribute to a new understanding of the impact of transportation infrastructure on individual TTEs.

This study estimates the impact of major road supply on individual TTEs using data on 30-year variations in transportation infrastructure and travel behavior in the Tokyo metropolitan area (TMA). Data from large-scale household travel surveys in the TMA are utilized. Repeated surveys, such as national, regional, or metropolitan household travel surveys, ask the same question to different samples of people at different points in time (*5*). This study then models individual TTEs for all travel purposes and for commuting by following the behavior of birth cohorts over time and

[1]Department of Civil Engineering, The University of Tokyo, Tokyo, Japan
[2]Institute of Transport Planning and Systems (IVT), ETH Zürich, Zürich, Switzerland

Corresponding Author:
Address correspondence to Ryosuke Abe: abe-r@ip.civil.t.u-tokyo.ac.jp

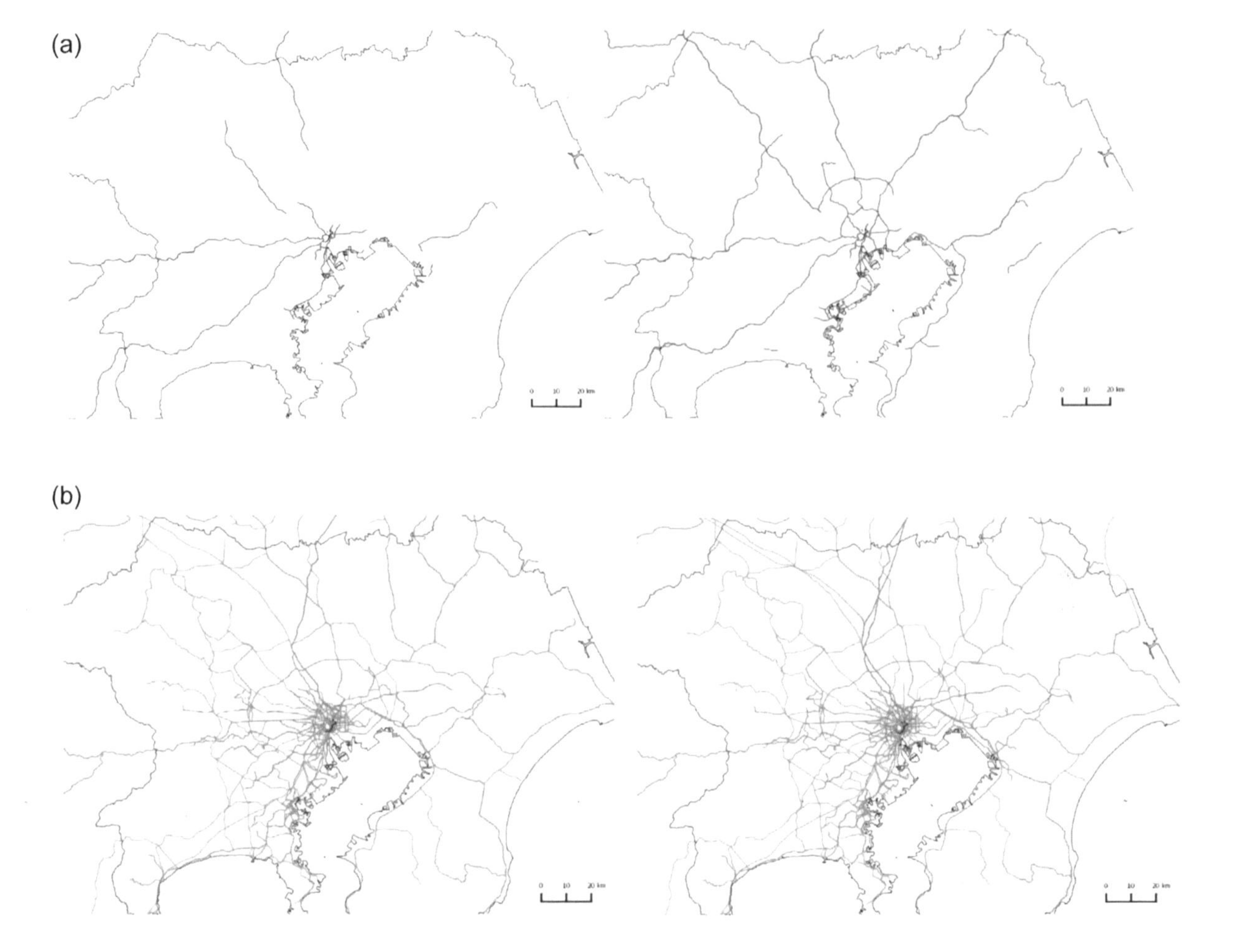

Figure 1. Transportation network in the Tokyo metropolitan area (TMA). (*a*) Panel A expressway network in 1978 (left) and 2008 (right) and (*b*) Panel B rail network in 1978 (left) and 2008 (right) (*4*).
Note: The expressway refers to "national expressways (Category A)," "national highways for motor vehicles only (Categories A' and B)," and the "Shuto expressway." The Tokyo Bay Aqua-Line is not depicted in 2008. The rail refers to urban rail (suburban railways, subways, monorails, and automated guideway transit) and high-speed rail.

incorporating the changes in transportation infrastructure. The next section reviews the literature related to this study. The section on models conceptualizes the impact of the supply of transportation infrastructure on individual TTEs and presents an empirical model. Then, data used in the empirical analysis are explained, and results of the model estimation are presented and discussed. The final section summarizes the findings.

Literature Review

Analyzing Long-Run Changes in Travel Behavior with Repeated Surveys Data

The basic method employed when using repeated surveys data to follow birth cohorts (i.e., people born in the same time interval) is to combine (pool) the surveys at the individual level (*5*). For example, in this way, the analysis of Scott et al (*6*) and Tilley and Houston (*7*) account for cohort fixed effects in a recent trend in daily travel mobility.

Meanwhile, long-run (20-year or more variation in data) studies based on repeated surveys data use pseudo-panel data, by grouping individuals into cohorts *and* constructing cohort variables (e.g., as the average of all individuals included in the cohort); in this case, they first aggregate the data at the cohort level (*8–10*). Although pseudo-panel data are useful for dynamic modeling, surveys need to be conducted in smaller intervals to make this form of model possible.

This study investigates the long-run changes in individual TTEs using repeated surveys data at the individual level, and following birth cohorts over time. Because repeated surveys of travel behavior typically record a large sample of individual TTEs, using individual variations could be highly beneficial for a robust statistical inference. In addition, the unique TTE characteristics of a cohort, which are constant over time, can be isolated by defining cohorts using repeated surveys data.

Impacts of the Supply of Transportation Infrastructure on Travel

The supply of transportation infrastructure that affects daily individual travel behavior typically refers to that which can be measured directly, such as lane kilometers of road supply. Although road traffic capacity may be the true output of road stock, it is not easily measurable on a larger geographical scale, such as a metropolitan, regional, or country level. At the same time, the (intermediate) output of public transit services is typically measured in vehicle kilometers or seat kilometers of service (*11*).

Past studies have intensively examined the impact of road supply on road traffic demand. Such studies have found significant effects of lane kilometers of road stock on average daily traffic and Vehicle Kilometers Traveled (VKT) at the metropolitan or regional level, termed induced travel demand, even after accounting for possible endogeneity in the supply (*12*, *13*). Duranton and Turner (*14*, pp. 2639–2641) examined the impact of interstate lane kilometers in a metropolitan area on individual VKT in the United States, using data from travel surveys conducted in 1995 and 2001, and found similar results, which generalizes the impact.

It still seems that few studies have empirically analyzed the impact of the supply of transportation infrastructure on individual travel behavior in general (e.g., covering all origin/destination pairs and modes of transportation). This is also true in the case of individual TTEs, although many studies have found that improvements to transportation infrastructure have reduced TTEs of the trips between specific places, for example, in the form of ex-post evaluations (*3*, *15*). A caveat is that measures of service levels in transportation networks, which are typically defined for a relatively small area (e.g., a neighborhood), do provide evidence of their impacts on individual travel behavior (e.g., *16*, *17*). This study prefers to use a direct measure of transportation infrastructure, or the supply of transportation infrastructure, which is typically measured for a larger area (e.g., a metropolitan area), to determine the relationship between individual TTEs and the extent of transportation infrastructure (and its cost of being built).

Model

Impacts of the Supply of Transportation Infrastructure on Individual TTEs

Past studies have analyzed the impact of the supply of transportation infrastructure by controlling for related exogenous factors. In this study, we express individual TTE y as a function of the stock of transportation infrastructure i, and the exogenous factors x as follows: $y = f(i, x)$. For example, x refers to all the other factors that affect TTEs, but that are not affected by TTEs and transportation infrastructure.

We then estimate the impact of the supply of transportation infrastructure on individual TTEs, $\partial f / \partial i$; we estimate the effect on individual TTEs, which works in the transportation network and, at best, works through land development activities, while controlling for the effects of individual socioeconomic conditions on individual TTEs; the relevant control x is chosen.

Further issues are discussed for clarity. The above effects on individual TTEs working through the factors outside the transportation network might take a positive sign. As a result, the supply of transportation infrastructure could have a strictly negative impact on individual TTEs with direct effects only, but a rather smaller negative impact with such indirect effects and direct effects.

Empirical Model

Two empirical TTE models are estimated. The first is a linear model, estimated by the ordinary least squares (OLS) approach. Here, the behavior of a cluster of individuals (based on year of birth) is followed over time. The model is specified as follows:

$$y_{i,j,t} = \alpha + \boldsymbol{i}'_{i,j,t}\boldsymbol{\beta} + \boldsymbol{x}'_{i,j,t}\boldsymbol{\gamma} + c_j + \varepsilon_{i,j,t} \quad (1)$$

where

$y_{i,j,t}$ is the TTE of the i th individual in cohort j in period t;

$\boldsymbol{i}_{i,j,t}$ is a vector of the stock of transportation infrastructure in the area where the i th individual in cohort j in period t lives;

$\boldsymbol{x}_{i,j,t}$ is a vector of control variables for the i th individual in cohort j in period t;

c_j is a fixed effect of cohort j;

and $\varepsilon_{i,j,t} \sim \text{iid } N(0, \sigma^2)$.

Then, α, $\boldsymbol{\beta}$, and $\boldsymbol{\gamma}$ are unknown parameters to be estimated. Transportation infrastructure is assumed to be predetermined in explaining individual TTEs; in other words, it is assumed to have lag in responding to a change in TTE (e.g., it requires substantial time to be built).

The second, and preferred, TTE model is developed using the standard Tobit model, which accounts for a dependent variable censored at zero. This application is appropriate because an individual could decide not to make any trips during a certain period and such individuals typically comprise a nontrivial fraction of the population. The Tobit specification for individual TTEs follows that of, for example, Chikaraishi et al (*18*). The Tobit TTE model is expressed as follows:

$$y^*_{i,j,t} = \alpha + \boldsymbol{i}'_{i,j,t}\boldsymbol{\beta} + \boldsymbol{x}'_{i,j,t}\boldsymbol{\gamma} + c_j + \varepsilon_{i,j,t}$$

$$y_{i,j,t} = \begin{cases} y^*_{i,j,t}, & \text{if } y^*_{i,j,t} > 0 \\ 0, & \text{if } y^*_{i,j,t} \leq 0 \end{cases}, \quad (2)$$

where $y^*_{i,j,t}$ is a latent variable for $y_{i,j,t}$. Equation 2 is a modification of Equation 1, and is estimated by the maximum likelihood (ML) approach.

Data

Repeated Surveys

The data set is constructed from the records of the TMA Person Trip Surveys (*19*), conducted in 1978, 1988, 1998, and 2008. The 1978 survey was the second Person Trip Survey covering this metropolitan area, after the first survey in 1968. The population aged 5 years or older in the TMA was 26.5, 30.7, 32.9, and 34.6 million in 1978, 1988, 1998, and 2008, respectively. The definitional expansion of the geographical area of the TMA was fairly minor for the period this study analyzes. The TMA population accounts for approximately a quarter of the nation's population in those years.

The survey details can be summarized as follows. Eligible participants were those who were 5 years old or older in the TMA with the average sampling rate of 2% to 3% in each survey. Face-to-face surveys were conducted in 1978, 1988, and 1998, whereas survey sheets were sent by post to participants and their responses were collected by post or online in 2008. Response rates were 84.9%, 81.5%, 71.5%, and 25.6% in 1978, 1988, 1998, and 2008, respectively (*20*). Respondents recorded their own travel behavior conducted on a typical weekday (i.e., weekdays excluding Monday and Friday) and adults compiled the records of individuals aged 15 or younger. Most of the items questioned and the way they are communicated have not changed, as documented by the survey forms used over the years (*19*).

Cohort-by-Period Data Array

Table 1 shows the TTEs of all travel purposes and commuting for TMA's cohorts, covering the period 1978–2008. The table shows the relative size (or the number of individual observations included in the cohort) and the average TTE for each cohort. Cohorts are defined by 10-year bands of the year of birth. The earliest of the 12 cohorts includes individuals born between 1884 and 1893, and the most recent cohort includes those born between 1994 and 2003. The variables for age groups

Table 1. Relative Sizes and the Average TTE (All Minutes and Commuting Minutes per Capita per Weekday) for Cohorts, 10-Year Intervals of 1978–2008, TMA

	1978 Survey (n = 587,695; N = 26.5 million)				1988 Survey (n = 667,863; N = 30.7 million)				1998 Survey (n = 882,924; N = 32.9 million)				2008 Survey (n = 697,997; N = 34.6 million)			
Cohort	Age	Size (%)	TTE	Com. TTE	Age	Size (%)	TTE	Com. TTE	Age	Size (%)	TTE	Com. TTE	Age	Size (%)	TTE	Com. TTE
1994–2003	—	—	—	—	—	—	—	—	—	—	—	—	5–14	8.5	45.1	—
1984–1993	—	—	—	—	—	—	—	—	5–14	11.1	43.4	—	15–24	8.4	92.0	9.7
1974–1983	—	—	—	—	5–14	15.2	42.3	—	15–24	13.4	89.6	11.1	25–34	12.4	85.5	27.9
1964–1973	5–14	19.7	43.2	—	15–24	16.8	90.9	12.5	25–34	15.7	84.4	25.0	35–44	16.7	85.2	25.7
1954–1963	15–24	14.3	95.5	13.1	25–34	13.8	81.8	23.1	35–44	14.0	82.8	22.9	45–54	12.9	84.4	25.9
1944–1953	25–34	20.0	80.5	19.5	35–44	18.6	77.7	21.0	45–54	16.5	80.3	22.2	55–64	17.0	75.9	18.4
1934–1943	35–44	18.7	82.0	20.1	45–54	15.7	78.1	21.8	55–64	14.2	72.7	16.5	65–74	15.0	59.5	5.6
1924–1933	45–54	12.8	75.0	20.0	55–64	10.5	63.9	15.4	65–74	9.5	47.7	4.7	75–84	7.2	38.4	1.2
1914–1923	55–64	7.4	59.8	13.6	65–74	6.0	38.5	4.5	75–84	4.3	26.8	0.9	85–94	1.8	16.0	0.2
1904–1913	65–74	4.9	40.5	5.7	75–84	2.8	23.0	1.1	85–94	1.2	11.9	0.2	—	—	—	—
1894–1903	75–84	2.0	18.9	1.1	85–94	0.6	8.8	0.2	—	—	—	—	—	—	—	—
1884–1893	85–94	0.3	6.8	0.3	—	—	—	—	—	—	—	—	—	—	—	—
Average (weighted by size)	—	—	70.0	16.7[a]	—	—	69.5	17.1[a]	—	—	71.1	17.0[a]	—	—	72.2	17.5[a]

Note: Com. = Commuting; — = not defined.
[a]Average of individuals aged 15 or older.

and cohorts in Table 2 are defined on the basis of the grouping in Table 1.

Three findings are summarized in Table 1. First, the relative sizes of the cohorts reflect the rapid aging of the population in this area. The share of the population aged 65 or older was (4.9 + 2.0 + 0.3 =)7.2% (of the population aged 5 +) in 1978, and increased considerably to (15.0 + 7.2 + 1.8 =)24% in 2008. Second, by tracing the TTE of a specific cohort over time, the effects of age (controlling for effects that arise from generations) on TTEs can be understood. Overall, this pattern is consistent over different time intervals, such that TTE (of all travel purposes) of a cohort peaks in young adulthood. Third, by comparing the TTEs of different cohorts in the same life stage, the effects of generations (controlling for effects that arise from age) on TTEs can be understood. Overall, recent generations have a higher TTE (of all travel purposes) than earlier generations.

Admittedly, Table 1 shows that the average TTE (of all travel purposes) has remained roughly the same at around 70 minutes over the 30 years, which is an example of the findings of Ahmed and Stopher (*1*) and Mokhtarian and Chen (*2*).

Descriptive Statistics

Table 2 shows the descriptive statistics of the variables used in the empirical analysis. A TTE of a linked trip is computed as the sum of reported TTEs of all stages (unlinked trips) of the trip. Then, the daily TTEs for all travel purposes and for commuting are computed for each individual. The commuting TTE is defined as daily TTEs for home-to-workplace trips and does not include TTEs for business/home-to-school trips. It can be safely assumed that the respondent's definition of commuting remained the same over the years.

Figure 2 shows the distributions of individual TTEs for all travel purposes and for commuting. The data set for the commuting TTE is a subset of the full data set, and excludes observations for individuals aged 5 to 14 because, in principle, legitimate employed work is not allowed for them in Japan. Censoring is evident in both distributions. The average TTE for all travel purposes is 71.4 min, and 83.5 min after excluding observations whose TTE is zero (13.9%, 14.7%, 15.3%, and 14.2% of the total in 1978, 1988, 1998, and 2008, respectively). The average commuting TTE is 17.1 min, and 41.9 min after excluding observations whose TTE is zero (60.1%, 57.9%, 59.3%, and 59.6% of the total of the commuting subset in 1978, 1988, 1998, and 2008, respectively).

Table 2 also shows the descriptive statistics of transportation infrastructure variables. This study measures major road supply on two levels. The higher-level major road supply is measured by lane kilometers of expressways including "national expressways (Category A)," "(designated) national highways for motor vehicles only (Categories A' and B)," and the "Shuto expressway" and those of other "designated national highways" in Japan. The expressways are in most cases the roads with full access control. While the roads with Categories A, A', and B comprise national arterial road networks in Japan, the Shuto (Metropolitan) expressway exclusively serves the central areas of Tokyo and their adjacent areas. The designated national highways that are not expressways date back to a national highways system before motorization and have been updated (e.g., adding bypass routes for some sections). The lower-level major road supply is measured by lane kilometers of "major prefectural highways," which in most cases are the major roads connecting places in a prefecture. The TMA consists of four prefectures (Chiba, Kanagawa, Saitama, and Tokyo) and the southern part of Ibaraki Prefecture.

Data on lane kilometers for the major roads are collected from the annual road statistics (*21*), in which original data are modified according to the two-level classification presented above. Lane kilometers are measured by the areas (in square kilometers) of lanes and their shoulders, excluding possible cycling lanes and sidewalks. By this measure, the road supply could be represented well in Japan and compared across different levels. Figure 3 shows trends in lane kilometers for the major roads by prefecture.

Although data for Ibaraki Prefecture can be collected from the annual road statistics, data for the southern part of Ibaraki Prefecture cannot be collected. This study collects expressways data for the southern part of Ibaraki Prefecture by examining the opening date of every section of each expressway that currently serves this region. In addition, this study collects data on other highways for this region in each year by weighting data for Ibaraki Prefecture by the fraction of the population in this region out of the population in Ibaraki Prefecture in the corresponding year.

Finally, data on vehicle kilometers for urban rail services are collected from the annual reports of urban transportation (*22*). These data cover vehicle kilometers at the metropolitan level (the population of the corresponding area is 24.9, 27.8, 29.6, and 32.0 million in 1978, 1988, 1998, and 2008, respectively). The urban rail considered in this study includes suburban railways, subways, monorails, and automated guideway transit, operated by private companies or public authorities. Vehicle kilometers of subways services account for 11% of the total in the TMA, while those of suburban railways services account for most of the rest, on average, for the period this study analyzes.

Table 2. Descriptive Statistics of Variables in All the Years (n = 2,836,479)

Variable	Mean	SD	Minimum	Maximum
TTE (minutes per weekday)	71.37	31.91	0	1,200
Non-zero TTE only (85.5% of total)	83.49	89.02	1	1,200
Commuting TTE (minutes per weekday) only for individuals aged 15 + (n = 2,462,065, commuting subset)	17.08	27.80	0	1,129
Non-zero commuting TTE only (40.8% of commuting subset)	41.90	29.26	1	1,129
Supply of transportation infrastructure				
Lane kilometer (km^2) of higher-level major road stock	5.75	1.60	1.34	8.35
Lane kilometer (km^2) of lower-level major road stock	7.72	3.04	3.02	12.28
(Per year) vehicle kilometer of urban rail service in 10^6	2.63	0.60	1.76	3.31
	Percentage			
Employment status (1 or 0)				
Production, manual, and sales	1.91	—	—	—
Service industry	4.67	—	—	—
Clerical	9.56	—	—	—
Professional and technical	9.54	—	—	—
Student	15.42	—	—	—
Homemaker	16.92	—	—	—
No job	12.38	—	—	—
Other jobs	29.60			
Household size				
1	8.73	—	—	—
2	18.82	—	—	—
3	21.04	—	—	—
4	30.74	—	—	—
5 +	20.67	—	—	—
Female (1 or 0)	50.96	—	—	—
Age (1 or 0)				
5–14	13.20	—	—	—
15–24	13.15	—	—	—
25–34	15.36	—	—	—
35–44	16.72	—	—	—
45–54	14.65	—	—	—
55–64	12.63	—	—	—
65–74	9.07	—	—	—
75–84	4.19	—	—	—
85 +	1.04	—	—	—
Cohort (1 or 0)				
1994–2003	2.09	—	—	—
1984–1993	5.52	—	—	—
1974–1983	10.78	—	—	—
1964–1973	17.05	—	—	—
1954–1963	13.75	—	—	—
1944–1953	17.88	—	—	—
1934–1943	15.68	—	—	—
1924–1933	9.85	—	—	—
1914–1923	4.73	—	—	—
1904–1913	2.03	—	—	—
1894–1903	0.57	—	—	—
1884–1893	0.06	—	—	—
Residence (1 or 0)				
Saitama Prefecture	20.00	—	—	—
Kanagawa Prefecture	27.12	—	—	—
Chiba Prefecture	18.03	—	—	—
Tokyo Prefecture	30.74	—	—	—
Southern part of Ibaraki Prefecture	4.15	—	—	—

Note: SD = standard deviation; — = not defined.

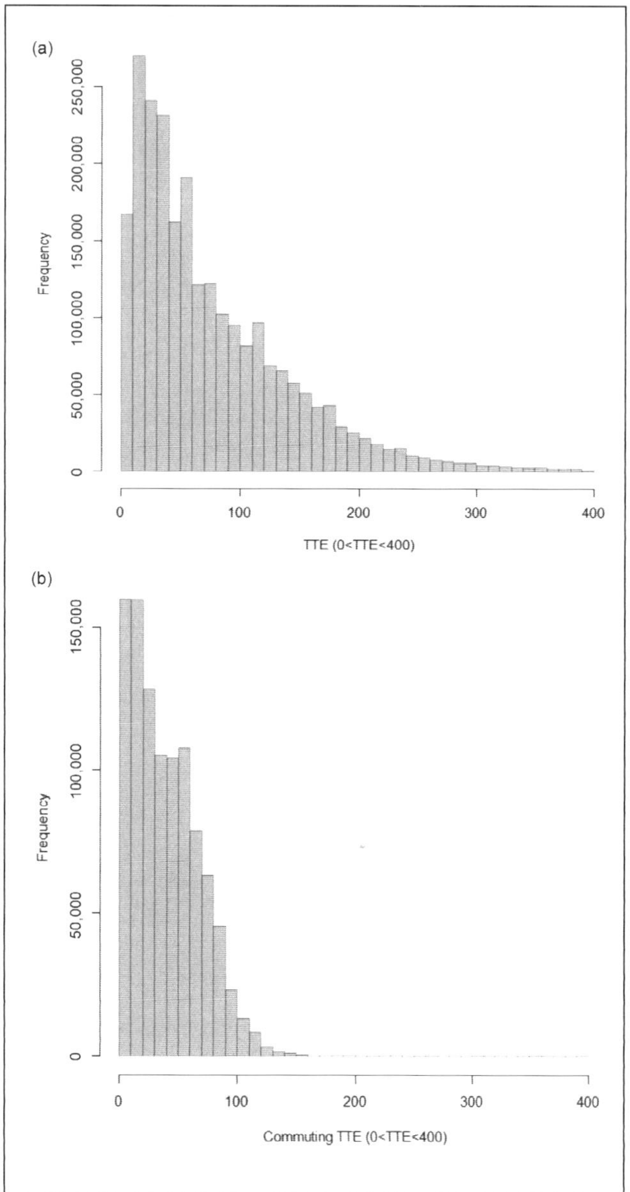

Figure 2. Distributions of individual TTEs (minutes per weekday) in the TMA.
(*a*) Panel A TTEs for all travel purposes (n = 2,836,479).
Note: Observations with a TTE of zero (n = 414,266) and 400 min or longer (n = 16,283) are not plotted.
(*b*) Panel B commuting TTEs (n = 2,462,065).
Note: A subset of the full data set. This subset only includes individuals aged 15 or older. Observations with a commuting TTE of zero (n = 1,458,201) and 400 min or longer (n = 75) are not plotted.

Results

Tables 3 and 4 show the estimation results of the empirical models for all travel purposes and for commuting, respectively. Each table shows both linear and Tobit model results. Estimated coefficients in the Tobit model are the partial effects of explanatory variables on the latent dependent variable, as shown in Equation 2, and so Tables 3 and 4 also report the partial effects on the observed dependent variable. Specifically, among the two common approaches used to compute an adjustment factor to obtain such partial effects, the average partial effect (APE) is computed, which is preferred in most cases (e.g., *23*, pp. 596–604). The fitness of all models seems sufficient, given the nature of the data set. All the estimated effects on the TTEs for all travel purposes (Table 3) are significant, with the exception of those of a worker in the service industry, the 5–14 age group, and some of the cohort fixed effects in the OLS model. This is also true of all the estimated effects on commuting TTEs (Table 4), except for the effects of vehicle kilometers of urban rail services and most of the cohort fixed effects.

Lane kilometers have a negative effect on the TTEs for all travel purposes and for commuting. Table 3 (all travel purposes) shows that the differences in the estimated effects of lane kilometers are minor between the linear and Tobit models. The results show that an increase in lane kilometers for higher-level major roads decreases the TTE for all travel purposes by 2.3 minutes and an increase for lower-level major roads does so by 1.1 minutes, after controlling for socioeconomic conditions and generations of individuals. In Table 4 (commuting purpose), our preferred Tobit results show that an increase in lane kilometers for higher-level major roads decreases the commuting TTE by 1 minutes and that an increase for lower-level major roads decreases the commuting TTE by 0.4 minutes, with the same controls.

Vehicle kilometers of urban rail services have a positive effect on the TTE for all travel purposes and do not have a significant effect on the commuting TTE.

Table 5 shows the results of a sensitivity analysis for the effects of infrastructure variables. First, it tests different combinations of infrastructure variables in the Tobit model using the same control variables as those used in Tables 3 and 4. Second, it estimates a Tobit model without incorporating cohort dummy variables. These results confirm our principal findings: road supply variables have a significant negative effect on the TTEs for all travel purposes (Panel A) and for commuting (Panel B). Finally, this study (in a separate analysis) confirms that our principal findings are also obtained by models estimated with random samples of the data set.

Discussion

The estimated effects of lane kilometers on TTEs can be regarded as travel time savings by the investments in major roads. The conversions of these savings to monetary terms are also possible. For example, the value of

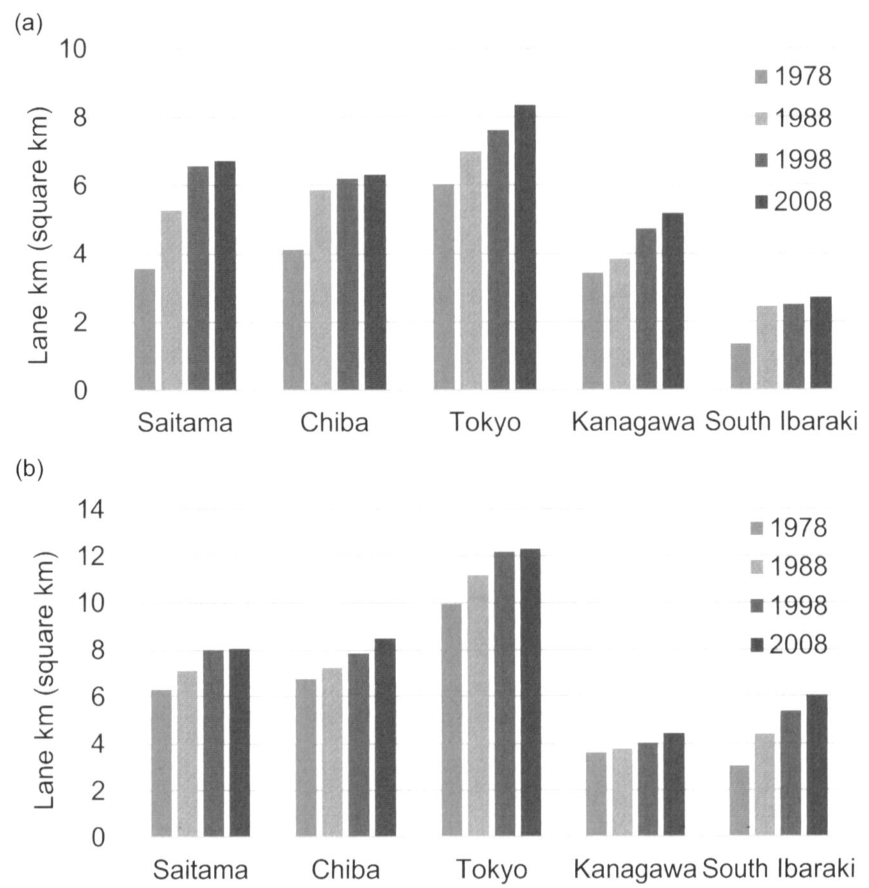

Figure 3. Trends in lane kilometers of major roads by prefecture. (*a*) Panel A higher-level major road and (*b*) Panel B lower-level major road.

travel time savings (VTTSs) in 2000 prices is JPY40.9/min for home-to-workplace travel, which are the averages taken from 261 VTTS estimates documented in 68 peer-reviewed papers in Japan between 1979 and 2003 (*24*).

The extent to which the estimated effect of transportation infrastructure reflects its direct effect can be discussed, because it also incorporates an indirect effect working through land development activities. This framework follows Giuliano (*25*) in general, Cervero (*26*) for road supply, and Abe and Kato (*27*) for urban rail supply.

The direct effect of transportation infrastructure is the one that is first translated into the condition (supply related or potential service level) in the transportation network, which affects individual travel behavior.

The indirect effect of transportation infrastructure is the one that first affects land development activities near major roads and urban rail stations which, in turn, affects individual travel behavior. This indirect effect may be the so-called induced travel demand by the supply of transportation infrastructure.

Our estimated effect of transportation infrastructure may reflect that the indirect effect is stronger for urban rail supply than it is for major road supply, for the following two reasons. First, the indirect effect of urban rail supply on individual TTEs could be significant and take a positive sign in the TMA. This is because a larger value of vehicle kilometers of urban rail services in the TMA could also imply higher housing development activities, particularly in the suburbs where residents travel to the central areas over a long distance, possibly resulting in a larger TTE. Note that the TMA has had a mono-centric urban form with a high concentration of office and commercial functions in the central areas and central business district, and the development of housing estates along rail lines in the suburbs (*28*). Note also that the 2008 TMA Person Trip Survey shows that urban rail is the most frequently used mode, both for all travel purposes (30%) and for commuting (53%), followed by cars (29% for all travel, 24% for commuting). Second, although major road supply could have an effect on land development activities along major roads in the TMA (e.g.,

Table 3. Estimation Results of All TTE (Minutes per Weekday) in the TMA (n = 2,836,479)

	Linear (OLS)		Tobit (ML)		
Explanatory variable	Coefficient	SD	Coefficient (APE in parenthesis)		SD
Constant	45.95	1.82	−31.18	(−23.92)	3.01
Lane km (km^2) of higher-level major road	−2.32	0.13	−3.03	(−2.32)	0.15
Lane km (km^2) of lower-level major road	−1.07	0.14	−1.52	(−1.17)	0.16
Vehicle km of urban rail service in 10^6	6.30	0.40	6.77	(5.19)	0.45
Employment status (other jobs as reference)					
Production, manual, and sales	6.82	0.31	8.83	(6.77)	0.35
Service industry	0.04[n.s]	0.21	−0.90	(−0.69)	0.24
Clerical	2.79	0.16	4.35	(3.34)	0.18
Professional and technical	4.00	0.16	4.26	(3.27)	0.18
Student	−10.23	0.31	−8.94	(−6.86)	0.35
Homemaker	−38.49	0.15	−46.64	(−35.77)	0.17
No job	−42.93	0.17	−61.42	(−47.11)	0.20
Household size (single family as reference)					
2	4.56	0.17	5.33	(4.09)	0.20
3	3.23	0.17	3.09	(2.37)	0.19
4	2.82	0.17	2.76	(2.11)	0.19
5 +	−3.07	0.17	−4.55	(−3.49)	0.20
Age (85 + as reference)					
5–14	−3.61[n.s]	1.96	18.32	(14.06)	2.28
15–24	36.30	1.70	57.82	(44.35)	1.99
25–34	30.62	1.47	52.30	(40.11)	1.73
35–44	27.69	1.26	50.34	(38.62)	1.49
45–54	24.25	1.05	46.86	(35.95)	1.26
55–64	21.80	0.85	44.97	(34.49)	1.04
65–74	16.43	0.66	40.72	(31.24)	0.84
75–84	8.88	0.53	28.15	(21.59)	0.71
Female	−22.89	0.09	−24.82	(−19.04)	0.11
Cohort (1884–1893 as reference)					
1994–2003	4.28[n.s.]	3.04	60.12	(46.12)	4.12
1984–1993	4.05[n.s.]	2.88	59.73	(45.81)	3.97
1974–1983	4.54[n.s.]	2.68	59.52	(45.65)	3.78
1964–1973	7.33	2.50	61.91	(47.49)	3.62
1954–1963	11.79	2.34	66.21	(50.79)	3.47
1944–1953	11.08	2.18	64.87	(49.76)	3.33
1934–1943	13.03	2.04	65.64	(50.35)	3.21
1924–1933	7.96	1.93	56.62	(43.43)	3.12
1914–1923	4.02[a]	1.82	45.43	(34.85)	3.03
1904–1913	3.42[n.s.]	1.75	36.99	(28.37)	2.98
1894–1903	0.35[n.s.]	1.76	18.21	(13.97)	3.02
Residence (southern part of Ibaraki Prefecture as reference)					
Saitama Prefecture	20.88	0.54	26.04	(19.97)	0.62
Kanagawa Prefecture	17.88	0.38	21.35	(16.38)	0.44
Chiba Prefecture	21.82	0.56	27.03	(20.74)	0.64
Tokyo Prefecture	28.89	1.02	37.80	(28.99)	1.17
R-squared	0.160		0.160[b]		
σ	69.33		77.63		
Log likelihood at zero	—		-1.46×10^7		
Log likelihood at convergence	—		-1.43×10^7		

Note: SD = standard deviation; — = not defined.
Statistical significance: [n.s.] = not significant; [a]$p < .05$; others $p < .01$.
[b]Square of the correlation between observed and predicted values of the dependent variable.

developments of freight distribution centers and large-scale shopping centers), whether such activities affect individual TTEs in a structured way might be less clear for a large part of the population.

Conclusion

In this study, we estimated the impact of major road supply on individual TTEs using data on 30-year variations

Table 4. Estimation Results of Commuting TTE (Minutes per Weekday) in the TMA (n = 2,462,065)

	Linear (OLS)		Tobit (ML)		
Explanatory variable	Coefficient	SD	Coefficient (APE in parenthesis)		SD
Constant	21.73	0.63	−35.36	(−14.36)	5.62
Lane km (km^2) of higher-level major road	−0.79	0.05	−2.57	(−1.04)	0.11
Lane km (km^2) of lower-level major road	−0.34	0.05	−0.99	(−0.40)	0.12
Vehicle km of urban rail service in 10^6	0.17[n.s.]	0.14	−0.05[n.s.]	(−0.02)	0.33
Employment status (other jobs as reference)					
Production, manual, and sales	6.64	0.11	12.29	(4.99)	0.19
Service industry	−1.08	0.07	−0.31[a]	(−0.13)	0.13
Clerical	12.65	0.05	20.48	(8.31)	0.10
Professional and technical	7.73	0.05	12.02	(4.88)	0.10
Student	−16.21	0.11	−102.80	(−41.75)	0.78
Homemaker	−19.70	0.05	−136.90	(−55.60)	0.63
No job	−19.46	0.06	−139.70	(−56.71)	1.03
Household size (single family as reference)					
2	0.98	0.06	1.92	(0.78)	0.13
3	1.20	0.06	1.45	(0.59)	0.13
4	0.62	0.06	−0.54	(−0.22)	0.12
5 +	−2.37	0.06	−8.25	(−3.35)	0.14
Age (85 + as reference)					
15–24	−7.65	0.61	32.83	(13.33)	2.06
25–34	3.07	0.53	57.40	(23.31)	1.94
35–44	2.37	0.45	56.23	(22.83)	1.83
45–54	3.09	0.37	58.26	(23.65)	1.73
55–64	1.85	0.30	56.58	(22.97)	1.65
65–74	−1.88	0.23	40.06	(16.27)	1.60
75–84	−1.56	0.18	17.76	(7.21)	1.60
Female	−7.83	0.03	−12.93	(−5.25)	0.08
Cohort (1894–1903 as reference)					
1984–1993	−0.64[n.s.]	1.01	−8.80[n.s.]	(−3.57)	6.09
1974–1983	2.42[a]	0.95	1.32[n.s.]	(0.54)	6.04
1964–1973	3.02	0.88	3.00[n.s.]	(1.22)	5.99
1954–1963	4.62	0.82	7.88[n.s.]	(3.20)	5.94
1944–1953	1.97	0.76	1.49[n.s.]	(0.60)	5.90
1934–1943	1.31[n.s.]	0.71	−0.97[n.s.]	(−0.39)	5.86
1924–1933	0.42[n.s.]	0.66	−5.45[n.s.]	(−2.21)	5.84
1914–1923	0.55[n.s.]	0.62	−9.99[n.s.]	(−4.06)	5.81
1904–1913	1.09[n.s.]	0.60	−7.27[n.s.]	(−2.95)	5.81
1884–1893	0.86[n.s.]	0.60	−6.69[n.s.]	(−2.71)	5.92
Residence (southern part of Ibaraki Prefecture as reference)					
Saitama Prefecture	8.19	0.20	21.01	(8.53)	0.48
Kanagawa Prefecture	7.13	0.14	16.67	(6.77)	0.33
Chiba Prefecture	8.83	0.21	22.03	(8.95)	0.49
Tokyo Prefecture	9.49	0.38	26.20	(10.64)	0.90
R-squared	0.285		0.293[b]		
σ	23.51		41.02		
Log likelihood at zero	—		-6.35×10^6		
Log likelihood at convergence	—		-5.58×10^6		

Note: Table 4 uses a subset of the full data set, which excludes observations for individuals aged 5–14. SD = standard deviation; — = not defined.
Statistical significance: [n.s.] = not significant; [a] $p < .05$; others $p < .01$.
[b] Square of the correlation between observed and predicted values of the dependent variable.

in transportation infrastructure and travel behavior in Tokyo by following the behavior of birth cohorts and incorporating the changes in transportation infrastructure. The results showed significant negative effects of lane kilometers for higher-level and lower-level major roads on the TTEs for all travel purposes and for commuting, after controlling for socioeconomic conditions and generations of individuals. In Tokyo, the estimated effect is more likely to reflect the effect of a major road network per se on individual TTEs than the (indirect)

Table 5. Sensitivity of Estimation Results

Panel A Results of All TTE, Tobit Models

Variable	Model 1a Coefficient	(APE)	SD	Model 1b Coefficient	(APE)	SD	Model 1c Coefficient	(APE)	SD	Model 1d Coefficient	(APE)	SD
Higher-level major road	−3.03	(−2.32)	0.15	−2.99	(−2.30)	0.15	−3.41	(−2.62)	0.15	−3.28	(−2.51)	0.14
Lower-level major road	−1.52	(−1.17)	0.16	−1.09	(−0.84)	0.16	—	—	—	—	—	—
Urban rail	6.77	(5.19)	0.45	—	—	—	5.97	(4.58)	0.45	—	—	—
	Model 1e			**Model 1f**			**Model 1g**			**Model 1a (without cohort dummies)**		
Higher-level major road	—	—	—	—	—	—	—	—	—	−3.00	(−2.30)	0.14
Lower-level major road	−1.97	(−1.51)	0.15	−2.42	(−1.85)	0.16	—	—	—	−1.15	(−0.89)	0.16
Urban rail	—	—	—	6.63	(5.08)	0.45	5.25	(4.03)	0.45	8.67	(6.65)	0.23

Panel B Results of Commuting TTE, Tobit Models

Variable	Model 2a Coefficient	(APE)	SD	Model 2b Coefficient	(APE)	SD	Model 2c Coefficient	(APE)	SD	Model 2d Coefficient	(APE)	SD
Higher-level major road	−2.57	(−1.04)	0.11	−2.57	(−1.04)	0.11	−2.80	(−1.14)	0.11	−2.82	(−1.14)	0.11
Lower-level major road	−0.99	(−0.40)	0.12	−1.00	(−0.41)	0.12	—	—	—	—	—	—
Urban rail	$-0.05^{n.s}$	−0.02	0.33	—	—	—	$-0.58^{n.s}$	(−0.23)	0.33	—	—	—
	Model 2e			**Model 2f**			**Model 2g**			**Model 2a (without cohort dummies)**		
Higher-level major road	—	—	—	—	—	—	—	—	—	−1.73	(−0.70)	0.11
Lower-level major road	−1.72	(−0.70)	0.12	−1.70	(−0.69)	0.12	—	—	—	−0.42	(−0.17)	0.12
Urban rail	—	—	—	$-0.28^{n.s}$	(−0.11)	0.33	−1.30	(−0.51)	0.32	$0.23^{n.s}$	(0.09)	0.17

Note: Only the estimated effects of infrastructure variables are shown in each model. Tables 4 and 5 present the results of Models 1a and 2a, respectively, with the estimated effects of other variables.
SD = standard deviation; — = the variable is not used in the model.
Statistical significance: $^{n.s.}$= not significant; $^{a}p < .05$; others $p < .01$.

effect of major road supply on individual TTEs working through land development activities (i.e., induced car travel demand). For example, the caveat is that actual road investment decisions still need to consider the induced component of road traffic in addition to the effect this study estimated.

Future areas of research are summarized. First, adding a variation regarding target urban areas and their developments would enable a robust estimation of different model specifications of the impact. For example, the effect that stems from a network aspect of transportation infrastructure on individual TTEs can be specified in detailed ways. Second, Tobit models use the same latent response to predict zero and positive reactions, whereas Hurdle models that allow the zero-TTE response to react differently from the positive reactions can also be estimated to better capture the unique features that lead people to stay at home. Finally, modeling techniques with long-run data from repeated surveys can be refined further. Future studies can more often utilize such long-run data to examine key issues with the assumption of the turnover of individuals in different generations that could have different mobility and residential preferences.

Acknowledgments

The authors thank the Ministry of Land, Infrastructure, Transport and Tourism for providing the records on the Person Trip Surveys in the Tokyo metropolitan area.

Author Contributions

The authors confirm contribution to the paper as follows: study conception and design: R. Abe and K. W. Axhausen; data collection: R. Abe; analysis and interpretation of results: R. Abe and K. W. Axhausen; draft manuscript preparation: R. Abe. All authors reviewed the results and approved the final version of the manuscript.

References

1. Ahmed, A., and P. Stopher. Seventy Minutes Plus or Minus 10: A Review of Travel Time Budget Studies. *Transport Reviews*, Vol. 34, 2014, pp. 607–625.
2. Mokhtarian, P. L., and C. Chen. TTB or Not TTB, That is the Question: A Review and Analysis of the Empirical Literature on Travel Time (and Money) Budgets. *Transportation Research Part A: Policy and Practice*, Vol. 38, 2004, pp. 643–675.
3. Metz, D. The Myth of Travel Time Saving. *Transport Reviews*, Vol. 28, 2008, pp. 321–336.
4. MLIT. *National Land Numerical Information Download Service*. Ministry of Land, Infrastructure, Transport and Tourism, Japan. http://nlftp.mlit.go.jp/ksj-e/index.html. Accessed November 15, 2017.
5. Firebaugh, G. *Analyzing Repeated Surveys*. Sage, Thousand Oaks, 1997.
6. Scott, D. M., K. B. Newbold, J. E. L. Spinney, R. Mercado, A. Páez, and P. S. Kanaroglou. New Insights into Senior Travel Behavior: The Canadian Experience. *Growth and Change*, Vol. 40, 2009, pp. 140–168.
7. Tilley, S., and D. Houston. The Gender Turnaround: Young Women Now Travelling More Than Young Men. *Journal of Transport Geography*, Vol. 54, 2016, pp. 349–358.
8. Dargay, J. M. The Effect of Income on Car Ownership: Evidence of Asymmetry. *Transportation Research Part A: Policy and Practice*, Vol. 35, 2001, pp. 807–821.
9. Dargay, J. M. The Effect of Prices and Income on Car Travel in the UK. *Transportation Research Part A: Policy and Practice*, Vol. 41, 2007, pp. 949–960.
10. Weis, C., and K. W. Axhausen. Induced Travel Demand: Evidence from a Pseudo Panel Data Based Structural Equations Model. *Research in Transportation Economics*, Vol. 25, 2009, pp. 8–18.
11. Small, K., and K. van Dender. *The Economics of Urban Transportation*, 2nd ed. Routledge, Abingdon, 2007.
12. Cervero, R. Induced Travel Demand: Research Design, Empirical Evidence, and Normative Policies. *Journal of Planning Literature*, Vol. 17, 2002, pp. 3–20.
13. Noland, R. B., and L. L. Lem. A Review of the Evidence for Induced Travel and Changes in Transportation and Environmental Policy in the US and the UK. *Transportation Research Part D: Transport and Environment*, Vol. 7, 2002, pp. 1–26.
14. Duranton, G., and M. A. Turner. The Fundamental Law of Road Congestion: Evidence from US Cities. *The American Economic Review*, Vol. 101, 2011, pp. 2616–2652.
15. Ironmonger, D., and P. Norman. Improvements in Transport Infrastructure are Designed to Increase Travel Speed: Comments on "the Myth of Travel Time Saving". *Transport Reviews*, Vol. 28, 2008, pp. 694–698.
16. Cao, X., P. L. Mokhtarian, and S. L. Handy. Examining the Impacts of Residential Self-Selection on Travel Behaviour: A Focus on Empirical Findings. *Transport Reviews*, Vol. 29, 2009, pp. 359–395.
17. Ewing, R., and R. Cervero. Travel and the Built Environment: A Meta-Analysis. *Journal of the American Planning Association*, Vol. 76, 2010, pp. 265–294.
18. Chikaraishi, M., A. Fujiwara, J. Zhang, K. W. Axhausen, and D. Zumkeller. Changes in Variations of Travel Time Expenditure: Some Methodological Considerations and Empirical Results from German Mobility Panel. *Transportation Research Record: Journal of the Transportation Research Board*, 2011. 2230: 121–131.
19. Transportation Planning Council in the Tokyo Metropolitan Area. *Person Trip Survey*. http://www.tokyo-pt.jp/person. Accessed July 2017. (In Japanese)
20. Mori, Y., and J. Morio. Tokyo Metropolitan Area: 50-Year Changes and a Future Vision. *IBS Annual Report*. The Institute of Behavioral Sciences, Tokyo, 2014. (In Japanese)
21. MLIT. *Annual Road Statistics*. Ministry of Land, Infrastructure, Transport and Tourism, Japan, 2017.
22. JTRI. *Annual Report of Urban Transportation*. Japan Transport Research Institute, Tokyo, 2017.

23. Wooldridge, J. M. *Introductory Econometrics: A Modern Approach*. South-Western, Cengage Learning, Boston, 2012.
24. Kato, H., M. Tanishita, and R. Abe. A Meta-Analysis of the Value of Travel Time Savings Incorporating GDP Per Capita: Evidence from Japanese Passenger Travels. Presented at 92nd Annual Meeting of the Transportation Research Board, Washington, D.C., 2013.
25. Giuliano, G. Land Use Impacts of Transportation Investments: Highway and Transit. In *The Geography of Urban Transportation*, 3rd ed (Hanson, S., and G. Giuliano, eds.). Guilford Press, New York, 2004, pp. 237–273.
26. Cervero, R. Road Expansion, Urban Growth, and Induced Travel: A Path Analysis. *Journal of the American Planning Association*, Vol. 69, 2003, pp. 145–163.
27. Abe, R., and H. Kato. What Led to the Establishment of a Rail-Oriented City? Determinants of Urban Rail Supply in Tokyo, Japan, 1950–2010. *Transport Policy*, Vol. 58, 2017, pp. 72–79.
28. Okata, J., and A. Murayama. Tokyo's Urban Growth, Urban Form and Sustainability. In *Megacities*, Springer Japan, Tokyo, 2011, pp. 15–41.

The Standing Committee on Transportation and Land Development (ADD30) peer-reviewed this paper (18-02249).

TRR
JOURNAL OF THE TRANSPORTATION RESEARCH BOARD

Article

Transportation Research Record
2018, Vol. 2672(3) 69–79

DOI: 10.1177/0361198118797203
journals.sagepub.com/home/trr
SAGE

Role of Spatial Configuration in Commuting Trade-Offs of Two-Worker Households: A Case Study in the Greater Toronto Area

Saidal Akbari[1] and Khandker Nurul Habib[1]

Abstract

This research investigates the commuting trade-offs between individuals in two-worker households with home and work locations in the Greater Toronto Area (GTA). A commuting trade-off occurs when a home relocation results in one worker incurring a higher commute distance for the other worker to work closer to the home location. This research uses stated preference data and multilevel modeling to demonstrate that two-worker households adjust their home–work spatial configuration which results in commuting trade-offs between individuals. This research uses the angle between the two workplaces, measured at the home location, as a variable in its empirical model. This variable is a descriptor of the home–work spatial configuration and a predictor of total household commute distance. The modeling results indicate an inverse relationship between total household commute distance and the difference between individual commutes. This suggests that individuals in two-worker households trade off their individual commute distances and, in that process, reduce total household commute distance. A key policy implication arising from this research relates to the jobs–housing balance within a catchment area. Two-worker households have been regarded as a hindrance to achieving jobs–housing balance as the two work locations present a constraint in relocating the home near both work locations. However, as this research shows, workers will trade off their individual commute distances such that a home relocation results in a shorter commute distance for one worker and longer commute distance for the other.

Where people live and work and how they travel between those destinations are among the most important considerations of land-use and transportation planning. It is therefore important to understand the behaviors of individuals and households whose decisions ultimately shape the urban form. One such important behavior is the commuting trade-off made between partners within a two-worker household in respect of their individual commute distances. Commuting trade-offs in this context occur when a home relocation results in one worker increasing their commute distance to reduce their partner's commute distance. Our research adds new findings to the existing body of research on two-worker households and reveals important aspects of commuting trade-offs that will be useful in long-term land-use and transportation planning.

This research investigates households of two out-of-home workers with home and work locations in the Greater Toronto Area (GTA). Two-worker households are an important group to study because in the City of Toronto alone, two-worker households comprise up to 28% of the population (*1*). Also, households where both workers travel outside of the home to two separate work locations are an interesting household type to examine as they face the dual constraint of considering two work locations when making commuting choices. That dual constraint can lead to commuting trade-offs between individuals in the household. Our research examines those commuting trade-offs using stated preference data and multilevel modeling.

This research relies on stated preference data, which—unlike revealed preference data—has the ability to find relationships and explain non-existing situations or choices (*2*). It also uses multilevel modeling techniques, which identify variation between households located in

[1]Department of Civil Engineering, University of Toronto, Toronto, Canada

Corresponding Author:
Address correspondence to Khandker Nurul Habib: khandker.nurulhabib@utoronto.ca

different geographic areas (regional municipalities), and provide significant results.

Literature Review

Understanding two-worker households is important for planning for future travel demand as some researchers estimate that the number of two-commuter households is growing and will make up a significant share of urban residents in the near future (*3–5*). This portion of the urban population is unique in that they face dual constraints with home and commuting choices because they have two workplace locations, which results in home–commuting trade-offs (*6*). These home–commuting trade-offs can have significant spatial effects on an urban region (i.e., where people work and live); therefore, to plan appropriately for the future, their behaviors and patterns should be considered in urban land-use and transportation policies. Tscharaktschiew and Hirte (*7*) explain that overlooking the influence of two-worker households could create an urban economy and spatial structure ill-suited for a significant part of the population.

Much of the literature on two-worker households has been dedicated to measures that capture the spatial configuration of the home and work locations. This includes creative measures, such as the use of angles, which have often been employed for understanding residential locations (*8*). Generally, there are two types of angles used throughout the existing literature: (1) the home–central business district (CBD)–workplace angle; and (2) the workplace 1–home–workplace 2 angle. The home–CBD–workplace angle, set out in Figure 1, is employed by several studies in the existing literature.

Vandersmissen et al. (*9*) used the home–CBD–workplace angle in their study, which examines spatial and social reasons for increased commuting distances over several decades in Quebec City, Canada. The authors explained the rationale of using this angle to measure the relationship between directional bias and work-trip length as follows: in a monocentric city, commuting distance should increase as the home–CBD–workplace angle increases; however, as a city evolves away from this city form, commuting time is less impacted by the home–CBD–workplace angle (p. 451). This relationship was true for their study of Quebec City, which has over time experienced a restructured transportation network and changes in home and work locations that reflect a polycentric urban form.

The use of this angle is most appropriate for a monocentric city, in which case a narrower angle is ideal as it would lead to faster travel. This is based on the theoretical assumption that it is generally faster to travel in the direction of the CBD, since transportation networks are designed to access the CBD. In the context of non-monocentric cities, the angle of workplace 1–home–workplace 2 has been used to describe the positioning of two-worker households relative to their associated workplaces (*8*). In using this approach, acute angles (less than 90°, the extreme being 0°) indicate that commuters travel in similar directions, while obtuse angles (greater than 90°, the extreme being 180°) indicate that commuters travel in dissimilar directions. Of these described angles, also identified in Figure 1, our study considers the angle of workplace 1–home–workplace 2.

Van Ommeren (*8*) evaluated the influence of the distance between workplaces as well as the angle created at the home location as factors influencing the job and residential search behavior of two-worker households. The

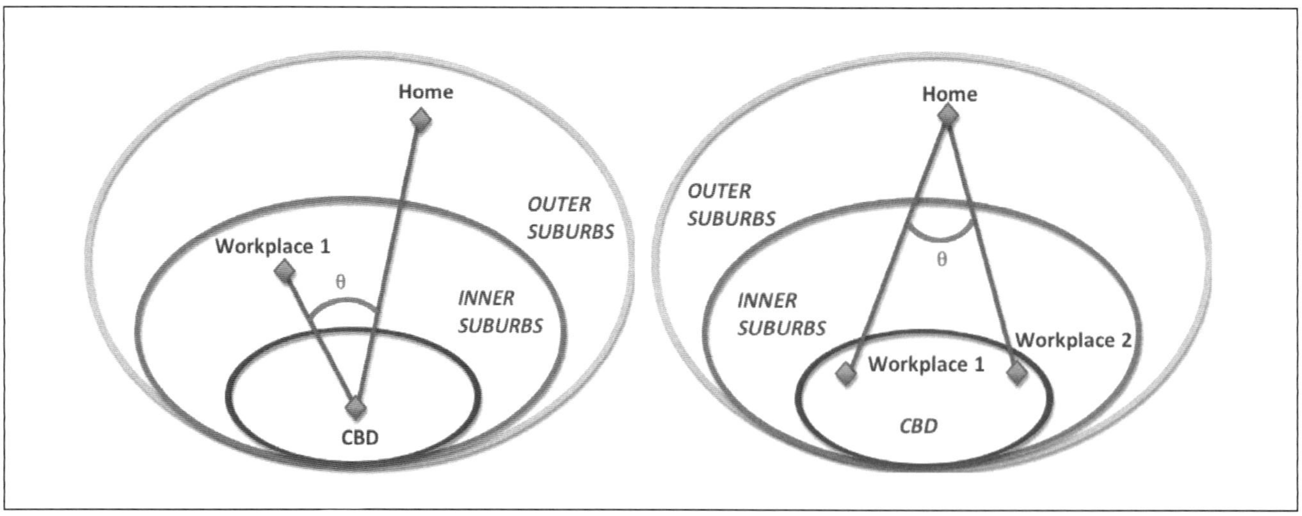

Figure 1. Home–central business district–workplace angle and workplace 1–home–workplace 2 angle.

author explains that in more than half of the households, the distance between the workplaces is less than 10 km, and in 85% of the households the angle is smaller than 90° (p. 385). Therefore, the typical household in this study is not located between the workplaces; instead, the wage earners travel to their work in the same direction (pp. 384–385). The deduction here is that acute angles indicate commuters traveling in the same direction, while obtuse angles indicate commuters traveling in opposite directions. Another study by Akbari and Habib (*10*) found a positive relationship between commuting modal accessibility and the workplace 1–home–workplace 2 angle, meaning that higher accessibility leads to a widening of this angle. In that study, modal accessibility was defined as a measure that considers the expected maximum utility that households derive from all modes of transportation.

There is much debate in existing research around whether two-worker households commute more or less than their single-worker counterparts. Intuitively, one would expect that two-worker households commute more than one-worker households simply because of the addition of a commute trip per household. In White's (*3*) study, this was cited as one factor that would likely explain why some urban workers choose to commute more than the average minimum commuting journey length. Since workers in two-worker households must commute to two different workplaces from a single residence, it is more likely that one or both workers will choose an outward or circumferential commute, which would then lead to relatively longer commute time and distance (excess commuting). Despite this intuitive reasoning, the existing literature on the subject has not been conclusive on that point and many, such as Kim (*6*), have reported the opposite findings.

Kim (*6*) analyzes the commuting behavior of two-worker households by accounting for the constraints they face, and then calculating that individuals in two-worker households in fact have, on average, shorter commute distances than those in one-worker households. Other studies, such as that of Sultana (*11*), reach similar conclusions, which ultimately lead us to consider the reasons or factors influencing this reported travel behavior. Kim (*6*) and Ma and Banister (*12*) make the same conclusion that two-worker households minimize commuting distance, but go further to say that the commuting distances of workers within two-worker household are complementary to one another.

While several studies investigate whether two-worker households commute more or less than their single-worker counterparts, their assertions require empirical investigations of the factors that support their findings. Some research has identified the trade-offs made by individuals in two-worker households as one such explanatory factor. Trade-offs arise out of the need to accommodate the home-to-work trips of two individuals residing in the same location but working in potentially very different locations. In a recent study, Surprenant-Legault et al. (*13*) consider commuting mode choice as an exogenous variable to the commuting distance regression model. Their study found that for every 1% increase in one partner's distance, total commute distance increased by less than 1% and therefore leads to the conclusion that individuals in two-worker households are in fact trading off commuting distance.

This empirical investigation in this research uses stated preference (SP) data, which are preferred over revealed preference (RP) data because of their ability to find relationships and explain non-existing situations or choices and their covariates. As Ben-Akiva (*2*) explains, SP data are based on respondents' intentions or choices within some hypothetical scenario. SP data have several advantages compared to RP data. Those advantages over RP data include using pre-specified choice sets and an extended range of attributes. Conversely, SP data do have some negative aspects, which researchers have tried to account for. Unlike RP data, SP data are not based on actual market behavior and therefore the validity and accuracy of their predictions of actual market behavior are often challenged (*2*). To account for these negative aspects, this research combines both RP and SP data. This has been an effective method used in previous research as it addresses the validity issue with SP data and improves the accuracy of the parameter estimates.

There is a gap in the existing literature for empirical analyses of the commuting trade-offs between individuals in a two-worker household using reliable SP data. Only by using SP data can one make conclusions about household decisions in respect of commuting trade-offs. Our research fills this gap by employing a multilevel modeling technique to examine commuting trade-offs based on SP data collected in the GTA.

Research Design

For this research, econometric models are developed for total household commuting distance. The models employ several explanatory variables that describe the spatial configuration of home and work locations in the urban environment, including *workplace 1–home–workplace 2 angle* and the *difference between longest and shortest Euclidean home-to-work distances*.

The variable *workplace 1–home–workplace 2 angle* is the angle between the two work locations measured from their common home location. This variable is a proxy for "optimality" of two-worker households' home location with respect to the two work locations, since any location on a line connecting and in between the two work locations results in a 180° angle and the shortest possible household commuting distance, provided the two work locations are fixed. Therefore, home locations that yield a larger (or wider) angle are considered more optimally located because they result in shorter household commuting distance. One can object that distance minimization is not the only factor that households consider. Other important factors include real estate prices, availability of housing stock, neighborhood characteristics, and lifestyle preferences, all of which may not be optimal at these locations from the household's perspective. The importance of these factors is acknowledged; however, once these are controlled for, a general tendency toward distance minimization should be observed.

In this research we wish to study the relationship between this angle and total household commuting distance. The sprawling urban environment of the GTA is such that if households living in outlying suburbs and working in the CBD moved closer to their workplaces, there would be a gradual reduction in household commute distance and a widening of their *workplace 1–home–workplace 2 angle* (refer to Figure 2). We therefore hypothesize that there is an inverse relationship between this angle and Euclidean household commute distance. There is no theoretical reason for this inverse relationship. Euclidean geometry permits having larger angles and larger distances at the same time. An inverse relationship would be an indication that households adjust their spatial configuration to reduce total Euclidean household commute distance.

The variable *difference between longest and shortest Euclidean home-to-work distances* is used in the household-level regression model to provide insights about commuting trade-offs. If the difference between individual commutes increases by a certain amount, and the total household commute distance increases by the same amount or greater, then it can be concluded that the individual commute distances are compliments (meaning that partners do not make trade-offs). If the difference between individual commutes increases by a certain amount and the total household distance does not increase by the same amount, or if it decreases, it can be concluded that individual commute distances are substitutes. However, a commuting trade-off can only occur when one worker increases their share of the total household commute distance to reduce their partner's commute distance (illustrated in Figure 2). Therefore, if an inverse relationship is observed between total household commute distance and the difference between individual commutes, then this is an indication that two-worker households make commuting trade-offs.

Study Area and Survey Design

The study area is the GTA, which comprises the census divisions of Toronto, Durham, Halton, Peel, and York. It covers 7,200 km^2 (*14*) and its population in 2016 was 6.4 million (*15*). Within the GTA, there is a notable concentration of residential and business areas in the City of Toronto, which is considered the most significant CBD in the region. Its significance is reflected in the relatively

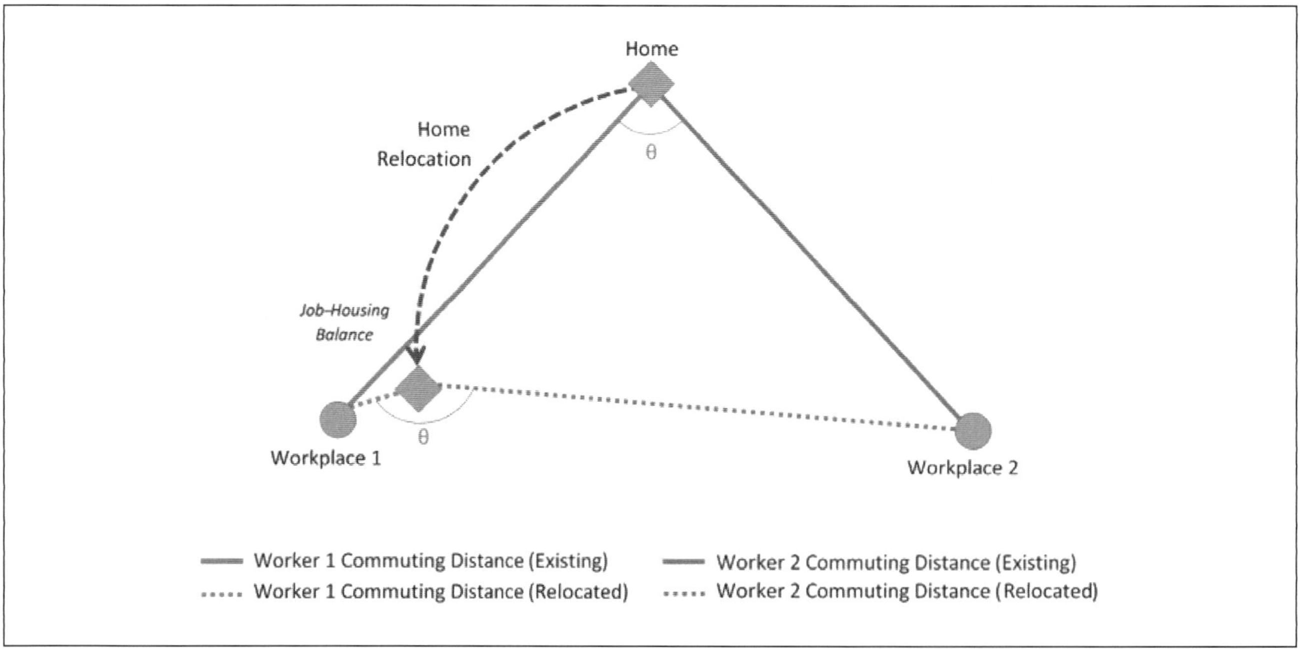

Figure 2. Commuting trade-offs.

higher-quality public transportation network that serves Toronto. In surrounding regions, there is notably less public transit and more auto-oriented infrastructure. There is a high number of cross-regional commuting trips (one in four trips in the GTA and Hamilton) made by GTA residents and the existing transportation network lacks the ability to support those trips, therefore making those trips unattractive and not financially feasible (*16*).

This study uses data from a stated-adaptation-based survey entitled *Car and Home Ownership Decisions in the Face of Increasing Commuting Expenses* (CHOICE) (*17*). The CHOICE survey collected mobility tool ownership, home relocation, and home location choices from GTA households with at least one cross-regional commuter. All interviews were conducted for this survey between March and April 2014. The survey consisted of an RP component, which collected sociodemographic information, characteristics of typical commuting trips, and retrospective housing and vehicle ownership information, as well as an SP component with two different experiments, namely SP1 and SP2.

In SP1, respondents were presented with a maximum of four scenarios of hypothetical commuting costs. In each scenario, respondents were requested to select a mobility tool "bundle" and/or relocate homes in response to the presented commuting costs. The mobility tool "bundle" included different combinations of car type, fuel technology, and transit pass ownership. Each bundle had an associated cost which respondents considered in their selection. Scenario 1 represented households' existing commuting costs based on the home and work locations and existing mobility tool ownership as reported in the RP component of this survey. In each subsequent scenario, commuting costs were significantly increased to induce households to adjust their mobility tool "bundle" as well as to change home location. If at any point in the scenarios commuting costs were considered excessive, respondents could choose to relocate their residence to mitigate that cost.

After voluntarily choosing to relocate, or after completing all four scenarios of SP1, respondents were asked to complete SP2. The purpose of SP2 was to have respondents choose a new residential location in reaction to the commuting cost increase in SP1. Respondents were asked to choose a new residential location by reviewing maps and selecting the desired region, city, or township, and neighborhood. Respondents were presented with qualitative neighborhood characteristics in addition to housing prices of typical dwellings in the neighborhood. Respondents were able to review these neighborhood and dwelling attributes and decide whether they met household needs—otherwise they may continue searching elsewhere.

Data Preparation and Statistics

The household location choices revealed in SP2 of the CHOICE survey provide the basis for investigating commuting trade-offs between individuals in two-worker households. A total of 413 households completed the full survey questionnaire. Of this sample, a total of 90 households had two out-of-home workers, meaning that their workplaces were somewhere other than their home, and thus this subset is used for empirical investigation. Given that two-worker households only make up about 28% of all households in Toronto, it is understandably difficult to obtain a large sample of survey responses for two-worker households, especially for an SP survey designed to collect detailed data. The central limit theorem requires a minimum of 30 samples for a normally distributed sample. For the purpose of this study, the sample size of 90 is considered adequate, especially considering that a stratified sampling strategy was utilized for collecting survey responses which minimize sampling error.

Some respondents reported that their households had more than two workers. In these cases, we defined the second worker in the respondent's household as the individual who made the highest income from the remaining workers (excluding the respondent). This was based on the rationale that the higher-income individual likely contributes the most to household expenditures and has a larger impact on household decision-making. In addition to the socioeconomic variables, this research also uses geometric variables to describe the spatial configuration of two-worker households (refer to Table 1). These variables were calculated using respondents' existing workplace locations, and the chosen home locations reported in SP2.

Methodology

The two models presented in this paper include a Cobb–Douglas log-linear model and a multilevel random intercept-only model. In both models, the dependent variable (*D*) is *total Euclidean household commute distance*, which is expressed as a function of housing attributes, household income, and the three geometrical variables that describe the spatial configuration of home and work locations. This methodology builds on the work by previous studies of two-worker households in two significant ways: (1) it uses SP data of people's home location choices, which has the advantage of identifying causal relationships among variables; and (2) it uses a multilevel modeling technique to improve group-level variation in the data.

The first model presented in this paper is the Cobb–Douglas log-linear model, whereby the natural logarithm of the dependent variable is expressed as a function of

Table 1. Definition and Numerical Summary of Variables in Models

	Definition	Minimum	Median	Mean	Maximum
Total household commute distance	This is the combined commute distance of both workers in the household in the SP scenario (km)	5.52	39.98	51.89	191.22
Household income	This is the combined income of both workers in the household ($)	57,500	137,500	139,194	250,000
Area of current home	This is the gross floor areas of a household's current home (square feet)	397	1,600	1,730	4,500
Owns current home	This is a dummy variable which takes the value of 1 if the household owns their current home	n/a	n/a	n/a	n/a
Workplace 1–home–workplace 2 angle	The angle between the two workplaces relative to the home location in the SP scenario (°)	0.342	33.30	49.94	147.74
Difference between longest and shortest commute distances	The difference between the longest and shortest home–work distances of two workers in the SP scenario (km)	0.197	9.72	12.86	71.1
Distance between workplace 1 and workplace 2	Distance between the two workers' workplaces in the SP scenario (km)	0.3	17.75	20.38	71.6

the natural logarithms of independent variables. Since the dependent variable is continuous, meaning that there are an infinite number of possible outcomes, the Cobb–Douglas log-linear model is appropriate for the dataset. Two major advantages of this model are that the estimated coefficients of variables represent their respective elasticities with respect to the dependent variable and that its formulation resembles the demand function associated with the Cobb–Douglas utility equation (*18*). The following is the formulation of said model:

$$\ln(D) = \beta_o + \sum \beta_i \ln(x_i) + \sum \gamma_j y_j, \tag{1}$$

where D is the dependent variable *total Euclidean household commute distance*, β_o is the intercept of the model, β_i refers to the vector of coefficients corresponding to the set of continuous independent variables x_i, and γ_j refers to the vector of coefficients corresponding to the set of dummy independent variables y_j. The final model is estimated using the ordinary least squares (OLS) regression method using the "lm" package in the statistical software R.

The first model used in this research is a classical regression model, which does not take into account repeated observations or variation across groups of data. The ability to model variation across groups is a critical feature that distinguishes multilevel modeling from classical regression (*19*). The SP data used in this study inherently has a multilevel structure, whereby households can be grouped into categories based on whether they live in urban or suburban locations. Since the classical regression model is unable to model the variation between these groups, a multilevel model is developed. For the multilevel model, the households in the data are grouped into five geographic areas delineated by municipal boundaries. These five groups include the City of Toronto and the suburban regional municipalities of Durham, York, Peel, and Halton. The rationale to model the variation across these groups is based on the concept that differences exist in commuting patterns between urban and suburban households because of factors such as lifestyle preferences and different land-use characteristics. To the authors' knowledge, there are no studies that use multilevel modeling to analyze commuting trade-offs of households.

A two-level model is adopted for this study. The upper level represents the geographic areas where households reside and the lower level corresponds to the households. The following is the multilevel random intercept model in which the only coefficient that varies across groups is the intercept α:

$$y_i = \alpha_{j[i]} + \beta_1 x_{i1} + \beta_2 x_{i2} + \varepsilon_i i = 1, .., N,\ j = 1, .., J \tag{2}$$

$$\alpha_j = \mu_\alpha + \eta_j, \text{with } \eta_j \sim N(0, \sigma^2{}_\alpha) j = 1, \ldots.., J, \tag{3}$$

where y_i is the observed value of total household commute distance corresponding to observation i, x_{i1} is the ith element of vector x_1, which represents the set of continuous independent variables, x_{i2} is the ith element of vector x_2, which represents the set of dummy independent variables, β_1 is a set of coefficients corresponding to x_1 and β_2 is a set of coefficients corresponding to x_2. Here α_j represents the model intercept of group j. The intercept α_j is constant for all observations within a group while varying between different groups (i.e., City of Toronto, Region of Durham). ε_i and η_j are independent households and group error terms that follow a normal distribution with mean 0 and variance σ^2.

To interpret the elasticities of the independent variables, Equation 2 is expressed in a log-linear format as follows:

$$\ln(y_i) = \alpha_{j[i]} + \sum \beta_1 \ln(x_{i1}) + \sum \beta_2(x_{i2}) i = 1, .., N, \; j = 1, .., J. \quad (4)$$

Development of the log-linear random intercept model follows a process whereby the model is built up gradually. First, a baseline generalized least squares model, denoted as Model 1, is developed. In Model 1, *total household commute distance* is expressed only as a function of an intercept. Second, we estimate a model denoted Model 2, which also expresses *total household commute distance* as a function of the intercept, but allows the intercept to vary across different groups in the data (i.e., City of Toronto, Region of Durham). The Akaike information criterion (AIC), Bayesian information criterion (BIC), and log-likelihood values are compared between the two models. Since the AIC, BIC, and log-likelihood values decreased from Model 1 to Model 2, this is an indication of a better fit to the model; therefore, a multilevel structure should be used in the model. Finally, the multilevel random intercept model, denoted as Model 3, is estimated, which expresses *total household commute distance* as a function of several independent variables. The AIC, BIC, and log-likelihood values for all three models are shown in Table 2. The log-likelihood values of the models are compared using a chi-square difference test. The chi-square value obtained has a reasonable level of significance. Based on the results, Model 3 is determined to be the best fit to the data and is therefore used in ensuing results discussion.

Empirical Results

The final models include variables that are statistically significant and have intuitive signs. The 95% confidence limit for one-tailed tests (critical t-statistics value of 1.64) was used to identify the estimated parameters' statistical significance. One-tailed tests are performed in accordance with our a priori intuition and theoretical understanding of the factors that affect household commute distance. The final models have the same independent variables; however, the variables have slightly greater coefficient values in the Cobb–Douglas model than in the multilevel random intercept model. The multilevel model accounts for group-level variation by allowing the model intercept to vary across the geographic areas where two-worker households reside. For example, the model intercept for the City of Toronto is 1.9342, while for the Region of Durham the intercept is 2.2742. The mean of all of the intercepts is 2.1132.

All variables related to housing or home location are based on the reported home location choices in the SP survey. The positive value of the coefficient for the logarithm of *household income* suggests a positive association between household income and total Euclidean household commute distance. This finding is perhaps intuitive as higher-income households have greater financial means to incur greater commuting costs resulting from living farther away from workplaces. This finding is also consistent with Alonso's (*20*) bid-rent theory, which states that higher-income households are more interested in maximizing utility gained from living in larger spaces of land and are less impacted by commuting costs.

The variable *area of current home* represents the total floor area of the respondents' current homes. The estimated coefficient of the logarithm of the *area of the current home* has a negative value, suggesting that households with larger homes chose to relocate in such a manner that reduces commute distance. The explanation for this finding is perhaps that respondents are choosing to relocate into smaller homes with lower maintenance costs. The variable *owns a current home* is a categorical variable that represents the tenure of a household's current residence. The estimated positive coefficient of this variable suggests a positive association between owning a home and higher total commute distance.

The following three geometrical variables are included in the presented models: (1) *workplace 1–home–workplace 2 angle*; (2) *Euclidean distance between workplace 1 and workplace 2*; and (3) *difference between longest and shortest Euclidean home-to-work distances*. These variables describe the spatial configuration of home and work locations of two-worker households based on the reported home location choices in the SP survey. It is important to study the effects of these variables as they provide an innovative way of assessing the influences that land use has on household commute distance and consequently travel demand. The modeling results indicate that the estimated coefficient for the logarithm of *workplace 1–home–workplace 2 angle* has a negative value, which confirms our a-priori hypothesis that an inverse relationship exists between this angle and total commute distance. There is no theoretical reason for the said angle to be inversely related to total distance; however, the explanation of the inverse relationship lies in the GTA's suburban sprawl, whereby we find a high portion of low-density residential areas in the urban periphery and work concentrated in the CBD. In this scenario, an inverse correlation between total commute distance and the said angle can be expected. It is also an indication that households adjust their spatial configuration to reduce total Euclidean household commute distance.

Table 2. Model Outputs

AIC, BIC, and log-likelihood ratio test results

	AIC	BIC	Log-likelihood	Test	Likelihood ratio	*p*-value
Model 1	199.359	204.36	−97.68			
Model 2	169.728	177.23	−81.86	Model 1 versus 2	31.63	<0.0001
Model 3	85.392	107.89	−33.7	Model 2 versus 3	96.34	<0.0001

Cobb–Douglas model

	Parameter estimate	*t*-stats
Intercept	2.5467	1.673
In household income ($)	0.2251	1.726
In area of current home (square feet)	−0.227	−2.379
Owns home (dummy)	0.2892	2.181
In workplace 1–home–workplace 2 angle	−0.5775	−14.5
In difference between longest and shortest commute distances (m)	−0.226	−4.522
In distance between workplace 1 and workplace 2 (m)	0.848	10.71
Multiple *R*-squared	0.7454	
Adjusted *R*-squared	0.727	

Multilevel random intercept

	Parameter estimate	*t*-stats
Fixed effects		
Intercept	2.1132	1.474
In household income ($)	0.2299	1.87
In area of current home (square feet)	−0.1676	−1.801
Own home (dummy)	0.233	1.85
In workplace 1–home–workplace 2 angle	−0.5169	−12.253
In difference between longest and shortest commute distances (m)	−0.2203	−4.695
In distance between workplace 1 and workplace 2 (m)	0.765	9.759
Random intercept		
Group	Parameter estimate	
City of Toronto	1.9342	
Region of Durham	2.2742	
Region of York	2.0969	
Region of Peel	2.0415	
Region of Halton	2.2194	

The estimated coefficient for the second geometrical variable *Euclidean distance between workplace 1 and workplace 2* has a positive value, suggesting a positive association between total commute distance and the physical distance between the two workplaces. This finding suggests that for households who have longer commute distances, workplaces are more "spread out."

The estimated coefficient for the variable *difference between longest and shortest commute distances* has a negative value, indicating an inverse relationship with *household commuting distance*. This suggests that commuters in two-worker households trade off their individual commute distances and, in that process, reduce total household commute distance. Recall that a commuting trade-off occurs when one worker incurs a longer proportion of the household commute distance to allow for the second worker to reduce their individual commute. However, an increase in the difference between individual commutes can also occur when total household commute distance increases. Therefore, a trade-off only occurs when the increase in difference between individual commutes leads to a reduction in total household commute distance.

In a recent study, Surprenant-Legault et al. (*13*) used a regression model and found that for every 1% increase in the difference between individual commutes, total commute distance increased by less than 1% and therefore concluded that individuals in two-worker households were trading off commuting distances. However, their result can also exist in many instances such as a household relocation that increases total commute distance. Furthermore, conclusions about trade-offs can only be made if households are observed to make commuting changes in this manner, therefore the need for SP data, which were not used in their study. Conversely, the modeling results of our research indicate that households will trade off individual commute distances such that a 1% increase in the difference between individual commutes results in a 0.22% reduction in total household commute distance. As the increase in the difference between individual commutes is positively associated with a reduction in household commute distance, it can be concluded that one worker increases their individual commute distance while the other worker reduces theirs.

Conclusions

This research made two significant advancements to the body of literature on two-worker households. Unlike previous studies, this research used SP data rather than RP data because of its ability to find relationships between variables based on hypothetical scenarios. Also, this research used multilevel modeling techniques, which allowed us to capture variation across different groups of households within the dataset.

In using the methods described above, two key conclusions were made. The model results showed that an inverse relationship exists between *workplace 1–home–workplace 2 angle* and total Euclidean household commute distance, whereby a smaller angle was associated with a longer total Euclidean household commute distance. The described scenario indicated the presence of a sprawling urban form whereby workers commute long distances into the CBD from a traditional suburban area with limited employment uses and high concentrations of residential land uses. Conversely, a larger angle was tied to shorter household commute distance, indicating that the home and work locations were in proximity to each other, which is a common characteristic of mixed-use urban environments.

The results from this research also revealed that as two-worker households trade off their commuting distances, such that the difference in their individual commutes increases, the total household commute distance decreases. An overarching policy direction in land-use and transportation planning is the reduction of commuting distances as it has a negative effect on sustainability, especially when commutes rely on the automobile. The results of this research lead us to conclude that the correlation between reducing total household commute and commuting trade-offs is significant and can form the basis of policy improvements.

Policy Implications

The movement of people in an urban environment is an important consideration for policy-makers, and the home-to-work commute contributes a significant proportion to that movement daily. Two-worker households comprise a significant and growing portion of GTA households and, as such, their home location decisions relative to where they commute to for work is an important consideration when planning for current and future urban activity systems. Two-worker households are unique in that they perform two home-to-work commutes from a single home location, and therefore, that home location choice can be seen as twice as important when planning for the movement of people in an urban environment.

A general and longstanding policy directive of transportation and land-use planners in urban regions such as the GTA is to reduce the amount of travel by automobiles and encourage the use of active transportation modes. Long commute distances, necessitated by isolated land uses, are a hindrance to that objective. By focusing on mixed-use development, urban regions can move away from the division of land uses, and for home-to-work commuting issue planners can look to achieve a "job–housing balance." The "job–housing balance" has been the topic of much research and is generally achieved when there is heterogeneity between employment locations and the home locations of workers. Many have argued that a "job–housing balance" is not achievable because of two-worker households since they are unable to locate near both workplaces (*21–24*).

Our research challenges that argument by revealing that two-worker households will adjust their home locations to be closer to one of the work locations. Workers in a two-worker household will make commuting tradeoffs such that one worker will incur a longer commute for the second worker to reduce their commute distance. Our study has found that by making commuting tradeoffs, two-worker households are achieving some level of job–housing balance by locating closer to one of the two work locations. In doing so, our research also reveals that two-worker household are achieving the overarching policy directive of reducing total household commute. Policy-makers can continue to promote "job–housing balance" in the urban environment as a method of reducing commuting distance and reducing the negative side effects associated with daily long-distance commutes by car (e.g., traffic congestion and pollution).

Acknowledgments

The authors acknowledge Elli Papaioannou for her work in designing and conducting the survey that produced data necessary for this study.

Author Contributions

The authors confirm contribution to the paper as follows: study conception and design: SA, KNH; data collection: KNH; analysis and interpretation of results: SA, KNH; draft manuscript preparation: SA, KNH. All authors reviewed the results and approved the final version of the manuscript.

References

1. Data Management Group. *Transportation Tomorrow Survey*, 2013. http://www.dmg.utoronto.ca/transportationtomorrowsurvey. Accessed July 2013.
2. Ben-Akiva, M. Combining Revealed and Stated Preferences. *Marketing Letters*, Vol. 5, No. 4, 1994, pp. 335–350.
3. White, M. J. Urban Commuting Journeys are not "Wasteful." *Journal of Political Economy*, Vol. 96, No. 5, 1988, pp. 1097–1110.
4. Hotchkiss, D., and M. J. White. A Simulation Model of a Decentralized Metropolitan Area with Two-Worker, Traditional, and Female-Headed Households. *Journal of Urban Economics*, Vol. 34, 1993, pp. 159–185.
5. Clark, W. A. V., Y. Q. Huang, and S. D. Withers. Does Commuting Distance Matter? Commuting Tolerance and Residential Change. *Regional Science and Urban Economics*, Vol. 33, No. 2, 2003, pp. 199–221.
6. Kim, S. Excess Commuting for Two-Worker Households in the Los Angeles Metropolitan Areas. *Journal of Urban Economics*, Vol. 38, 1995, pp. 166–182.
7. Tscharaktschiew, S., and G. Hirte. How Does the Household Structure Shape the Urban Economy?. *Regional Science and Urban Economy*, Vol. 40, 2010, pp. 498–516.
8. Van Ommeren, J. Job and Residential Search Behaviour of Two-Earner Households. *Papers in Regional Science*, Vol. 79, No. 4, 2000, pp. 375–391.
9. Vandersmissen, M. H., P. Villeneuve, and M. Thériault. Analyzing Changes in Urban Form and Commuting Time. *Professional Geographer*, Vol. 55, No. 4, 2003, pp. 446–463.
10. Akbari, S., and K. M. N. Habib. An Econometric Investigation on the Relationship between Modal Accessibility and the Home–Work Spatial Configuration of Two-Commuter Households: The Case of the National Capital Region (NCR) of Canada. *Transportmetrica A*, Vol. 11, No. 10, 2015, pp. 897–917.
11. Sultana, S. Effects of Married-Couple Dual-Earner Households on Metropolitan Commuting: Evidence from the Atlanta Metropolitan Area. *Urban Geography*, Vol. 26, No. 4, 2005, pp. 328–352.
12. Ma, K.-R., and D. Banister. Excess Commuting: A Critical Review. *Transport Reviews*, Vol. 26, No. 6, 2006, pp. 749–767.
13. Surprenant-Legault, J., Z. Patterson, and A. M. El-Geneidy. Commuting Tradeoffs and Distance Reduction in Two-Worker Households. *Transportation Research Part A*, Vol. 51, 2013, pp. 12–28.
14. Hodge, G., and I. M. Robinson. *Planning Canadian Regions*. UBC Press, Vancouver, Canada, 2001.
15. Statistics Canada. *2016 Census of Canada Population and Dwelling Count Highlight Tables, for Census Divisions, 2016 Census*. Catalogue no. 98-316-X2016001. Statistics Canada, Ottawa, Canada, 2017. http://www12.statcan.gc.ca/census-recensement/2016/dp-pd/hlt-fst/pd-pl/Table.cfm?Lang=Eng&T=701&S=3. Accessed July 1, 2017.
16. Metrolinx. *The Big Move: Transforming Transportation in the Greater Toronto and Hamilton Area*. Queen's Printer for Ontario, Ontario, Canada, 2008.
17. Papaioannou, E., and K. M. N. Habib. *A Study of Car and Home Ownership Decisions in the Face of Increasing Commuting Expenses (CHOICE) in the Greater Toronto Area (GTA)*. University of Toronto, Toronto, Canada, 2014.
18. Oum, T. H. Alternative Demand Models and their Elasticity Estimates. *Journal of Transport Economics and Policy*, 1989, pp. 163–187.

19. Gelman, A., and J. Hill. *Data Analysis Using Regression and Multilevel/Hierarchical Models*. Cambridge University Press, Cambridge, United Kingdom, 2006.
20. Alonso, W. A Theory of the Urban Land Market. *Papers in Regional Science*, Vol. 6, No. 1, 1960, pp. 149–157.
21. Giuliano, G. Is Job Housing Balance a Transportation Issue? *Transportation Research Record: Journal of the Transportation Research Board*, 1991. 1305: 305–312.
22. Giuliano, G., and K. A. Small. Is Journey to Work Explained by Urban Structure? *Urban Studies*, Vol. 30, 1993, pp. 1485–1500.
23. Levine, J. Rethinking Accessibility and Jobs–Housing Balance. *Journal of the American Planning Association*, Vol. 64, 1998, pp. 133–150.
24. Dieleman, F. M., M. Dist, and G. Burghouwt. Urban Form and Travel Behavior: Micro-Level Households Attributes and Residential Context. *Urban Studies*, Vol. 39, 2002, pp. 507–527.

The Standing Committee on Transportation and Land Development (ADD30) peer-reviewed this paper (18-04752).

All views and interpretations made in the paper belong only to the authors.

Article

TRR
JOURNAL OF THE TRANSPORTATION RESEARCH BOARD

Transportation Research Record
2018, Vol. 2672(3) 80–91

DOI: 10.1177/0361198118797459
journals.sagepub.com/home/trr
SAGE

Relationship between Neighborhood Characteristics and Demand for Emergency Health Service Vehicles: A Poisson Hurdle Regression Modeling Approach

Muhammad Ahsanul Habib[1], Babatope Olajide[2], Mikiko Terashima[3], and Sara Campbell[2]

Abstract

The main objective of this study is to explore the spatial and temporal variability of demand for emergency health service vehicles, measured at the 1 km-by-1 km grid level in Halifax, Nova Scotia, Canada. This study utilizes and compares a Poisson regression and Poisson hurdle regression model that examine the effects of neighborhood characteristics on emergency health service vehicle demand. It also develops a time-segmented model to investigate the temporal variability of the effects of factors considered in relation to demand. It analyzes the Nova Scotia emergency health service administrative database for the period from January 2012 to December 2012. A comprehensive set of socio-demographic attributes, land use characteristics, and measures of accessibility to services are used to achieve the objectives of this research. Results found that demand for emergency health service vehicles was higher in areas where there was higher population density, more heterogeneous land uses, and a larger proportion of the population aged 40 years and above. The time-segmented model shows demand was highest during morning peak periods for residential areas, afternoon peak for commercial areas, and morning peak and mid-day for members of the population aged over 75 years. An assessment of the time-segmented model suggests that the generic model is sufficient for predicting how neighborhood characteristics relate to the demand for emergency health service vehicles. The findings of this study will be beneficial for urban planners and health professionals in designing healthy cities and targeting health promotions to reduce the need for emergency services.

Emergency health services (EHS) are often the first responders to provide medical attention and transportation during emergencies. In recent years, there has been a considerable increase in the demand for EHS vehicles in the United States and other countries, for example, Australia (*1–3*). The primary causes of this trend appear to be: recurrent need for doctor visits, domestic accidents, automobile collisions, heart attacks, and many other unexpected illnesses suffered by increasingly aging populations. The built environment can also have an effect on population health, including risk of exposure to traffic and air pollution, and physical inactivity (*4, 5*). Understanding how the continuous increase in the rate of demand for EHS vehicles affects different population groups is essential to improve emergency healthcare planning and reduce operational costs as well as implement necessary actions to combat potential deterioration in EHS quality. This paper utilizes and compares Poisson regression and Poisson hurdle regression models to explore the neighborhood characteristics affecting demand for EHS vehicles, which receives limited attention in the existing literature.

This study investigates the relationships between EHS vehicle demand and socio-demographic attributes, land use characteristics, and accessibility measures, at the 1 km-by-1 km grid level in Halifax, Nova Scotia,

[1]School of Planning and Department of Civil and Resource Engineering, Dalhousie University, Halifax, Canada
[2]Dalhousie Transportation Collaboratory, Dalhousie University, Halifax, Canada
[3]School of Planning and Department of Community Health and Epidemiology, Dalhousie University, Halifax, Canada

Corresponding Author:
Address correspondence to Muhammad Ahsanul Habib: ahsan.habib@dal.ca

Canada. A Poisson hurdle regression model is employed to explore the geographic and temporal variation in demand for EHS vehicles. The hurdle model provides on and off states for requesting emergency response. The Poisson hurdle model was preferred over other count models (e.g., zero-inflated Poisson) as it assumes that the existence of excess zeros in the grids is from a structural source. As the zero and non-zero counts are modeled uniquely, the hurdle model seems advantageous in accommodating the large proportion of zeros and a right-skewed distribution for the positive counts. This paper utilizes the Nova Scotia Emergency Health Service (NS EHS) administrative database for the period from January 2012 to December 2012. The dataset consists of 24,403 cases of emergency transportation to hospitals in Halifax. Neighborhood attributes at the grid level are derived using ArcGIS functions. A comprehensive set of socio-demographic attributes (e.g., population density, average household income, and employment rate), land use characteristics (e.g., land use type and dwelling characteristics) and accessibility measures (e.g., distance to park and hospital) are used to achieve the objectives of this research. The study first examines spatial factors and their influence on EHS vehicle demand using all observations. Then, it develops five time-segmented models: AM (morning) peak, midday, PM (afternoon) peak, evening, and overnight, to determine the effects of the factors considered with respect to temporal variation in demand.

Literature Review

Recent studies suggest that the growth in emergency vehicle use can be attributed to demographic changes, public expectations, and issues of accessibility of medical care (*1, 2*). Other researchers report that government policies, the cost of medical insurance, and the regional health management process have an influence on demand (*3*). In certain regions, emergency services have come to be considered an attractive healthcare option, due to the ease of accessing medical services without having to make an appointment (*6–8*). According to the 2011/12 annual report from the NS EHS (*9*), 137,374 requests were reported for ground ambulances, and just 54% of were emergency or urgent (*9*). Similarly, a study in the United States showed that up to 27% of emergency calls can be handled without the need for an emergency vehicle to respond, saving approximately $4.4 billion in healthcare expenses annually (*10*). Additional research has shown that older citizens are more frequent users of emergency health services (*11*), which is a major concern in Nova Scotia, as the province has the highest proportion of elderly population (over the age of 65 years) in Canada (*12*).

Studies have shown that city planning plays an important role in promoting healthy living and preventing diseases (*4, 5*). Giles-Corti et al. (*5*) emphasize the need to explore both transport opportunities and urban design further in relation to demand for EHS vehicles. The researchers acknowledge that urban designers and planners still do not fully comprehend the consequences of built environment for public health (*5*). Reed and Bendall (*13*) studied the effect of living outside the urban center on emergency service demand. Although a few studies have documented an increase in demand for emergency health services (*1–3*), Reed and Bendall (*13*) report underutilization for those living outside a major city. These studies indicate the need for further exploration of the influence of land use characteristics and accessibility measures on EHS vehicle demand.

There have been a considerable number of studies looking at EHS vehicle demand in North America, Australia, and other parts of the world (see summary in Table 1). These studies can be grouped into three categories: (a) spatial and temporal descriptive analysis (*14*), (b) diagnostic and predictive modeling (*15–17*), and (c) location coverage and service optimization (*18–20*). For example, Munjal et al. (*14*) conducted a spatial and temporal analysis on requests for EHS vehicles in relation to population size and medical complaints. Alternatively, Siler (*16*) and Zhou and Matteson (*17*) developed predictive demand models for ambulance services in Los Angeles County, United States, and Melbourne, Australia, respectively. Furthermore, Ong et al. (*19*) used a geo-spatial analysis to optimize service and reduce ambulance response times. The category of study most frequently seen within this field of research is the empirical data analysis of demand for EHS vehicles. Recent studies include rate of use of emergency medical services (*21*), retrospective study of "superusers" of emergency medical services (*22*), and pattern analysis of visits to hospital emergency departments (*2*). Additionally, Aldrich et al. (*23*) used a linear regression model to analyze the demographics of emergency department users. Using the count of demand as a linear function of socio-economic characteristics of the census tract within a community, a linear regression model was developed to find the association between users' socio-economic status and their frequency of emergency department visits (*23*).

Furthermore, studies that model demand for EHS vehicles utilize alternative count data modeling methods, including multiple regression analysis (*16*), auto regressive models (*15*), Gaussian mixture models (*20*), kernel warping method (*17*) and Poisson modeling (*6*). Spatial demand models inherently consist of a large proportion of zeros, due to areas where no emergency calls are observed. To tackle this challenge, researchers have developed alternative Poisson regression models that account

Table 1. Examples of Major Studies on EHS Vehicle Demand Analysis

Reference	Objective	Data used	Methods	Major findings
(*14*)	To understand and analyze the patterns of EHS utilization and trends over time	New York City 911 EHS calls between 1999 and 2007	Retrospective epidemiologic survey using Spearman's rho (p) to examine temporal trends	1.16% annual rise in EHS utilization rates
(*17*)	To provide spatial density predictions for ambulance demand during any hour in Melbourne, Australia	2011–2012 ambulance demand from Melbourne, Australia	Spatio-temporal kernel warping method	Proposed method gives more accurate predictions compared with current industry practice
(*16*)	To predict demand for ambulance service in Los Angeles County	Los Angeles County 1970 census data and 1973 data on utilization of emergency ambulances	Regression models analysis	Demand for public emergency medical vehicles appears to be highly predictable using a nonlinear per capital model using socio-economic, land use, and emergency services variables
(*20*)	Estimating the spatial distribution of ambulance demand in Toronto, Canada, as it changed over 2-hour intervals, to reduce response times	2007–2008 Toronto EMS performance data	Time-varying finite mixture model	This proposed method predicts the EMS operational performance more accurately, reducing error by as much as two-thirds compared with industry practice
(*6*)	Identify spatial patterns in emergency department use within Durham County, North Carolina, and examine patient and neighborhood level factors	2009 Duke University Health System hospitals and clinics decision support repository (DSR) patient records	Spatial Poisson hurdle model with Bayesian estimation approach	The spatial Poisson hurdle model should prove useful to investigators confronting spatially dependent count data that are characterized by an abundance of zeros
(*15*)	To create models that could be used for simulation of the EMS system, and for forecasting daily and hourly call volumes	2000–2004 EMS call data from Calgary, Alberta	Estimated models of daily volume via two approaches: 1. Autoregressive models of data obtained after eliminating trend, seasonality, and special-day effects, 2. Doubly-seasonal ARIMA models with special-day effects	The doubly-seasonal ARIMA model performed poorly when forecasting more than a week into the future

Note: EMS = emergency medical services.

for excess zeros (*6, 24–26*). The Poisson regression model assumes the mean and variance of the dataset are equal, unlike the linear regression model that assumes the true values are normally distributed around an expected mean value. However, the standard Poisson formulation is likely to produce a greater error due to larger variance in the dataset as a result of excess zero count for regions where there is no demand. Hence, models have been further developed using a Poisson hurdle regression model which can address both the abundance of zeros and the non-zero counts. The Poisson hurdle regression modeling approach in this study builds on recent studies such as Neelon et al. (*6*), which used this model specifically to explore the geographic variation of emergency department visits. However, this study differs significantly as the research focuses on the effects of the neighborhood characteristics on EHS vehicle demand. Previous studies mostly examined socio-demographic characteristics (e.g., race and ethnicity) and health related attributes (e.g., ability to pay for medical insurance). In contrast, this research develops finer-grained land use measures and accessibility variables that are common in travel behavior research (for instance, land use mix index in Ferdous and Bhat [*27*]) for better understanding of their impacts on EHS demand. Moreover, time-segmented models are developed to observe relative differences among factors in multiple traffic peak and off-peak periods. Thus, the specific objectives of the research include: (1) to understand better the spatial and temporal variability of EHS vehicle demand in Nova Scotia, (2) to assess various spatial characteristics (e.g., land use characteristics and accessibility measures) that may influence the demand, and (c) to compare the Poisson regression and Poisson hurdle regression models for generic and time-segmented observations of vehicle requests.

Data Used

The data source used in this study is the NS EHS administrative database. In the Province of Nova Scotia, NS EHS provides the dispatch of emergency vehicles through a single access centralized computer system. It has a fleet of 139 ground ambulances, each staffed by a team of two paramedics. In 2011, NS EHS received 137,374 requests for ground ambulance service, of which 74,101 (54%) were considered emergency or urgent, while the rest were inter-facility transfers (*9*). For administrative purposes, NS EHS requires that all paramedic responses be documented. Each response is assigned a unique identification case number and uploaded to a central server. Records were obtained for a period of one year from January 2012 to December 2012. The dataset comprised of 24,403 cases of emergency transportation to hospitals in Halifax, representing approximately 40% of the total EHS vehicle demand during this period. This dataset includes the coordinates of pick-up locations, time of pick-up and drop off, age, and gender of the service user. All neighborhood characteristics data were obtained from the 2011 Canadian Census from Statistics Canada at the Dissemination Area (DA) level. Accessibility and land use measures were generated using ArcGIS data obtained from DMTI Spatial.

Methodology

This paper uses and compares a Poisson regression and Poisson hurdle regression model at the 1 km-by-1 km grid level in Halifax, Nova Scotia. This study expands the recent efforts on spatial modeling with excessive zero counts (*6, 28*). The study uses multiple fitness tests to compare a Poisson regression and a Poisson hurdle regression models.

Data Preparation

To prepare the data, the accessibility measures are generated by utilizing the Network Analyst tool in ArcGIS platform. These measures include distances from the emergency service users' pick-up locations to the central business district (CBD), to the nearest hospital, and to parks, among others. Land use characteristics including commercial, residential, and industrial land uses were determined using data obtained from DMTI Spatial. Land use index was derived according to the methodology outlined by Ferdous and Bhat (*27*). To analyze how emergency demand varies across space, a spatial join was performed in ArcGIS to merge the attributes of the emergency requests to each 1 km-by-1 km polygon. The sum of these points was calculated as the total demand in a particular cell. This method generated 9,512 cells with neighborhood level attributes. Of those, about 59% of cells had no demand due to no requests being made within that area. Using a similar approach, the 2011 census data for the Province of Nova Scotia was used to generate geographical attributes of the cells. The percentage of the DA area that fell inside the 1 km-by-1 km cell was calculated to determine the weighted average area of each DA within the cell. The adjusted attribute values calculated from the proportion of the DA present in the polygon form the basis of the variables used for the econometric modeling. Analysis and modeling were conducted using the LIMDEP 11 software package (*29*). A brief description of the modeling approaches for the Poisson regression and Poisson hurdle regression models is presented below.

Poisson Regression Model

A Poisson regression model assumes that each observed count is drawn from a Poisson distribution. This model restricts the conditional variance to equal the conditional mean, a term commonly referred to as equidispersion. The dependent variable in a Poisson regression model is assumed to follow a standard Poisson distribution, as expressed below:

$$Y \sim \text{Poisson}(\mu), \text{then } \mathbb{E}(Y) = Var(Y) = \mu \qquad (1)$$

where Y is a count-valued response (dependent variable), $\mathbb{E}(Y)$ is the mean of Y, and $Var(Y)$ is the variance of Y. The Poisson regression model can be expressed as:

$$\Pr(Y = y|X_1, X_2, \ldots, X_m) = \left(\frac{e^{-\lambda}\lambda^y}{y!}\right), y = 0, 1, 2, \ldots, n \qquad (2)$$

where the parameter λ is the mean rate of occurrence; particularly, the expected frequency of occurrence of an event per unit time. λ is an exponential function of the independent variables X_1, X_2, X_3,..., X_m, which can be written as:

$$\lambda = \exp(\beta_0 + \beta_1 X_1 + \beta_2 X_2 + \cdots + \beta_m X_m) \qquad (3)$$

The log-likelihood of the Poisson model can be written as:

$$LL_{Poisson} = \sum_{i=1}^{n} [-\lambda + yLog(\lambda) - Log(y!)] \qquad (4)$$

If the count data are equidispersed, then as the value of λ increases the mass of the distribution shifts to the right, specifically, $\mathbb{E}(Y) = Var(Y) = \lambda$. However, in practice, the count variables often have a variance that is greater than the mean, which is known as overdispersed, and if the variance is less than the mean then it is known as underdispersed. One of the major reasons for such overdispersed and underdispersed distribution is the existence of excess zeros. When this occurs, the Poisson regression model will likely produce a greater standard error, indicating the model is not sufficiently fit to describe the variables. Hence an adjustment needs to be made to the model. The development of a Poisson hurdle regression model is an attempt to account for, and minimize, the effect of overdispersion.

Poisson Hurdle Regression Model

The Poisson hurdle regression model is a two-stage model that consists of a point mass at zero and a truncated Poisson distribution for the non-zero observation, such that it does not distinguish between zeros from the degenerate distribution versus those from the Poisson distribution (*6*). The Poisson hurdle regression model, therefore, can be used to handle both overdispersed and underdispersed count variables. Generally, the structure of the Poisson hurdle regression model is given by:

$$Pr(Y = 0) = (1 - p),\ 0 \leq p \leq 1 \qquad (5)$$

$$Pr(Y = y) = (p)\frac{e^{-\lambda}\lambda^{y}}{y!(1 - e^{-\lambda})} \qquad (6)$$

where p is the utilization probability. The following can be deduced from the model: when $(1 - p) > e^{-\lambda}$, the zeros are inflated relative to the base distribution; and when $(1 - p) < e^{-\lambda}$, there is a zero deflation. In extreme cases, when $p = 0$, the model reduces to a truncated Poisson distribution; and when $p = 1$ the model degenerates at zero. Typically, one assumes that p is precisely between 0 and 1, so there is a non-zero utilization probability. The log-likelihood of the Poisson hurdle model can be written as:

$$LL_{Hurdle} = \sum_{1}^{n}\begin{bmatrix} I(Y = 0)Log(1 - p) \\ I(Y>0)Log(p) - \lambda + yLog(\lambda) - Log(1 - e^{-\lambda}) - Log(y!)) \end{bmatrix} \qquad (7)$$

Model Fitness Test

The goodness-of-fit of the models are tested using the log-likelihood ratio (LR), Akaike information criterion (AIC) and Bayesian information criterion (BIC) measures. The model fitness test shows that the Poisson hurdle regression model outperforms the traditional Poisson regression model. The Poisson regression model displayed a log-likelihood of −83,923.87, with an AIC value of 17.65 and BIC value of 17.67. The Poisson hurdle regression model reported a log-likelihood value of −72,160.75, with an AIC value of 15.18 and BIC value of 15.20. Moreover, the results of the LR test exceed the critical value for the chi-squared distribution with two degrees of freedom, which allows the rejection of the Poisson model in favor of the Poisson hurdle model (LR stat 23526.23). Table 2 shows the summary statistics of the variables used in the final model specification.

Time-Segmented Models

From the Poisson hurdle regression model, five time-segmented models were developed to analyze the temporal variations in the relationship of EHS vehicle demand. These time periods were defined as AM (morning) peak (6.00 a.m.–9.00 a.m.), midday (9.00 a.m.–3.00 p.m.), PM (afternoon) peak (3.00 p.m.–6.00 p.m.), evening (6.00 p.m.–midnight) and overnight (midnight–6.00 a.m.).

Model Results and Discussion

Exploratory Analysis

An exploratory analysis was conducted on the NS EHS database, including analysis by age and gender; distribution of transport by hour, day of the week, month, and season; travel time including pick-up demand per day; and spatial distribution of demand. The exploratory analysis shows that EHS vehicle demand varies considerably across age and gender (see Figure 1*a*), with more pick-up requests being submitted by the group over 75 years old than any other age group. Although users within the age bracket 25 to 34 years had a relatively lower demand compared with age group of 15 to 24 years, the general trend reveals that emergency vehicle demand increases with the increase in age, which is consistent with the finding of other studies (*11, 21*).

The demand for EHS vehicles increases gradually during the early hours of the day (see Figure 1*b*), reaching a peak at 10.00 a.m. after which demand begins to decline throughout the day. The monthly variation shows the highest requests during March, October, and December. As for seasonality, EHS vehicle demand is highest in the winter months, with the lowest demand in the summer (see Figure 1*c* and *d*).

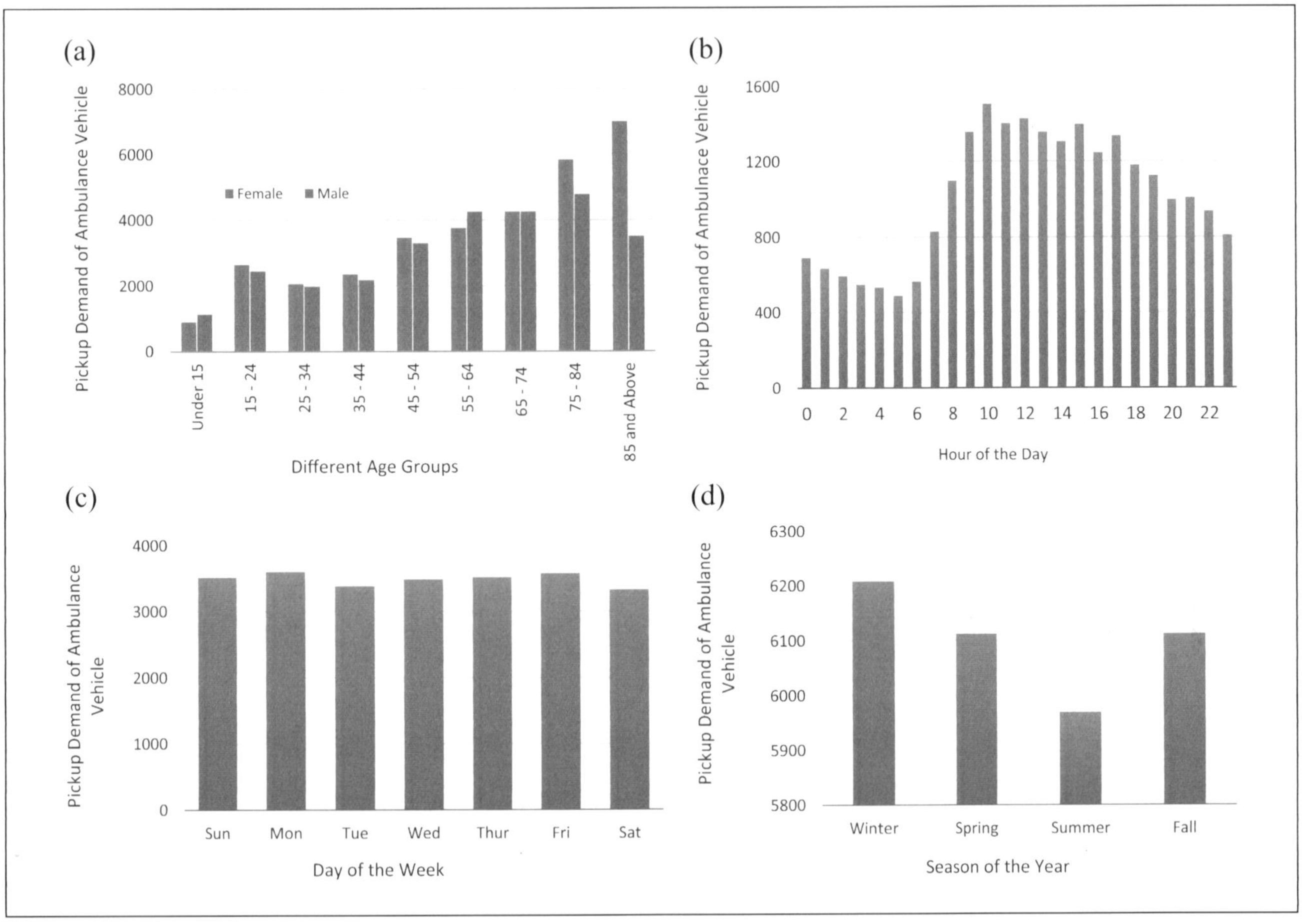

Figure 1. Distribution of demand for EHS vehicles in Halifax by (*a*) age group and gender, (*b*) hour of day, (*c*) day of the week, and (*d*) season of the year.

The demand for the whole province is also analyzed spatially by creating a heat map using the Kernel Density tool in ArcGIS ArcMap 10.2 (see Figure 2). In comparison with other counties in Nova Scotia, Halifax had the highest demand with a cumulative percentage of approximately 42%. This was expected considering approximately 44% of the province's population reside in the Halifax Regional Municipality (HRM). Demand hotspots were distributed through the entire region, with higher proportion in high-density areas. Most of the demand in HRM was located in the urban core area, specifically Halifax peninsula, which is not surprising given the concentration of population in these areas.

Poisson Hurdle Regression Model

The parameter estimates of the Poisson hurdle regression model are reported in Table 3. The model suggests that higher population density will lead to higher demand for EHS vehicles. For instance, a larger population size, typically in urban core areas, will tend to increase the demand. Furthermore, average household income exhibits a negative relationship with EHS vehicle demand, suggesting that lower income households are more susceptible to a higher demand for EHS vehicles than are higher income households. Age is also one of the major predictors of EHS vehicle demand. The model results exhibit a negative relationship for EHS vehicle demand in areas with higher numbers of younger individuals (up to 39 years), while areas with an older population, including workers (40 years and above), show higher demand for EHS vehicles. A closer observation of the model parameters shows that, among these older age groups, the age group of 40 to 64 years demonstrates a lesser need for EHS vehicles than the group aged 65 years and above. The results are in alignment with healthcare literature which indicates that demand for emergency care tends to increase with age (*21*).

In terms of land use characteristics, the model demonstrates a negative relationship with land use index, indicating that demand EHS vehicles will decrease if the land use is more heterogeneous. This result supports

Table 2. Summary Statistics and Description of Variables

Variable	Description	Mean	Standard deviation
Demand count	Number of requests for EHS vehicles in the grid	21.6003	79.8934
Socio-demographic attributes			
Population density	Population density (people/km^2) in the grid	358.294	1452.73
HH average income	Household average income ($ CAD) in the grid	80740.9	30346.9
Employment rate (%)	Percentage of employment in the grid	58.2968	14.2089
Population under 15 years	Number of population aged 15 years or less in the grid	3.10529	10.8883
Population age 16–39 years	Number of population aged between 16 and 39 years in the grid	11.4162	44.0286
Population age 40–64 years	Number of population aged between 40 and 64 years in the grid	11.3457	37.4017
Population age 65–75 years	Number of population aged between 65 and 75 years in the grid	2.28158	8.64036
Population over 75 years	Number of population aged 75 or more in the grid	1.96013	11.7724
Land use characteristics			
Single detached house	Number of single detached houses in the grid	6.10457	20.4864
Apt 5 or more	Number of apartment buildings with 5 or more storeys in the grid	1.88082	18.3662
Apt 5 few	Number of apartment buildings with less than 5 storeys in the grid	3.52731	22.8446
Apt duplex	Number of duplex apartments in the grid	0.576289	3.73888
Land use residential	Proportion of land use for residential purpose	2.72258	12.6472
Land use commercial	Proportion of land use for commercial purpose	0.06084	1.65322
Land use park	Proportion of land use for park purpose	0.238869	1.88266
Land use industrial	Proportion of land use for industrial purpose	0.108898	1.50782
Land use index	The land use mix index[a]	0.03942	0.09851
Accessibility measures			
Distance to CBD	Distance to the Central Business District (m) from user's location	40795.2	30261.4
Distance to hospital	Distance to the nearest hospital (m) from user's location	17981.7	9970.00
Distance to park	Distance to the nearest park (m)[b] from user's location	4329.72	5047.64
Summary statistics and description of temporal variables based on different time segments			
AM_PEAK	Number of requests for EHS vehicles during morning peak	2.10765	7.81499
MIDDAY	Number of requests for EHS vehicles midday	7.32328	27.0228
PM_PEAK	Number of requests for EHS vehicles	3.52986	13.3531
EVENING	Number of requests for EHS vehicles in the evening	5.39319	19.7789
OVERNIGHT	Number of requests for EHS vehicles overnight	3.24632	13.4297

[a]The land use mix is more homogenous if the value is closer to 0 and more heterogeneous if the value is closer to 1.
[b]Proportion of distance to park within 2 km = 47.67%.

other studies on the impact of land use mix, for example, Bramley et al. (*30*), which concludes that better designed neighborhoods with mixed land uses enable better access to and increased use of local facilities. The land use index effects can also be established with the individual components of the land uses. For instance, the model demonstrates a positive association with land use for industrial, residential, and commercial areas. This means that EHS vehicle demand will increase in areas with a higher proportion of land with major land use types, such as apartments, shops, or industries. Among these three land use measures that positively affect EHS vehicle demand, the land use for industrial purposes exhibits the highest rate, followed by residential and commercial land uses. This suggests that demand for EHS vehicles is more likely to come from people at work than people at home or in commercial areas. In addition, in the model, neighborhoods of single detached homes show less demand than those with apartments. As shown in the spatial distribution map (Figure 2), demand was very high in the urban core compared with the suburban and rural areas of the province.

Finally, the analysis reveals that demand for EHS vehicles can also be influenced by accessibility measures. The model demonstrates a negative relationship between location of pick-up and distance to the nearest park. The negative relationship rate is higher when the park is within 2 km of the user's location, compared with the overall effect for the park distance, meaning the farther the distance from the park the more likely the demand for EHS vehicles will increase. Additionally, the model suggests a positive relationship between the distance between user's location and the nearest hospital locations, as well as a negative relationship for the distance from the user's location to the CBD.

Time-Segmented Model

As indicated earlier, time-segmented models are developed in this study to investigate temporal variation of demand for EHS vehicles. A preliminary exploratory analysis finds that the most requests are made during the midday period (9.00 a.m.–3.00 p.m.), followed by the

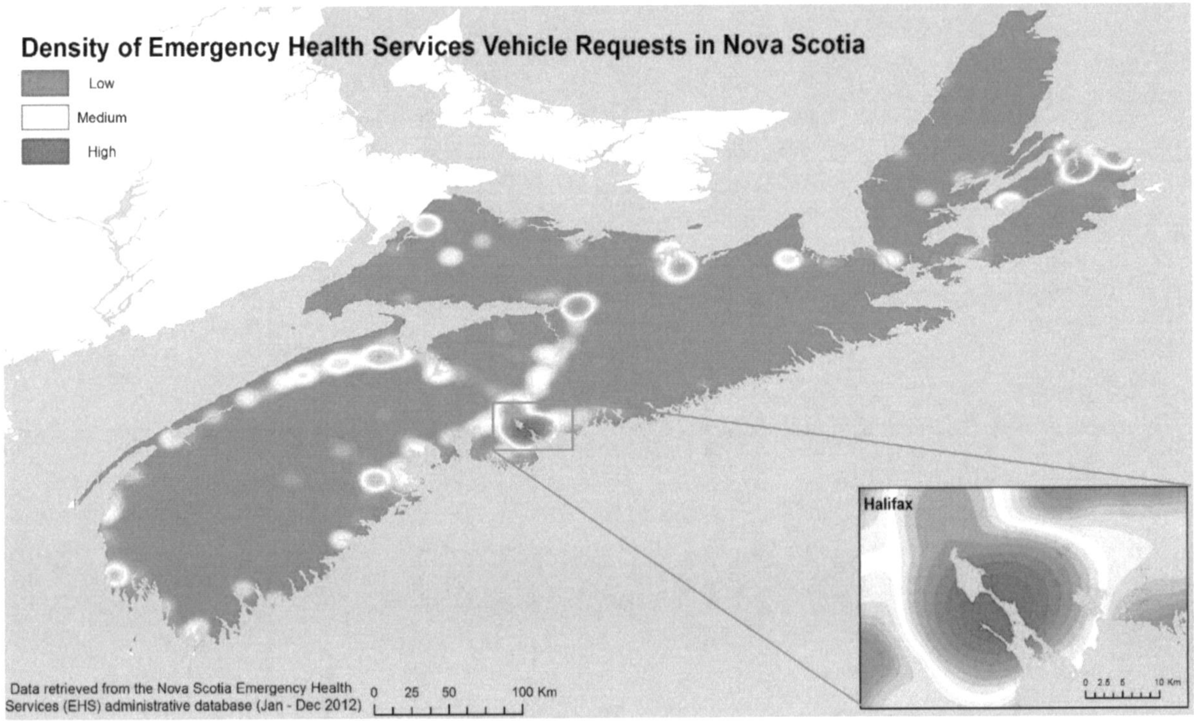

Figure 2. Spatial distribution of EHS vehicle requests in Nova Scotia.

Table 3. Parameter Estimation Results of EHS Vehicle Demand Models

	Poisson regression model		Poisson Hurdle regression model	
Variable	Coefficient	*t*-stats	Coefficient	*t*-stats
Constant	8.14956528	167.256	8.3935567	1056.733
Socio-demographic attributes				
Population density in the grid	0.00021574	41.530	0.0002111	560.668
HH average income in the grid	−0.00003921	−54.970	−0.3610020	−473.538
Employment rate (%) in the grid	0.00014811	0.757	−0.0005152	−29.229
Population under 15 years (# in the grid)	−0.00175525	−7.661	−0.0021606	−79.129
Population age 16–39 years (# in the grid)	−0.00067826	−11.523	−0.0003834	−59.875
Population age 40–64 years (# in the grid)	−0.00015939	−1.423	0.0002212	15.680
Population age 65–75 years (# in the grid)	0.00414220	14.781	0.0036248	87.534
Population over 75 years (# in the grid)	0.00000203	0.002	0.0001163	8.846
Land use characteristics				
Single detached house (# in the grid)	−0.0055933	−42.021	−0.0054228	−360.643
Apt 5 or more (# in the grid)	0.0040086	4.386	0.0001573	14.379
Apt 5 few (# in the grid)	0.0008606	11.438	0.0005952	69.206
Apt duplex (# in the grid)	0.0053623	20.023	0.0047964	133.120
Land use residential	0.0059498	50.693	0.0052695	349.418
Land use commercial	0.0049592	11.415	0.0050075	64.717
Land use park	−0.0054597	−11.170	−0.0031739	−44.410
Land use industrial	0.0060779	13.568	0.0060783	123.640
Land use index	0.1221007	7.867	−0.0042806	−2.767
Accessibility measures				
Distance to CBD	−0.0002431	−90.441	−0.0026459	−980.024
Distance to hospital	0.0007073	26.670	0.0010347	373.780
Distance to park	−0.0017767	−137.309	−0.0012631	−650.402
Distance to park within 2 km	−1.9121892	−41.586	−2.1911864	−283.070
Hurdle binary parameters				
Constant	–	–	−2.4550654	−17.707
Land use park	–	–	0.29078660	8.518
Distance to hospital	–	–	−0.00010004	−32.929

Table 4. Parameter Estimation Results of Time-Segmented EHS Vehicle Demand Models: Poisson Hurdle Regression Model

	Parameter coefficients									
	Morning peak		Midday		Afternoon peak		Evening		Overnight	
Variables	Coefficient	*t*-stats	Coefficient	*t*-stats	Coefficient	*t*-stats	Coefficient	*t*-stats	Coefficient	*t*-stats
Constant	5.723537	93.254	7.45782203	354.073	6.44238302	153.632	6.76426132	213.014	6.5252226	130.754
Socio-demographic attributes in the grid										
Population density	0.000159	39.454	0.202625	167.218	0.172570	71.488	0.186969	116.099	0.183753	88.921
HH average income (CAD)	−0.000031	−45.153	−0.254820	−114.466	−0.245930	−51.709	−0.376012	−117.146	−0.325342	−74.917
Employment rate	−0.000275	−1.536	−0.001083	−19.580	−0.001237	−12.163	−0.000974	−13.311	0.0006727	0.688
Population under 15 years	−0.005164	−17.651	−0.002618	−32.416	−0.003535	−23.415	−0.001739	−14.482	−0.002934	−19.170
Population age 16–39 years	−0.000958	−1.587	−0.000116	−5.860	0.917867	2.465	−0.000720	−27.079	0.460131	1.330
Population age 40–64 years	0.000182	1.381	−0.000117	−2.709	0.001010	12.189	0.000939	15.047	0.001226	15.008
Population age 65–75 years	0.003061	7.972	0.003424	26.731	0.002248	9.924	0.001327	7.813	0.002658	12.393
Population over 75 years	0.001381	8.795	0.000709	14.738	−0.000223	−2.652	−0.000409	−8.787	−0.000741	−13.913
Land use characteristics in the grid										
Single detached house	−0.004605	−32.781	−0.004960	−113.336	−0.005258	−59.999	−0.004953	−77.740	−0.006499	−72.628
Apt 5 or more	−0.000256	−2.406	0.821261	2.544	0.000110	1.800	0.000622	13.428	0.000357	6.384
Apt 5 few	0.000223	2.631	0.000429	16.166	−0.329628	−0.610	0.000661	18.169	0.000860	1.834
Apt duplex	0.004236	13.238	0.005431	54.206	0.005307	29.481	0.002931	17.423	0.000779	3.589
Land use residential	0.005260	39.203	0.004167	95.290	0.003421	37.938	0.004890	75.868	0.004440	57.557
Land use commercial	0.005953	5.662	0.004360	16.486	0.006006	11.699	0.005820	21.028	0.004423	13.477
Land use park	−0.000991	−1.458	−0.003537	−16.677	−0.002584	−6.196	−0.002078	−6.957	−0.000431	−1.155
Land use industrial	0.005275	11.099	0.006276	39.568	0.004601	15.033	0.004932	26.799	0.005207	20.383
Land use index	−0.152648	−9.931	−0.060100	−12.951	−0.267190	−28.838	−0.109525	−17.032	−0.287887	−33.005
Accessibility measures in the grid										
Distance to CBD	−0.000250	−99.492	−0.000279	−370.322	−0.000279	−168.437	−0.000328	−314.838	−0.000351	−228.152
Distance to hospital	0.000115	45.176	0.000121	159.903	0.000126	72.385	0.000181	167.721	0.000202	125.700
Distance to park	−0.000803	−42.296	−0.001112	−192.568	−0.000850	−76.670	−0.000912	−119.223	−0.000829	−69.859
Distance to park within 2 km	−1.921463	−32.774	−2.336409	−115.122	−2.022091	−49.889	−1.888890	−60.998	−2.195808	−44.943
Hurdle binary parameters										
Constant	−2.274733	−13.516	−2.393873	−15.822	−2.319227	−14.095	−2.708938	−17.325	−2.492333	−14.914
Land use park	0.209776	10.554	0.302084	8.918	0.274782	9.080	0.294200	9.317	0.237153	9.828
Distance to hospital	−0.000143	−36.821	−0.000118	−34.413	−0.000135	−35.448	−0.000118	−33.623	−0.000134	−34.854
Employment rate	0.047259	17.829	0.050811	21.455	0.048291	18.646	0.054008	21.985	0.050258	18.977

Note: All p values ~ 0.000 and t-stats values ≥ 3.2 which indicates a significance level of 99.99%.

afternoon peak period (3.00 p.m.–6.00 p.m.). The number of requests gradually declines from the evening (6.00 p.m.–midnight) to the overnight period (midnight–6.00 a.m.).

The results of the Poisson hurdle model (Table 4) reveal that there are slight variations in the proportion contributed by each coefficient of variables of interest for different time segments. For instance, the age group over 75 years old has higher demand during the morning peak period with demand declining throughout the day. Interestingly, the age group between 16 and 39 years, which showed a negative relationship with demand in the general model, showed a positive relationship with demand during the afternoon peak and overnight periods.

In addition, among the three land use measures that positively affect demand for EHS vehicles, the coefficient for residential land use exhibits a higher demand during the morning peak period in contrast to that of the afternoon peak period. On the other hand, the coefficient for commercial land use exhibits the highest demand during the afternoon peak period. Industrial land use shows the highest demand during the midday period. This signifies that demand for EHS vehicles will vary according to activity purposes and location. The higher demand in industrial areas during the day coincides with the period when people are at work, and demand in commercial areas during afternoon peak coincides with the period when many of the users are most likely running errands after work.

Conclusion

The objectives outlined for this study were achieved by utilizing and comparing a Poisson regression and Poisson hurdle regression model to examine the effects of neighborhood characteristics on EHS vehicle demand. It also developed a time-segmented model to investigate the temporal variability of demand in Halifax, Nova Scotia. The general and time-segmented analyses have shown that the Poisson hurdle model has provided better model fits and useful insights into EHS vehicle demand, and are therefore suitable methods for the analysis of overdispersed count data.

The challenges presented by the increased proportion of senior citizens and pressures on emergency health services are growing. This study can be used to assist decision makers to develop better health care services for communities. Understanding how spatial and temporal factors affect EHS vehicle demand can assist with resource allocation of personnel and vehicles. More importantly, it can be used to help determine a cost-effective method of deploying health resources to underserved communities in order to relieve access to emergency medical services, particularly by addressing neighborhoods that are prone to emergency need. For instance, if demand for EHS vehicles is higher in commercial areas during the afternoon peak period, then the vehicles could be strategically deployed within such areas during this time of day, thereby reducing resource allocation for certain periods of the day as well as response travel times. Making these directed efforts to address user needs will be crucial to reduce the burden placed on EHS vehicles, subsequently reducing overall costs associated with healthcare within the province.

Improvements can be made to the predictive modeling tools and techniques used in this study by introducing a model that can accommodate random effects and spatial–temporal smoothing in order to reduce bias in the spatial covariance parameters and the intercept of the Poisson component (*6, 31*). Future work could also include investigating the spatial–temporal demand for EHS vehicles among smaller groups of patients with specific medical diagnoses to understand the causes of demand among specific communities. For instance, according to Neelon et al. (*6*), if the increase in EHS vehicle usage is mainly due to mental health related issues, then policy makers and health officials could focus on ways of improving the community mental health services as opposed to optimizing the use of emergency service resources. Nevertheless, the models and analysis of this study offer newer insights for understanding how neighborhood factors affect the demand for EHS vehicles. This study will guide transportation planners and healthcare professionals to develop plans for transportation and health service planning and resource allocation.

Acknowledgments

The authors thank the Natural Sciences and Engineering Research Council (NSERC) of Canada and Nova Scotia Department of Health and Wellness for their contribution in supporting this research. Special thanks to Stephen McCarthy and Nazmul Arefin Khan for proof-reading and formatting.

Author Contributions

The authors confirm contribution to the paper as follows: study conception and design: MAH, and BO; data collection: BO, MT, and MAH; analysis and interpretation of results: MAH, BO, and SC; draft manuscript preparation: MAH, SC, and BO. All authors reviewed the results and approved the final version of the manuscript.

References

1. Lowthian, J. A., P. A. Cameron, J. U. Stoelwinder, A. Curtis, A. Currell, M. W. Cooke, and J. McNeil. Increasing Utilisation of Emergency Ambulances. *Australian Health Review*, Vol. 35, No. 1, 2011, pp. 63–69.

2. Weber, E. J., J. A. Showstack, K. A. Hunt, D. C. Colby, B. Grimes, P. Bacchetti, and M. L. Callaham. Are the Uninsured Responsible for the Increase in Emergency Department Visits in the United States? *Annals of Emergency Medicine*, Vol. 52, No. 2, 2008, pp. 108–115.
3. McCaig, L. F., and C. W. Burt. National Hospital Ambulatory Medical Care Survey: 2003 Emergency Department Summary. *Advance Data from Vital and Health Statistics*, No. 358, May 26, 2005. U.S. Department of Health & Human Services, Centers for Disease Control and Prevention, National Centre for Health Statistics, Hyattsville, MD, 2005. https://www.cdc.gov/nchs/data/ad/ad358.pdf.
4. Giles-Corti, B., A. Vernez-Moudon, R. Reis, G. Turrell, A. L. Dannenberg, H. Badland, and N. Owen. City Planning and Population Health: A Global Challenge. *The Lancet*, Vol. 388, No. 10062, 2016, pp. 2912–2924.
5. Srinivasan, S., L. R. O'Fallon, and A. Dearry. Creating Healthy Communities, Healthy Homes, Healthy People: Initiating a Research Agenda on the Built Environment and Public Health. *American Journal of Public Health*, Vol. 93, No. 9, 2003, pp. 1446–1450.
6. Neelon, B., P. Ghosh, and P. F. Loebs. A Spatial Poisson Hurdle Model for Exploring Geographic Variation in Emergency Department Visits. *Journal of Royal Statistics Society A*, Vol. 176, No. 2, 2013, pp. 389–413.
7. Guttman, N., M. S. Nelson, and D. R. Zimmerman. The Many Faces of Access: Reasons for Medically Nonurgent Emergency Department Visits. *Journal of Health Politics, Policy and Law*, Vol. 28, No. 6, 2003, pp. 1089–1120.
8. Cunningham, P. J. What Accounts for Differences in the Use of Hospital Emergency Departments across U.S. Communities? *Health Affairs*, Vol. 25, No. 5, 2006, pp. w324–w336.
9. Emergency Health Services. *Emergency Health Services Annual Report 2011-2012*. Halifax, NS, Canada, 2017. https://novascotia.ca/dhw/ehs/documents/EHS-Annual-Report-2011-2012.pdf.
10. Weinick, R. M., R. M. Burns, and A. Mehrotra. Many Emergency Department Visits Could Be Managed at Urgent Care Centers and Retail Clinics. *Health Affairs*, Vol. 29, No. 9, 2010, pp. 1630–1636.
11. Clark, M. J., and G. FitzGerald. Older People's Use of Ambulance Services: A Population Based-Analysis. *Emergency Medicine Journal*, Vol. 16, 1999, pp. 108–111.
12. Institute of Public Administrators of Canada. *Healthcare Governance Models in Canada: A Provincial and Territorial Perspective*, Toronto, ON, Canada, November 2013. http://neltoolkit.rnao.ca/sites/default/files/Healthcare%20Governance%20Models%20in%20Canada_A%20Provincial%20Perspective_Pre-Summit%20Disscussion%20Paper%20March%202013.pdf.
13. Reed, B., and J. C. Bendall. Rurality as a Factor in Ambulance Use in Health Emergencies. *Australasian Journal of Paramedicine*, Vol. 12, No. 1, 2015, pp. 1–6.
14. Munjal, K. G., R. A. Silverman, J. Freese, J. D. Braun, B. J. Kaufman, D. Isaacs, and D. J. Prezant. Utilization of Emergency Medical Services in a Large Urban Area: Description of Call Types and Temporal Trends. *Prehospital Emergency Care*, Vol. 15, No. 3, 2011, pp. 371–380.
15. Channouf, N., P. L'Ecuyer, A. Ingolfsson, and A. N. Avramidis. The Application of Forecasting Techniques to Modeling Emergency Medical System Calls in Calgary, Alberta. *Health Care Management Science*, Vol. 10, No. 1, 2007, pp. 25–45.
16. Siler, K. F. Predicting Demand for Publicly Dispatched Ambulances in a Metropolitan Area. *Health Services Research*, Vol. 10, No. 3, 1975, pp. 254–263.
17. Zhou, Z., and D. S. Matteson. Predicting Melbourne Ambulance Demand using Kernel Warping. *The Annals of Applied Statistics*, Vol. 10, No. 4, 2016, pp. 1977–1996.
18. Das, A., and F. Martin. An Econometric Analysis of the US Health Care Expenditure. *Global Journal of Health Science*, Vol. 2, No. 1, 2010, p. 150.
19. Ong, M. E. H., T. F. Chiam, F. S. P. Ng, P. Sultana, S. H. Lim, B. S. H. Leong, and V. Anantharaman. Reducing Ambulance Response Times using Geospatial-Time Analysis of Ambulance Deployment. *Academic Emergency Medicine*, Vol. 17, No. 9, 2010, pp. 951–957.
20. Zhou, Z., D. S. Matteson, D. B. Woodard, S. G. Henderson, and A. C. Micheas. A Spatio-Temporal Point Process Model for Ambulance Demand. *Journal of the American Statistical Association*, Vol. 101, No. 509, 2015, pp. 6–15.
21. Goldstein, J., J. L. Jensen, A. J. E. Carter, A. H. Travers, and K. Rockwood. The Epidemiology of Prehospital Emergency Responses for Older Adults in a Provincial EMS System. *Canadian Journal of Emergency Medicine*, Vol. 17, No. 5, 2015, pp. 491–496.
22. Hall, M. K., M. C. Raven, J. Hall, C. Yeh, E. Allen, R. M. Rodriguez, and J. F. Brown. EMS-STARS: Emergency Medical Services "Superuser" Transport Associations: An Adult Retrospective Study. *Prehospital Emergency Care*, Vol. 19, No. 1, 2015, pp. 61–67.
23. Aldrich, C. A., J. C. Hisserich, and L. B. Lave. An Analysis of the Demand for Emergency Ambulance Service in an Urban Area. *American Journal of Public Health*, Vol. 61, No. 6, 1971, pp. 1156–1169.
24. Agarwal, D. K., A. E. Gelfand, and S. Citron-Pousty. Zero-Inflated Models with Application to Spatial Count Data. *Environmental and Ecological Statistics*, Vol. 9, No. 4, 2002, pp. 341–355.
25. Rathbun, S. L., and S. Fei. A Spatial Zero-Inflated Poisson Regression Model for Oak Regeneration. *Environmental and Ecological Statistics*, Vol. 13, No. 409, 2006, pp. 406–426.
26. Ver Hoef, J. M., and J. K. Jansen. Space-Time Zero-Inflated Count Models of Harbor Seals. *Environmetrics*, Vol. 18, No. 7, 2007, pp. 697–712.
27. Ferdous, N., and C. R. Bhat. A Spatial Panel Ordered-Response Model with Applications to the Analysis of Urban Land-Use Development Intensity Patterns. *Journal of Geographical Systems*, Vol. 15, No 1, 2013, pp. 1–29.
28. Arab, A. Spatial and Spatio-Temporal Models for Modeling Epidemiological Data with Excess Zeros. *International Journal of Environmental Research and Public Health*, Vol. 12, No. 9, 2015, pp. 10536–10548.

29. Greene, W. H. *LIMDEP Version 11.0 Econometric Modeling Guide*. Econometric Software, Inc., New York, NY, 2016.
30. Bramley, G., N. Dempsey, S. Power, C. Brown, and D. Watkins. Social Sustainability and Urban Form: Evidence from Five British Cities. *Environment and Planning A*, Vol. 41, No. 9, 2009, pp. 2125–2142.
31. Su, L., B. D. Tom, and V. T. Farewell. Bias in 2-Part Mixed Models for Longitudinal Semicontinuous Data. *Biostatistics*, Vol. 10, No. 2, 2009, pp. 374–389.

The Standing Committee on Transportation and Land Development (ADD30) peer-reviewed this paper (18-05049).

Article

TRR
JOURNAL OF THE TRANSPORTATION RESEARCH BOARD

Transportation Research Record
2018, Vol. 2672(3) 92–103

DOI: 10.1177/0361198118794543
journals.sagepub.com/home/trr
SAGE

Using Indicators to Assess Sustainable Transportation and Related Concepts

Tara Ramani[1]

Abstract

The overall goal of this study is to assess the concept of sustainability in relation to the related concepts of "health" and "livability" that have emerged in transportation planning discourse. This study achieves the goal using an indicator-based case study, conducted for the El Paso metropolitan area in the United States. Data from the regional travel demand model and other sources were used to quantify a sustainability index, livability index, and health index for individual traffic analysis zones in the region, for four analysis years over a 30-year planning horizon. Each index was comprised of representative indicators, which were normalized and aggregated in accordance with common multi-criteria decision-making methods. The analysis results demonstrated little correlation between the quantified livability, sustainability, and health indices developed for the El Paso region. The indices also showed relatively low levels of change over time for a location. That is, the relative performance of a traffic analysis zone tended to stay the same, despite the modeled changes to the transportation system, demographics, and land use. The main implication of the research findings is that despite overlaps at a theoretical level, concepts such as livability and health cannot necessarily serve as proxies for sustainability when implemented in practice. The study also provides insight into the challenges of making meaningful change in the area of sustainability over time and highlights the influence of factors beyond transportation, such as land use and socio-economic issues.

The term "sustainable transportation" is broad and overarching. In the context of transportation planning, it is generally understood as covering the "triple bottom line," that is, the environmental, economic, and social dimensions, in the present and in the future. In recent years, other similarly broad concepts have emerged in transportation planning discourse, most notably concepts of livability, public health, and resilience.

At a qualitative level, these concepts tend to address various aspects of sustainable transportation, often implicitly. These theoretical overlaps in discourse raise the question of whether using a framework oriented toward "livability" or "health" in transportation planning could serve as a proxy for promoting sustainable transportation, by addressing complementary goals. However, there is the potential for certain aspects of sustainability to be neglected when promoting these alternatives. It important to understand the practical implications of this, especially in the United States, where sustainable transportation is not necessarily seen as an overarching goal or an organizing principle for the sector. Rather, it is seen as a concept that "competes" as an alternative to other policy goals or frameworks, including concepts such as resilience, livability, or health.

The aim of this paper is to provide a quantitative, indicator-based case study contrasting the concept of sustainable transportation against the concepts of livability and health.

The findings will provide further insight into how the overall framing of sustainability issues affects the outcomes observed from indicators, and how concepts related to sustainable transportation complement or conflict with each other.

Background and Literature Review

Sustainable Transportation and Related Concepts

As mentioned in the introduction, discussions of sustainability in the context of transportation (often termed as "sustainable transportation") converge on the consideration of the economic, social, and environmental dimensions in the present and in the future (*1–4*). Authors distinguish "holistic" and "transportation-centric"

[1]Texas A&M University/Texas A&M Transportation Institute, College Station, TX

Corresponding Author:
Address correspondence to Tara Ramani: t-ramani@tti.tamu.edu

perspectives on sustainable transportation (*2*), and also discuss "weak" or "strong" approaches, depending on the extent to which the environmental dimension or natural resources can be "traded off" for economic or social gains (*1*, *5*, *6*).

The broad remit covered by the prevailing view on sustainable transportation, leads it to have several overlaps with transportation planning discourse oriented toward concepts such as livability, health or "public health," and "resilience." As argued by the author in previous work, a case can be made for viewing this concept through a sustainability lens, potentially as "alternative conceptualizations" of sustainable transportation (*7–10*). The concepts or discourse related to sustainability in the transportation planning sector can broadly be viewed as (*9*, *10*):

- Sustainability-based discourse
  - Eco-centric sustainability: Discussions of sustainability in transportation that are centered on long-term environmental and ecological considerations, most notably climate change.
  - Anthropocentric sustainability: Discussions of sustainability in transportation that reference the environmental, economic, and social dimensions, but do not specify a hierarchy between the various dimensions or explicitly reference equity issues.
  - Holistic sustainability: Discussions of sustainability in transportation that reference the environmental, economic, and social dimensions, and also place constraints or limitations in terms of distributional effects/equity, and an acknowledgment of limits to natural capital.
- Alternative discourse related to sustainability
  - Livability: Discourse that is generally concerned with transportation as it relates to community-scale impacts, primarily on human well-being.
  - Health: Discourse organized around transportation's relationship to human health, especially in relation to four key elements – safety, air quality, active living opportunities, and access to critical destinations.
  - Resilience: Discourse related to transportation's ability to continue functioning when faced with disruptive events.

Measuring and Evaluating Sustainability: Indicators

Measuring and evaluating sustainability (as well as related concepts such as livability, resilience, and health) is a key part of operationalizing the concepts, and allowing for them to be appropriately considered in transportation decision-making. This is especially true as sustainability and related constructs tend to be broad, and are not always clearly defined. In terms of technical approaches or analysis tools, methods for assessing sustainable transportation range from using descriptive statistics to spatial maps and statistics, to mode advanced modeling (*11*). At the very basic level however, most qualitative, analytical, or modeling approaches still utilize the concept of indicators. Indicators are a very useful tool for the understanding and assessment of broad, multi-dimensional concepts such as sustainability, and can range from simple values to complex indices/functions, and can be non-numerical. The term indicator is often used interchangeably with other terms such as parameter, measure, proxy, value, variable, performance measure, and so forth (*2*, *12*).

Selecting and Using Indicators for Sustainability

As indicators have a wide range of uses outside of the context of sustainability, a question arises about what makes an indicator an indicator of sustainability. Maclaren distinguishes sustainability indicators from regular planning indicators as being 1) integrating, 2) forward looking, 3) distributional, and 4) developed with input from multiple stakeholders in the community (*13*). Others emphasize the need to focus on outcomes, as opposed to inputs/processes and outputs for sustainability indicators (*1*, *14*). There are several compilations of sustainable transportation indicators available in the literature, the majority of which identify and classify indicators based on specific goals or areas of interest relevant to sustainability (*1*, *3*, *4*, *15*, *16*). Others also focus on the urban scale, quantifying sustainable transportation indicators for the comparison of cities (*17*, *18*). The actual application or implementation of these indicators generally follows approaches based in the field of multi-criteria decision making (MCDM) or multi-criteria decision analysis (MCDA), including approaches such as the analytic hierarchy process and pairwise comparisons to elicit priorities and values. Examples include the use of MCDM to obtain priority weights for environmental outcomes (*19*), to select appropriate measures from a long-list, and for identifying desirable attributes of an indicator (*15*), and for developing composite sustainability indices (*20*, *21*). Other approaches to assess sustainability in the transportation context include sustainability rating systems (*22*, *23*), and monetization of dimensions into a single value, similar to a benefit–cost analysis (*24*).

Spatial Data and Sustainability Indicators

Spatial mapping and spatial statistics were previously discussed as methods of assessing sustainable transportation (*11*). With advancements in geographic information systems (GIS), the use of spatial data to quantify sustainability-oriented transportation and land use indicators at the disaggregate level has become increasingly common. These analyses often make use of data from travel demand models, censuses, travel surveys, transit providers, and similar sources. Applications found in the literature range from understanding accessibility (*25*, *26*), pedestrian connectivity (*27*), equity and environmental justice analyses (*28*, *29*), and emissions estimation (*30*). Increasingly, GIS-based analyses are also used in combination with MCDA methods to reflect composite results for multiple indicators or attributes (*31*). The term GIS-MCDA is used broadly to cover these types of applications. Transportation-related examples include computation of sustainability indices in the Atlanta metropolitan region (*21*, *32*), and work on the ranking of zones in a city on the basis of transportation sustainability criteria (*33*).

Research Approach and Scope

This research effort is driven by the premise that the root conceptualization of sustainability (or any similar broad concept) can be used to identify key goals and representative indicators that reflect the overarching construct. The aim of this study is therefore to assess how conceptualizations of sustainable transportation differ when translated into quantified indicators applied to a common test case. It is acknowledged that indicators have different definitions and applications (*34*, *35*), but for the purposes of this study, the term is used to denote "a variable, or a combination of variables, selected to represent a certain wider issue or characteristic of interest" (*2*).

Study Area

The indicator-based analysis was conducted for the El Paso metropolitan area in the United States. It is located in west Texas, at the U.S.–Mexico border, across from the city of Ciudad Juarez in Mexico. El Paso is also adjacent to the city of Las Cruces, New Mexico. The area is currently in nonattainment of federal ambient air quality standards. It also serves as a gateway for large amounts of cross-border freight and passenger movement through the border crossings in the area. As a border community, the area also has unique demographics and equity considerations. The El Paso Metropolitan Planning Organization (MPO) is responsible for transportation planning in this region, and it covers a regional planning area that includes El Paso County in Texas and portions of Dona Ana and Otero counties in New Mexico.

Analysis Approach

A subset of the alternative conceptualizations of sustainability discussed in the Background section was selected for the analysis, namely:

- Anthropocentric sustainability (termed in this analysis as sustainability) – to be representative of "baseline" sustainability considerations as commonly applied in the area of transportation planning.
- Livability and health – as the alternative conceptualizations of sustainability, that represent emerging alternative discourse in transportation planning that sometimes replace sustainability considerations.

The overall analysis approach is consistent with what are broadly viewed as GIS-MCDA analyses, and is as follows:

- Identify representative indicators for the three conceptualizations being compared, namely, sustainability, livability, and health.
- Quantify these indicators using spatially disaggregated data for areas within the study region.
- Combine the indicators into composite sustainability, livability, and health indices, using a simplified aggregation method based on MCDA methods.

Data Sources and Unit of Analysis

The primary data source for this analysis is the El Paso MPO's regional travel demand model (TDM). As part of the federally mandated transportation planning process, MPOs use travel demand modeling (or travel demand forecasting) to estimate future travel demand and assess planning requirements (*36*). The most common model used for this purpose is the traditional four-step model, in which traffic analysis zones, or transportation analysis zones (TAZs) are the building blocks or unit of analysis. They contain demographic and land use data that are used in the estimation of travel between the zones (*36*–*38*). The design of TAZs is not an exact science, though existing literature discusses different approaches to TAZ development, and rules of thumb for appropriate TAZ size and scope (*36*, *38*, *39*). In general, TAZs are constructed to be relatively homogenous in terms of land use and demographic characteristics, and the density of TAZs (number of zones per unit area) increases in areas of greater population density such as city centers or highly urbanized areas. In line with several examples from the literature that use TAZs as the unit of analysis in quantifying sustainability-oriented transportation and land use indicators (*25*, *28*, *32*, *33*), this study also uses

Table 1. Summary of Indicator Selection Process

Discourse	Considerations for indicator set	Indicators selected	References/sources and additional notes
Sustainability (anthropocentric)	Indicators reflecting environmental, economic, and social factors	• Mode split • Greenhouse gas emissions • Criteria pollutant emissions • Land use mix • Safety	• General sustainability indicators: Jeon and Amekudzi (*4*); Litman (*3*); Zietsman et al. (*1*); Haghshenas and Vaziri (*17*); World Business Council on Sustainable Development, (*18*) • Perspectives on strong vs weak sustainability (anthropocentric vs holistic): Gudmundsson and Höjer (*6*); Black, (*53*) • Equity element: Karner (*54*); Golub and Martens (*28*) • Eco-centric perspective: Amekudzi et al. (*55*); Black (*53*); Wackernagel and Rees (*56*)
Livability	Indicators concerned with transportation as it relates to community-scale impacts, primarily on human well-being/ quality of life	• Employment–population balance • Proximity to bicycle routes • Proximity to transit facilities • Proximity to parks and recreation	• Livability indicators: Gough (*41*); Litman (*42*); United States Department of Transportation (USDOT) (*43*) • Employment–population balance as a variation of jobs–housing balance: California Planning Roundtable (*44*); Horner and Murray (*45*) • Limitations of measure discussed by Guiliano (*46*); and Cervero (*47*), but was selected due to extensive use in literature as a livability measure
Health	Indicators reflecting transportation's relationship to human health, especially in relation to four key elements: safety, air quality, active living opportunities, and access to critical destinations	• Traffic density • Safety • Proximity to clinics and hospitals • Proximity to parks and recreation	• Health indicators: California DPH (*48*); USDOT (*49*); and Lyons et al.; (*50, 51*) • Traffic density or intensity measure, defined as vehicle miles of travel per unit area, is used in literature as a proxy for exposure to traffic-related emissions: Rioux et al. (*52*); and Rowangould (*29*)

TAZs as the basic unit of analysis in the indicator and composite index calculations.

The El Paso MPO provided data from their current version of the TDM for this study. The data included transportation network data (network files) and demographics and land use information (TAZ files), with spatially linked attributes covering four analysis years: 2010, 2020, 2030, and 2040. Other data such as GIS maps containing locations of sites such as transit stops and bicycle lanes were also used in the analysis, obtained from a web repository of GIS datasets maintained by the University of Texas at El Paso, in collaboration with public agencies in the region (*40*).

Indicator Selection

Given that concepts such as sustainability, livability, and health are broad and overarching, there are inherent complexities with selecting appropriate indicators to represent them. In the absence of a universal consensus on indicators, the approach to indicator selection was to identify those that could reasonably reflect the issue at hand. Thus, the selection of indicators for analysis was conducted taking into account examples from literature, to reflect the discourse under consideration, and of appropriate scale for quantification using transportation planning-level data. Table 1 summarizes the indicators selected, along with considerations that governed the indicator selection, relevant literature sources reviewed, and additional clarifying notes. As seen in the table, the sustainability indicators selected were reflective of what could be considered an anthropocentric or weak approach to sustainability, and included a set of indicators reflecting social, economic, and environmental aspects. The greenhouse gas and criteria pollutant emissions cover the environmental aspects; the safety measures address social considerations, whereas the land use measure and mode split measure can be viewed as addressing both the social and economic dimensions. This is consistent with several sources that discuss sustainability indicators (*1*, *3*, *4*, *17*, *18*). The livability indicators were selected to reflect community-scale impacts, primarily on human well-being/quality of life, consistent with livability indicators from the literature (*41–47*). These include measures of proximity to bicycle routes, transit facilities, and parks and recreation, along with the employment–population balance, which is a common measure akin to the jobs–housing balance measure, and aims to assess how close people are to jobs and economic opportunity. Although there are limitations to this indicator (*46*, *47*), it was selected over alternatives such as commute times or distances due to considerations of data availability for future years. The health indicators were selected to reflect four key elements – safety, air quality, active living opportunities, and access to critical destinations – consistent with literature on the subject (*48–51*). This included a measure of traffic intensity used as a proxy for exposure to traffic-related emissions (*29*, *52*), as well as the safety indicator, and proximity to parks and recreational facilities (active living opportunities), and to clinics and hospitals (critical destinations).

Overlaps in Indicator Sets

As seen in Table 1 one indicator (proximity to park and recreation facilities) overlaps with the health and livability indicator sets, and one indicator (safety) overlaps with the health and sustainability indicator sets. This is expected, and representative of reality, given the overlaps between the concepts seen in the literature review and in the qualitative analysis. At the same time, the number of overlapping measures was intentionally limited to a single measure between any two indicator sets.

It is also evident that some of the measures between the indicator sets represent similar considerations, but at a different scope/scale. For example, the traffic density measure looks at exposure impacts of traffic emissions, from an air quality and health perspective. The criteria pollutant emissions measure in the sustainability indicator set also addresses the air quality issue, though at a different scale, focused on absolute emissions levels for a TAZ. Similarly, mode split in the sustainability indicator set, and transit or bike lane access in the livability indicator set, each address similar issues (of alternative transportation modes) at different levels. Again, these conceptual overlaps do not detract from the main study objective, which is to investigate whether the prevailing notions of sustainability, livability, and health differ significantly when quantified using indicators. Though a case can be made for some of the non-overlapping indicators in one indicator set to also be included in the other, this study is focused on identifying a limited set of the most representative indicators for each concept. It does not imply that the indicators represent the entirety of important sustainability, livability, and health considerations, they instead signify a representation of key considerations when quantified as indicators at the transportation planning level.

Indicator Definitions, Computations, and Desired Direction

Table 2 provides a summary of the indicators used, specific definition and computation, desired direction of the indicator, and the discourse (sustainability/livability/health) that the indicator is applicable to.

Development of Sustainability, Livability, and Health Indices

The eleven indicators quantified for each TAZ in this study are expressed in different units, and have different desired directions, as shown in Table 2. Further, individual subsets of these indicators are applicable to the sustainability, livability, and health discourse. In order to assess the overall differences between the indicator sets, it is necessary to express relative performance of individual measures and aggregate them at the TAZ level. The process is a simplified MCDA approach, similar to other examples in the literature (*17*, *21*, *57*), and is described briefly below. Further details of the process are provided in research by Ramani (*10*):

- Quantified indicators for each TAZ were scaled (expressed on a 0–1 scale) to allow for easy comparison of low-performing and high-performing TAZs.
- TAZs were ranked into quintiles, based on their indicator values and desired direction of the indicator. For example, in the case of the greenhouse gas emissions measure, TAZs with highest emissions will rank in the lowest quintile, and vice-versa.
- The bottom- to top-performing quintiles were then assigned scores of 0, 0.25, 0.5, 0.75, and 1. This approach to scaling allowed for an even distribution of scaled indicator values among the TAZs It also ensured that TAZs with extreme indicator values did not disproportionately affect the scores of other TAZs, which is possible with other scaling approaches that were based on normalization of values to highest and lowest observed values in the dataset.

Table 2. Summary of Indicators

Measure	Computation	Desired direction	Relevant discourse: Sustainability	Livability	Health
1. Mode split	Ratio of non-SOV trips to SOV trips originating from TAZ	Higher	■		
2. Emissions: greenhouse gases	Total CO_2 emissions from TAZ, tons per day	Lower	■		
3. Emissions: criteria pollutants	Total NOx and PM_{10} emissions from TAZ, in tons of NOx equivalent weighted by damage costs	Lower	■		
4. Land use mix	Aggregated area type for TAZ from TDM (rated from urbanized to rural land use types)	Closer to 1 on 1–5 scale	■		
5. Employment–Population Ratio	Ratio of employment to population in a TAZ	As close to 1 as possible		■	
6. Proximity to bicycle routes	Distance of TAZ centroid to nearest bicycle route	Lower		■	
7. Proximity to parks and recreation	Distance of TAZ centroid to nearest park or public recreation facility	Lower		■	■
8. Proximity to transit facilities	Distance of TAZ centroid to nearest transit facility	Lower		■	
9. Proximity to clinics and hospitals	Distance of TAZ centroid to nearest clinic or hospital	Lower			■
10. Safety	Uncalibrated total annual crashes per TAZ using PLANSAFE model estimate	Lower	■		■
11. Traffic density	VMT/mi^2	Lower			■

Note: SOV = single occupant vehicle; TAZ = traffic analysis zone; CO_2 = carbon dioxide; NOx = oxides of nitrogen; PM_{10} = particulate matter of 10 micrometers or below; TDM = travel demand model; VMT = vehicle miles of travel.

- The scaled indicator values corresponding to the sustainability, livability, and health datasets were then aggregated as simple averages to obtain composite sustainability index, livability index, and health index values for each TAZ. The use of average values is equivalent to assigning equal weights to all measures comprising the respective index; this was determined to be appropriate for the purposes of this analysis.

Analysis Results

Sustainability, Livability, and Health Indices

Using the methods described in previous sections, TAZ-level indicators, scaled indicators, and sustainability, livability, and health indices were computed for four analysis years (2010, 2020, 2030, and 2040) based on the El Paso TDM data. The results were also linked to spatial data to allow for visualization of results for better interpretation. Figures 1–3 show maps of the calculated sustainability, livability, and health index values, respectively, for each analysis year, between the different TAZs. Two main observations can be drawn from the results:

- There is very little change between the relative distribution of scores for the sustainability, livability, and health indices over time (i.e., TAZs scoring well in one particular analysis year, tend to perform well in other analysis years);
- There do not appear to be high correlations or clear trends between the livability, health, and sustainability indices for any given year.

These findings were confirmed by a correlation analysis of the index values. The findings show that there are high correlations between the same index for different analysis years. For example, the 2010 sustainability index values are seen to have correlation coefficients of 0.93, 0.89, and 0.87, with the 2020, 2030, and 2040 sustainability index values respectively. There are similarly high correlations for the livability and health indices. However, for a given analysis year, the correlations between the sustainability, livability, and health indices are found to be very low. For example, in 2010, the correlation between the sustainability and livability index pairs, the livability and health index pairs, and the health and sustainability index pairs, are 0.11, 0.33, and 0.48 respectively.

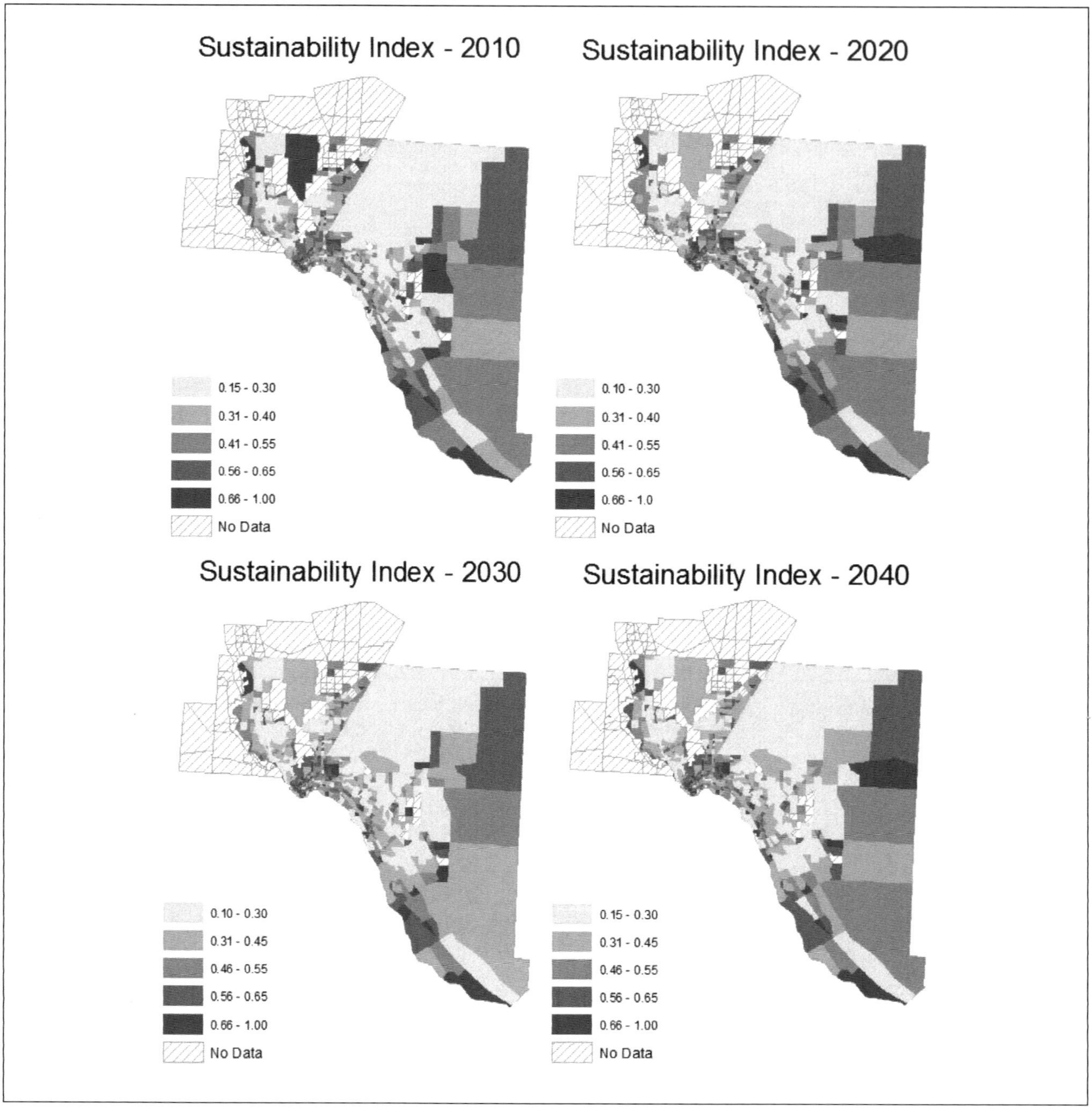

Figure 1. Sustainability index values by TAZ: all analysis years.

Additional Observations and Analysis

From the results shown in figures 1–3, it was seen that areas scoring well from a livability and health perspective tend to be located closer to downtown and urbanized areas, compared with performance from a sustainability perspective, which seems to be more widely dispersed. Additional analyses were conducted to assess:

- The use of CO_2 emissions as a proxy for the sustainability index;
- Assessment of equity implications in terms of index values for low- and high-income areas.

A detailed discussion of these analyses and results is beyond the scope of this paper, however, it was seen that sustainability index values and the scaled CO_2 emissions measure were highly correlated (correlation coefficients in the range of 0.9 for all analysis years), despite being only one of five measures comprising the sustainability index. The presence of other measures in the index that also correlate with vehicle miles of travel (VMT) could

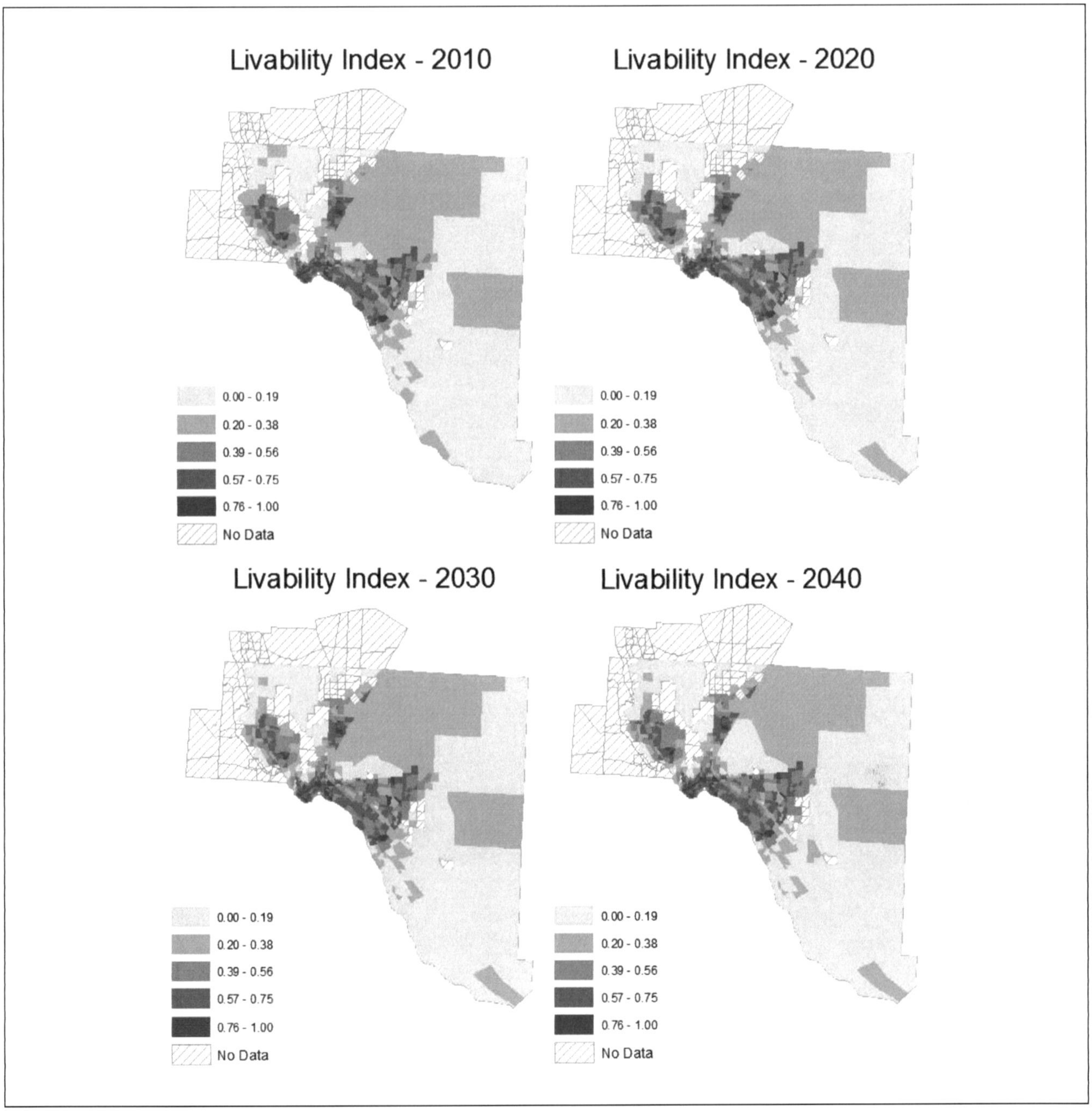

Figure 2. Livability index values by TAZ: all analysis years.

partially explain this observation. The equity analysis results showed that the average sustainability, livability, and health index values for the low-income zones were actually higher than the average for all zones in the analysis – which is an interesting finding with regards to potential equity considerations. This could be explained due to the location of the low-income TAZs being close to the densely populated urbanized areas, leading to better overall performance on some indicators such as proximity to various services. For the higher-income TAZs, it was found that livability and sustainability index values tend to be lower than average, whereas the health index values remain close to average values.

Summary and Conclusions

This paper describes an analysis conducted to investigate how alternative discourse related to sustainable transportation can affect outcomes measured using indicators. This was done for a test case in the El Paso metropolitan

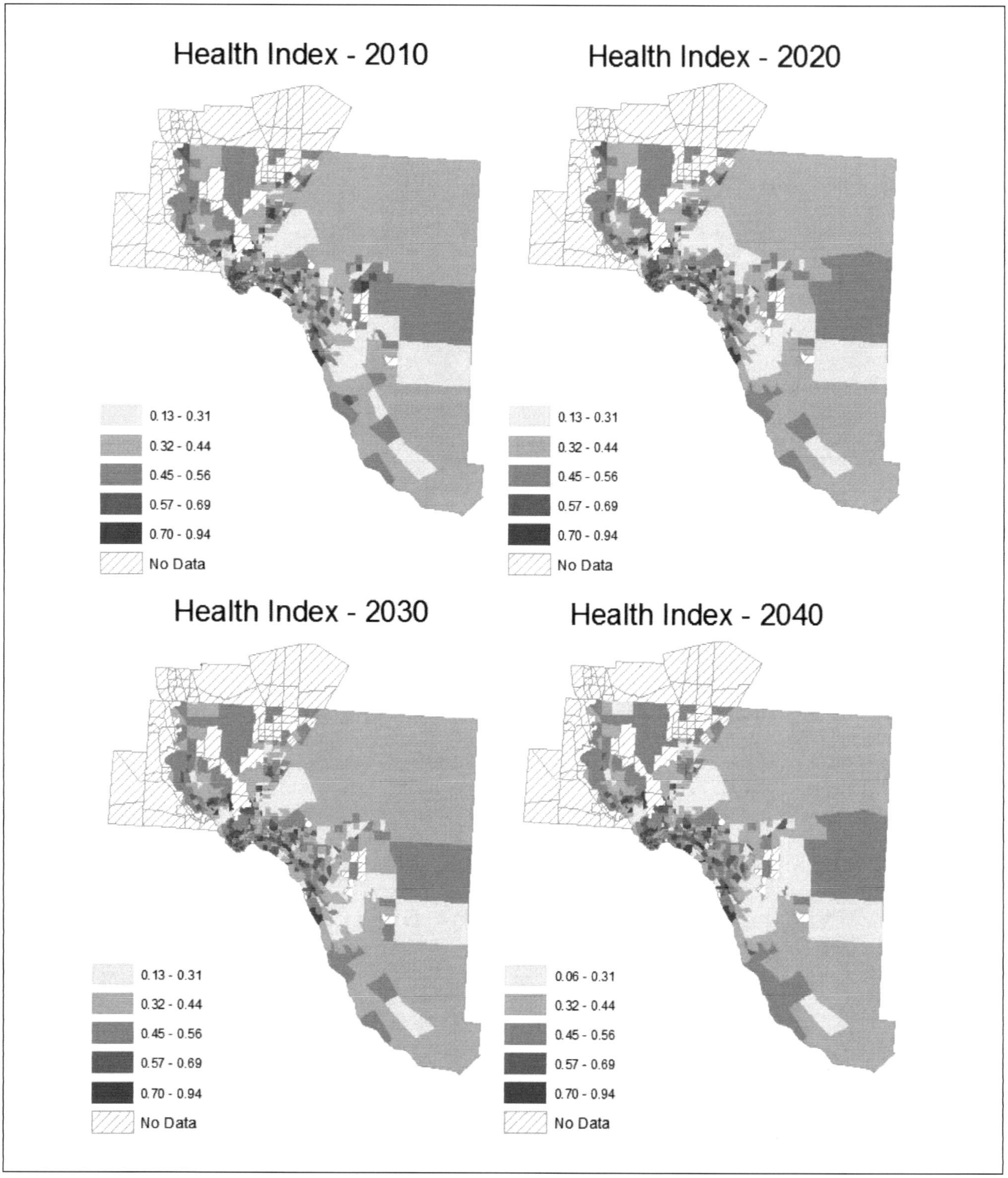

Figure 3. Health index values by TAZ: all analysis years.

region. Local data were used to quantify indicators that were aggregated to a sustainability index, livability index, and health index for traffic analysis zones in the region, for four analysis years. The analysis results demonstrated little correlation between the quantified livability, sustainability, and health indices for a specific analysis year. It was also found that the indices tended to stay the same over time, indicating that the relative performance of a zone remained unchanged despite the modeled changes to the transportation system, demographics, and land use. Additional analyses of CO_2 emissions compared with the sustainability index as a whole indicated the possibility that an eco-centric perspective could serve as a reasonable proxy for a triple-bottom-line sustainability perspective. An equity analysis showed lower-income zones having higher index values compared with overall averages, with high-income zones faring worse than average in terms of the sustainability and livability indices. These findings are somewhat counterintuitive, and indicate that areas that may be considered sustainable or livable from a transportation and land use perspective do not necessarily correlate to having a better quality of life from the perspectives of overall income or poverty levels.

Although the El Paso region is unique in terms of its geography and location, its transportation system is similar to that of many large cities in the United States, and the study findings could be expected to be generalizable to other regions. However, further application of this research to other areas is required to conclusively determine the generalizability of the findings.

This study indicates that the concept of sustainability cannot be replaced by alternative discourse such as livability, health, or resilience in the transportation sector, despite theoretical overlaps that exist between the concepts. That is, despite speaking to similar themes, the concepts are not necessarily interchangeable. At a practical level, this is a mixed outcome for sustainability advocates. On the one hand, it demonstrates the value of sustainability as a concept that cannot be replaced by other discourse or organizing principles. On the other hand, this can be viewed as a negative outcome, as it indicates that alternative approaches such as livability and health cannot serve as proxies for sustainability, which may be useful in advancing sustainability programs in politically challenging climates.

The study also reiterates the inherent complexity of sustainability issues, and brings to light the importance of indicators used for measuring concepts. Though this study only used selected indicators to test and compare the concepts, they were chosen to reflect prevailing notions of sustainability, health, and livability. Therefore, the approach to indicator selection was to rely on examples from the literature and the stated goals of the relevant discourse. It can be viewed as representative of a good-faith attempt by a transportation practitioner to quantify a concept (such as sustainability, livability, or health) that has been deemed important, and to presumably make planning decisions accordingly. However, the results also show the influence that indicator selection can have on overall results, and raises the possibility of indicators being "cherry-picked" to produce desired outcomes.

Further, the study also confirms what has been addressed at length in literature and research – that transportation cannot be made sustainable in isolation. The results also indicate very small changes in relative performance of TAZs over time, for the sustainability, livability, as well as health indices – pointing to the relative inelasticity of sustainability indicators, and the importance of a strategic, long-term view on sustainability issues.

There are several limitations to the study, stemming from the limitations of the available data and the simplified MCDA process used to aggregate index values. Some areas for future research to build on the study findings and address these limitations include:

- Similar studies with expanded indicator sets or different indicators to test the concepts.
- Comparison of multiple transportation and land use scenarios.
- Implementing the study approach in other areas to determine whether the findings are generalizable.
- Use of further disaggregated data with applicable measures to test the concepts at a finer scale than TDM data.
- Research into the impact of indicator selection, indicator weighting, and examination of influential indicators that can serve as proxies for sustainability.
- Use of Delphi methods or other approaches to identify and prioritize indicators to reflect the different frameworks of interest.

Acknowledgments

The data used for the analysis was provided by the El Paso Metropolitan Planning Organization, and the emissions estimation tool used to compute transportation emissions was developed as part of a research project conducted by the Texas A&M Transportation Institute (TTI) for the El Paso Metropolitan Planning Organization. The author thanks Michael Medina, Salvador Gonzalez-Ayala, and Sonia Perez of the El Paso Metropolitan Planning Organization for providing data and input to this work. The author also acknowledges the contributions of Josias Zietsman, Reza Farzaneh, Madhusudan Venugopal, Jeremy Johnson, and L.D. White of TTI to the development of the emissions estimation methodology used in this paper.

Author Contributions

The author confirms sole contribution to the paper as follows: study conception and design; analysis and interpretation of results; and draft manuscript preparation. The author reviewed the results and approved the final version of the manuscript.

References

1. Zietsman, J., T. Ramani, V. Reeder, J. Potter, and J. DeFlorio. *NCHRP Report 708: A Guidebook for Sustainability Performance Measurement for Transportation Agencies*. Transportation Research Board, of the National Academies, Washington, D.C., 2011.
2. Gudmundsson, H., R. P. Hall, G. Marsden, and J. Zietsman. *Sustainable Transportation: Indicators, Frameworks, and Performance Management*. Springer-Verlag Berlin Heidelberg, 2015.
3. Litman, T. Developing Indicators for Comprehensive and Sustainable Transport Planning. *Transportation Research Record: Journal of the Transportation Research Board*, 2007. 2017: 10–15.
4. Jeon, C. M., and A. Amekudzi. Addressing Sustainability in Transportation Systems: Definitions, Indicators, and Metrics. *Journal of Infrastructure Systems*, Vol. 11, No. 1, 2005, pp. 31–50.
5. Ramani, T., J. Zietsman, H. Gudmundsson, R. Hall, and G. Marsden. Framework for Sustainability Assessment by Transportation Agencies. *Transportation Research Record: Journal of the Transportation Research Board*, 2011. 2242: 9–18.
6. Gudmundsson, H., and M. Höjer. Sustainable Development Principles and Their Implications for Transport. *Ecological Economics*, Vol. 19, No. 3, 1996, pp. 269–282.
7. Ramani, T., and J. Zietsman. Sustainable Transportation – Alternative Perspectives and Enduring Challenges *International Journal of Urban Sciences*, 2016.
8. Ramani, T., J. Zietsman, and M. R. Pryn. Towards Sustainable Transport Planning in the United States. *European Journal of Transportation and Infrastructure Research*, 2017. Forthcoming.
9. Ramani, T. Assessing Alternative Conceptualizations of Sustainable Transportation. Presented at 96th Annual Meeting of the Transportation Research Board, Washington, D.C., 2017.
10. Ramani, T. *Validating Alternative Conceptualizations of Sustainable Transportation*. PhD dissertation. Landscape Architecture and Urban Planning, Texas A&M University, College Station, 2017.
11. Black, J. A., A. Paez, and P. A. Suthanaya. Sustainable Urban Transportation: Performance Indicators and Some Analytical Approaches. *Journal of Urban Planning and Development*, Vol. 128, No. 4, 2002, pp. 184–209.
12. Gallopín, G. C. Environmental and Sustainability Indicators and the Concept of Situational Indicators. A Systems Approach. *Environmental Modeling & Assessment*, Vol. 1, No. 3, 1996, pp. 101–117.
13. Maclaren, V. Urban Sustainability Reporting. *Journal of the American Planning Association*, Vol. 62, No. 2, 1996, pp. 184–202.
14. Marsden, G., C. Kelly, C. Snell, and J. Forrester. Improved Indicators for Sustainable Transport and Planning. *DISTILLATE Deliverable C1*, 2005.
15. Castillo, H., and D. E. Pitfield. ELASTIC – A Methodological Framework for Identifying and Selecting Sustainable Transport Indicators. *Transportation Research Part D: Transport and Environment*, Vol. 15, No. 4, 2010, pp. 179–188.
16. Lautso, K., and S. Toivanen. SPARTACUS System for Analyzing Urban Sustainability. *Transportation Research Record: Journal of the Transportation Research Board*, 1999. 1670: 35–46.
17. Haghshenas, H., and M. Vaziri. Urban Sustainable Transportation Indicators for Global Comparison. *Ecological Indicators*, Vol. 15, No. 1, 2012, pp. 115–121.
18. World Business Council on Sustainable Development. *Sustainable Urban Mobility - Methodology and Indicator Calculation Method.* World Business Council on Sustainable Development, 2015.
19. Accorsi, R., E. Zio, and G. E. Apostolakis. Developing Utility Functions for Environmental Decision Making. *Progress in Nuclear Energy*, Vol. 34, No. 4, 1999, pp. 387–411.
20. Ramani, T., J. Zietsman, W. L. Eisele, D. Rosa, D. Spillane, and B. Bochner. *Developing Sustainable Transportation Performance Measures for TXDOT's Strategic Plan: Technical Report.* http://tti.tamu.edu/documents/0-5541-1.pdf
21. Jeon, C. M., A. A. Amekudzi, and R. L. Guensler. Evaluating Plan Alternatives for Transportation System Sustainability: Atlanta Metropolitan Region. *International Journal of Sustainable Transportation*, Vol. 4, No. 4, 2010, pp. 227–247.
22. Muench, S. T., J. Anderson, and T. Bevan. Greenroads: A Sustainability Rating System for Roadways. *International Journal of Pavement Research and Technology*, Vol. 3, No. 5, 2010, pp. 270–279.
23. Oswald, M., and S. McNeil. Rating Sustainability: Transportation Investments in Urban Corridors as a Case Study. *Journal of Urban Planning and Development*, Vol. 136, No. 3, 2009, pp. 177–185.
24. McVoy, G., K. Gunasekera, L. R. Sousa, and P. Schaffner. An Analytical Framework for Sustainability Analysis of Transportation Investments Across the Triple Bottom Line Using a Common Metric. *2013 International Conference on Ecology and Transportation (ICOET 2013)*, 2013.
25. Benenson, I., K. Martens, Y. Rofé, and A. Kwartler. Public Transport versus Private Car GIS-Based Estimation of Accessibility Applied to the Tel Aviv Metropolitan Area. *The Annals of Regional Science*, Vol. 47, No. 3, 2011, pp. 499–515.
26. Yigitcanlar, T., N. Sipe, R. Evans, and M. Pitot. A GIS-based Land Use and Public Transport Accessibility Indexing Model. *Australian Planner*, Vol. 44, No. 3, 2007, pp. 30–37.
27. Randall, T. A., and B. W. Baetz. Evaluating Pedestrian Connectivity for Suburban Sustainability. *Journal of Urban Planning and Development*, Vol. 127, No. 1, 2001, pp. 1–15.
28. Golub, A., and K. Martens. Using Principles of Justice to Assess the Modal Equity of Regional Transportation Plans. *Journal of Transport Geography*, Vol. 41, 2014, pp. 10–20.
29. Rowangould, G. M. A Census of the US Near-roadway Population: Public Health and Environmental Justice Considerations. *Transportation Research Part D: Transport and Environment*, Vol. 25, 2013, pp. 59–67.

30. Brown, A. L., and J. K. Affum. A GIS-based Environmental Modelling System for Transportation Planners. *Computers, Environment and Urban Systems*, Vol. 26, No. 6, 2002, pp. 577–590.
31. Malczewski, J. GIS-based Multicriteria Decision Analysis: A Survey of the Literature. *International Journal of Geographical Information Science*, Vol. 20, No. 7, 2006, pp. 703–726.
32. Jeon, C. M., A. A. Amekudzi, and R. L. Guensler. Sustainability Assessment at the Transportation Planning Level: Performance Measures and Indexes. *Transport Policy*, Vol. 25, 2013, pp. 10–21.
33. Jakimavičius, M., and M. Burinskiene. A GIS and Multicriteria-based Analysis and Ranking of Transportation Zones of Vilnius City. *Ukio Technologinis ir Ekonominis Vystymas*, Vol. 15, No. 1, 2009, pp. 39–48.
34. Joumard, R., and H. Gudmundsson. Indicators of Environmental Sustainability in Transport. 2010.
35. Gilbert, R., N. Irwin, B. Hollingworth, and P. Blais. Sustainable Transportation Performance Indicators (STPI). *Transportation Research Board (TRB), CD ROM*, 2003.
36. Cambridge Systematics, VHB Inc, Gallop Corporation, C.R. Bat, Shapiro Transportation Consulting, Martin/Alexiou/Bryson. *NCHRP Report 716: Travel Demand Forecasting: Parameters and Techniques*. Transportation Research Board, of the National Academies, Washington, D.C., 2012.
37. Spielberg, F., and P. Shapiro. *Special Report 288: Determination of the State of the Practice in Metropolitan Area Travel Forecasting: Findings of the Surveys of Metropolitan Planning Organizations*. Transportation Research Board of the National Academies, Washington, D.C., 2007.
38. Federal Highway Administration. *Defining Traffic Analysis Zones*. https://www.fhwa.dot.gov/planning/tmip/publications/other_reports/technical_synthesis_report/page01.cfm
39. You, J., Z. Nedović-Budić, and T. J. Kim. A GIS-based Traffic Analysis Zone Design: Technique. *Transportation Planning and Technology*, Vol. 21, No. 1–2, 1998, pp. 45–68.
40. *Paso Del Norte Map for Public Access*. http://www.pdnmapa.org/
41. Gough, M. Z. Reconciling Livability and Sustainability: Conceptual and Practical Implications for Planning. *Journal of Planning Education and Research*, 2015.
42. Litman, T. *Sustainability and Livability: Summary of Definitions, Goals, Objectives and Performance Indicators*. Victoria Transport Policy Institute, Canada, 2010.
43. USDOT. *Livability in Transportation Guidebook: Planning Approaches that Promote Livability*, 2010.
44. California Planning Roundtable. *Deconstructing Jobs-Housing Balance*, 2008.
45. Horner, M., and A. Murray. A Multi-objective Approach to Improving Regional Jobs-Housing Balance. *Regional Studies*, Vol. 37, No. 2, 2003, pp. 135–146.
46. Giuliano, G. *Is Jobs-housing Balance a Transportation Issue?* University of California Transportation Center, 1991.
47. Cervero, R. Jobs-housing Balance Revisited: Trends and Impacts in the San Francisco Bay Area. *Journal of the American Planning Association*, Vol. 62, No. 4, 1996, pp. 492–511.
48. California Department of Public Health. *A Primer for California's Public Health Community on Regional Transportation Plans and Sustainable Communities Strategies*. California Department of Public Health, 2013.
49. United States Department of Transportation. *Transportation and Health Tool*. United States Department of Transportation. http://www.transportation.gov/transportation-health-tool. Accessed November 5, 2015.
50. Lyons, W., L. Morse, L. Nash, and R. Strauss. *Statewide Transportation Planning for Healthy Communities*, 2014.
51. Lyons, W., H. Peckett, L. Morse, M. Khurana, and L. Nash. *Metropolitan Area Transportation Planning for Healthy Communities*, 2012.
52. Rioux, C. L., D. M. Gute, D. Brugge, S. Peterson, and B. Parmenter. Characterizing Urban Traffic Exposures Using Transportation Planning Tools: An Illustrated Methodology for Health Researchers. *Journal of Urban Health*, Vol. 87, No. 2, 2010, pp. 167–188.
53. Black, W. R. Sustainable Transportation: A US perspective. *Journal of Transport Geography*, Vol. 4, No. 3, 1996, pp. 151–159.
54. Karner, A. Planning for Transportation Equity in Small Regions: Towards Meaningful Performance Assessment. *Transport Policy*, Vol. 52, 2016, pp. 46–54.
55. Amekudzi, A. A., C. Jotin Khisty, and M. Khayesi. Using the Sustainability Footprint Model to Assess Development Impacts of Transportation Systems. *Transportation Research Part A: Policy and Practice*, Vol. 43, No. 4, 2009, pp. 339–348.
56. Wackernagel, M., and W. E. Rees. Perceptual and Structural Barriers to Investing in Natural Capital: Economics from an Ecological Footprint Perspective. *Ecological Economics*, Vol. 20, No. 1, 1997, pp. 3–24.
57. Ramani, T., J. Zietsman, W. Knowles, and L. Quadrifoglio. Sustainability Enhancement Tool for State Departments of Transportation Using Performance Measurement. *Journal of Transportation Engineering*, Vol. 137, No. 6, 2011, pp. 404–415.

The Standing Committee on Transportation and Sustainability (ADD40) peer-reviewed this paper (18-03007).

Article

TRR
JOURNAL OF THE TRANSPORTATION RESEARCH BOARD

Transportation Research Record
2018, Vol. 2672(3) 104–117

DOI: 10.1177/0361198118799033
journals.sagepub.com/home/trr
SAGE

Developing Assessment Criteria for Sustainable Transport Appraisal

Merrill Jones Barradale[1] and Yannick Cornet[2]

Abstract

Defining a clear, comprehensive, and non-overlapping set of assessment criteria is a crucial step in multi-criteria analysis processes. Although criteria are typically developed based on planning objectives, there currently exists no standard procedure for conducting this exercise when both transportation and wider sustainable development goals need to be considered jointly. Based on research applying the multi-actor multi-criteria analysis approach to the case of a high-speed rail project in the UK, this paper presents a framework and a replicable stakeholder involvement process for selecting, defining, and naming assessment criteria to be used in sustainable transport appraisal (STA). The framework consists of a set of systematic guidelines and considerations for producing a coherent list that addresses known cognitive biases and can be understood by a broad range of stakeholders. A further practicable outcome of the research is a comprehensive list, developed using this framework, of 28 transport assessment criteria encompassing direct project impacts, indirect societal impacts, and environmental impacts. This list can serve as a starting point for planners wishing to conduct MCA-based STA. One key take-away from this research is the emphasis on an inclusive, interactive, and iterative approach for defining sustainable transport assessment criteria.

Multi-criteria analysis (MCA) has gained significant traction in the field of transport appraisal and is also considered useful for addressing sustainability issues because of its ability to accommodate impacts that are difficult to monetize (*1–3*).

By its very definition, MCA requires a list of multiple criteria to be drawn up as the basis for conducting the analysis. Defining the list of criteria to be assessed is in itself a crucial task that sets the stage for what is to be included and made explicit in assessment, and what is considered less relevant and can therefore be ignored. Criteria in standard MCA are typically defined based on planning objectives. But this limited set of criteria is insufficient to address the complex needs of sustainable transport appraisal (STA), for which the literature has highlighted the need for "new and if possible, standard set of criteria for assessing sustainable transportation altogether" (*4*).

This paper presents a framework for developing precisely such a set of criteria aimed at enabling the appraisal of large-scale transport projects from a sustainability perspective. The paper goes on to describe how this framework was used in practice to come up with a comprehensive list of 28 criteria encompassing conventional transport assessment goals as well as broader impacts on society and the environment. The framework presented in this paper was designed and applied as part of a research project on improving MCA-based appraisal processes for the case of the High Speed 2 (HS2) rail project in the UK (*5*, *6*).

Literature

MCA literature proposes a wide range of techniques to assess impacts that are currently not feasible or practical to monetize (*7*, *8*). The approach comprises eight standard steps (*9*):

1. Identify the objectives within the decision context.
2. Identify the stakeholders, decision makers and other key players.
3. Identify options for achieving the objectives.
4. Identify and organize the criteria to be used for comparing options.

[1]Renewable and Sustainable Energy Institute (RASEI), University of Colorado, Boulder, CO
[2]Department of Planning, Aalborg University, Copenhagen, Denmark

Corresponding Author:
Address correspondence to Merrill Jones Barradale: merrill.j.barradale@colorado.edu

5. Weigh the criteria to reflect their relative importance to the decision.
6. Score the expected performance of each option against the criteria.
7. Combine weights and scores to derive an overall value for each option.
8. Conduct a sensitivity analysis of the results to changes in weights and scores.

These steps are still today the basis for MCA subtechniques such as multi-actor multi-criteria analysis (MAMCA), the approach selected for the case to which this research contributes. MAMCA emphasizes the inclusion of multiple viewpoints not just through the incorporation of multiple criteria but through the involvement of multiple actors (*10*, *11*).

This paper focuses on Step 4 of the appraisal process: the identification and organization of criteria. By itself, this step does not guarantee the adoption of more sustainable options in transport project appraisal. It is, however, essential: if the criteria do not incorporate sustainability (and incorporate it adequately), then the subsequent appraisal steps based on those criteria cannot and will not take sustainability into consideration. Steps for incorporating these criteria in a MAMCA-style appraisal highlighting a sustainability perspective have been proposed elsewhere (*5*, *6*). This paper focuses on the prerequisite to those steps—drawing up a viable list of sustainability assessment criteria—and describes in detail both the process through which the set of criteria was established and the resulting list of criteria.

A key contribution to criteria selection is the existing (and quite rich) literature on indicator selection frameworks (*12*, *13*). In the broad sense, indicators are appraisal criteria seeking to represent operationally and scientifically one or more attributes (qualities, characteristics, properties) of the system under study (*14*). However, because indicators also serve to summarize and simplify, they can never fully represent or explain a phenomenon of interest. In many disciplines quantitative indicators are often preferred "because of the potential precision and reproducibility provided by standard numerical metrics" (*15*). According to Gallopín, qualitative indicators can be preferable when "the attribute of interest is not inherently quantifiable" (*14*). Finally, performance indicators provide a means to monitor progress and assess the effectiveness in reaching transport policy goals (*16*, *17*). In recent years, the operationalization of sustainability principles into performance measurement systems has resulted in a number of transportation rating systems (e.g., INVEST) and criteria-based appraisal processes (e.g., Transport Analysis Guidance [TAG]), which this paper uses as a departure point and also contributes to in relation to the criteria definition process (*18*, *19*).

In this paper, the conceptual unit behind performance indicators is understood to be assessment criteria for STA. In STA the overarching planning goals are determined by both conventional transport objectives (such as mobility and capacity) and long-term sustainable development goals (such as biodiversity or accessibility for the wider population). This paper does not define the measurable unit, whether that may be based on quantitative or qualitative indicators, as this will depend on the context, available data, and the chosen approach for implementing the MCA process. Limiting the scope to criteria definition is well-suited to MCA, which enables non-quantifiable criteria to be assessed on a relative basis (as contrasted with benefit–cost analysis [BCA], in which the scope of analysis is limited to criteria that can be quantified in absolute terms and monetized). However, criteria must be in their definition categorical and non-overlapping in a way that they can feed into steps 5 and 6 of the MCA process above.

Methods

The goal was to develop a set of assessment criteria to be used in transport appraisal that encompasses all three dimensions of sustainability (economic, social, environmental). The assessment criteria are defined in terms of impacts, in which impacts are understood to be effects that are exceptionally important for management and decision making, thus emphasizing the consequences of human action. Impacts can be positive (benefits), negative (costs), or both.

Social cost theory provides the basis for categorizing impacts, distinguishing between internal costs and benefits (direct project impacts) and external costs and benefits (indirect impacts or externalities). The latter was then subdivided into societal impacts and environmental impacts, resulting in the following three categories of appraisal criteria:

1. Direct project impacts (internal costs and benefits)
2. Indirect societal impacts (externalities: people)
3. Environmental impacts (externalities: planet)

To be practicable as well as compatible with a MAMCA appraisal approach, the set of assessment criteria should be:

- Sufficiently comprehensive to address the concerns of a broad range of stakeholder groups (including future generations)
- Short enough to be tractable
- Defined in terms that can be operationalized in MCA steps 5 (criteria prioritization) and 6 (performance assessment)

- Understandable for people and professionals with varied backgrounds and perspectives
- Coherent and well-reasoned (avoiding gaps and overlaps and addressing relevant sources of potential cognitive bias)

The rest of this section describes the process for developing a set of assessment criteria that meets these objectives. The purpose of describing the process in detail is two-fold: 1) the reader may be more confident in the quality of the resulting list of assessment criteria; and 2) this process can be replicated in other cases to modify existing criteria or develop new ones. The overall steps for developing the list of criteria were:

1. Start with existing lists.
2. Interview transport professionals and sustainability experts.
3. Develop a framework to guide defining and naming of criteria.
4. Engage in vigorous and detailed discussions to define and name each criterion, taking into consideration both the input of interviewees and the framework.

Steps 2, 3, and 4 were conducted in parallel rather than sequentially, yielding a process that was both iterative and interactive.

The starting point was the already comprehensive list contained in the official *Transport Analysis Guidance* (WebTAG) by the UK government (*20*) and the (older but still relevant) environmental assessment guidance for roads and bridges (*21*). These were complemented by the impacts assessed in the HS2 appraisal of sustainability (*22*) and environmental statement documents (*23*). These two sets of reports were helpful in pointing out a number of more detailed HS2 and rail-specific criteria (e.g., "solid waste and disposal" associated with tunneling).

Transport professionals and sustainability experts, as well as non-experts with an interest in transport or sustainability issues, were then involved for fine-tuning the list. This was undertaken in connection with developing an assessment questionnaire for HS2 (*5*) to help ensure the criteria would be applicable in the context of a MAMCA-style appraisal. The feedback received in interviews and discussions on each draft of the criteria was helpful in augmenting the list of relevant impacts and refining their names and definitions.

Finally, the internal process required interactive, in-depth discussions among three individuals, who among them cover a broad range of expertise in transport, energy, economics, environment, and sustainability.

In the process of discussing the criteria, it was helpful to distinguish between defining the criteria and choosing names for them. For both tasks, checklists were developed to guide the deliberations. These were then formalized into the framework described below.

Defining Criteria

To guide discussions for defining criteria, determining which impacts belong where, and delineating the boundaries between related criteria, a list was developed of different dimensions for categorizing criteria (Table 1). This (non-exhaustive) list of dimensions includes:

- Internal vs. External Impact (which might or might not coincide with Direct vs. Indirect and User vs. Non-User Impacts)
- Quantity vs. Quality
- Magnitude vs. Uncertainty
- Outcome vs. Mechanism
- Environmental Medium vs. Type of Damage
- Construction vs. Operating Phase
- Passenger vs. Freight Transport.

Keeping the criteria to a reasonable number requires grouping impacts. There are multiple dimensions along which impacts can be grouped (or separated), and these are often in conflict. For each dimension, there are not necessarily definitive conclusions, but many valid considerations (Table 1). Although consistency was one of the design goals, it did not always take precedence. Consistency was achieved where possible, but differential treatment was sometimes necessary (see examples in Table 1).

One of the challenges faced when defining the boundaries of impacts was the appropriate choice of dimension for delineation in cases in which multiple dimensions were deemed useful and reasonable. For example, instead of distinguishing between users and non-users, it is possible to distinguish along dimensions of time and space. In general, impacts are assumed to apply to all phases of the project's life cycle, but some impacts are more relevant during the construction phase (e.g., solid waste disposal and therefore road traffic disruption), whereas some impacts differ in quality between phases (e.g., noise from construction is not comparable to noise from operation). Where relevant, these considerations are included in the impact descriptions. In relation to the spatial dimension, geographic scope is assumed to be the entire railway line. Yet some impacts are more relevant in an urban context (e.g., housing in the London area) and others in rural areas. One could consider deconstructing the problem into phases and into the official 26 line segments (Community Forum Areas, see [*24*]). But that would also result in an exponential number of assessments and would defeat the purpose of comparing

Table 1. Dimensions for Defining Assessment Criteria

Dimension	Considerations and general guidelines	Noteworthy examples
Internal vs. External (Direct vs. Indirect) Impact	• User vs. Non-User Impacts is often a useful proxy for the Internal/Direct vs. External/Indirect dimension. • Sometimes "Intended vs. Unintended" or "Expected vs. Unexpected" may also be useful guidelines. • Impacts that affect users only (e.g., journey cost; journey time) clearly fall under direct project impacts. • Some impacts that affect both users and non-users are considered direct impacts (e.g., *Project cost* is paid by taxpayers, which include far more non-users than users, because it is such a core component of the project itself); whereas others are considered indirect impacts (e.g., *Accidents*, even if it affects users more than non-users, because the impact is unintended or inadvertent).	Affordability is typically grouped together with equity and distributional effects. However, the distinction between internal and external costs was prioritized by establishing a distinction between direct/user affordability and indirect/non-user affordability, putting the former under *Journey cost & affordability* and the latter under *Equity & distributional effects*. Both descriptions explain the distinction carefully.
Quantity vs. Quality	• Some impacts are clearly quantitative (e.g., journey time), some are primarily qualitative (e.g., journey experience), and some have both quantitative and qualitative aspects. • Are those criteria that refer exclusively to one or the other clear about that? • For those impacts that involve both aspects, is it best to group them together or separately?	• Wilderness areas can be measured both quantitatively (e.g., number of hectares) and qualitatively (e.g., location and distribution of land). In this case, the impacts are grouped together (under *Biodiversity & nature*). • *Agriculture, forestry & soils* is a measure of soil quality only, not acreage (for reasons explained in the description).
Magnitude vs. Uncertainty	• Impacts can be quantified in terms of magnitude (expected value, mean) and in terms of uncertainty (risk, variance). Both aspects are relevant, though one is sometimes considered more salient than the other. • Are both aspects of a particular impact included in the same criterion or separated out (e.g., journey time and journey reliability)? • When both aspects are included in the same criterion, should this be made explicit or is it sufficient to leave it implicit (e.g., CO_2 emissions)? • In an effort to condense the criteria, is it better to group by type of impact (e.g., cost vs. performance) or to group various risks together (e.g., project delays, cost overruns, underperformance)?	• Combining *Journey time* and *Journey reliability* into a single criterion was considered, as they are not really separate impacts but, rather, the mean and variance of the same impact. However, in the transport field, time and reliability are considered sufficiently important as separate criteria that they have been kept separate. • *Project delivery risks* is a variance-only impact, grouping together the risks associated with several factors: project cost, construction time, and technology performance. An earlier version of this criterion ("Risks & uncertainty"), a cross-cutting criterion covering variance of all impacts, was considered too broad to be meaningful (as well as inconsistent, as variance of some impacts is clearly included elsewhere).
Outcome vs. Mechanism	• Should criteria focus on outcome only (e.g., tons of CO_2 emitted) or on causal mechanisms (e.g., modal shift)? • Focusing on outcomes is more consistent with describing criteria in terms of impacts; however there is a "marketing" issue here: people often do not think about impacts without reference to the causal mechanism.	Passenger modal shift is often considered to be a significant environmental benefit of high-speed rail and could therefore constitute a criterion. Modal shift is also closely related to rail capacity and could therefore be grouped together with freight capacity. Considering why modal shift matters helped to determine where modal shift belongs. It was decided its significance lies in the potential for reducing the carbon intensity of transportation. In this context, modal shift is a mechanism, not a goal in and of itself. Furthermore, the primary impact is CO_2 emissions and therefore belongs under *Carbon footprint*. However, removing explicit references to "modal shift" from the criteria names may reduce people's consideration of this important mechanism. The descriptions therefore emphasize modal shift.

(continued)

Table 1. (continued)

Dimension	Considerations and general guidelines	Noteworthy examples
Environmental Medium vs. Type of Damage	• Environmental impacts can be categorized by medium (air, water, soil) or by type of damage (contamination, resource depletion, etc.). • Should all water impacts be grouped together? • Should water contamination be grouped with soil contamination? • Should CO_2 be grouped together with other emissions or treated separately?	• *Water resources & flood risk*: As implied by the name, this criterion refers primarily to impacts on the structure and flow of waterways (e.g., changing the course of streams or channels; emptying lakes or reservoirs) and on floodwater drainage networks. The official HS2 environmental statement (*23*) also includes liquid waste disposal as part of this category. It was decided to separate out the quality aspect of water (i.e., water pollution) from the flow aspect (how much and where) by removing liquid waste disposal from this category. • *Water & land contamination*: Here it was decided to combine impacts on water and soil, because contamination is interconnected (soil contaminants often leach into surface and ground water). Disposal of liquid waste is another source of water contamination and so is included under this category. • *Carbon footprint groups* together all greenhouse gas (GHG) emissions expressed as carbon dioxide equivalent (CO2e). • *Air quality* groups together all non-GHG air pollutants.
Construction vs. Operating Phase	• Most impacts are relevant during both construction and operating phases (e.g., costs, accidents, emissions, etc.). • Should these be grouped together, and if so, does the description make this explicit? • Are there some impacts that are associated exclusively or primarily with one phase, and is this clear?	• *Traffic & transport disruption* refers to construction phase only, to avoid overlap with *Community severance* and *Land-use & transport interaction*. Descriptions guide respondents toward appropriate criteria, depending on underlying concerns. • *Project delivery risks* is primarily a construction-phase impact: once the project has been "delivered" (i.e., is operating), construction is no longer delayed or continuing to incur costs. However, underperformance or operations and maintenance cost overruns may be ongoing. • By including both embedded carbon (CO_2 emitted during project construction as well as in connection with producing the steel, cement, and materials used in infrastructure) and fuel carbon (CO_2 emitted during transport operations), *Carbon footprint* covers both construction and operating (as well as pre-construction) phases.
Passenger vs. Freight Transport	• Some impacts are relevant only for passenger transport (e.g., *Journey experience*), and some are relevant for both. • For those that are relevant for both, does it make sense to group them together (e.g., accidents, innovation, landscape) or separately? • In what cases might components of an impact diverge when assessing options, thus necessitating the designation of two separate impacts?	• For most impacts associated with both passenger and freight transport (e.g., accidents, innovation, landscape), assessments and preferences are driven only by total magnitude and not by the passenger-freight distinction. • Only in in the case of capacity were assessments expected to diverge, thus requiring passenger capacity and freight capacity to be designated as separate impacts.

full alternatives with similar geographical footprints. These were therefore left to the deliberation phases with interviewees.

Naming Criteria

To make assessment as "user-friendly" as possible, criteria names had to be easily understood by respondents, be consistent with names of other impacts (e.g., carbon footprint and material footprint), and be as succinct as possible without losing accuracy with regards to what they covered in the detailed descriptions.

In determining what to call each criterion, a number of considerations were deemed useful (Table 2), ranging from whether to phrase a concept in positive or negative terms, as an absolute or a relative quantity, or using standard, established terms or new but more applicable terms. Table 2 also includes examples of how these considerations were applied to specific cases.

Addressing Cognitive Bias

The literature has highlighted a number of potential risks with the elicitation and quantification of human judgments. Based on the work of Montibeller and von Winterfeldt (*25*), the process addressed three biases that can arise in developing a list of criteria.

Omission bias is when an important variable is overlooked, which limits respondents' answers (e.g., a missing criterion). In terms of scope, we sought to include all potentially relevant impacts—in other words, not just those that are specifically known or agreed upon, but any that might be of interest to various stakeholders, including future generations. The purpose of comprehensiveness is to address omission bias and ensure conceptual validity in relation to sustainability (for which comprehensiveness is key, see [*12*]). Macharis et al. suggest MAMCA would naturally cover all the effects, as they would be "reflected in the goals of the stakeholders (if all relevant stakeholders are included)" (*10*, p. 188).

Proxy bias is when a proxy is used and receives a higher weight than what it intends to measure, for example, due to different positioning in a long cause-effect chain. Proxy bias was addressed by adopting a uniform definition of "impact" inspired by the "Drivers, Pressures, State, Impact and Response" framework as a means of structuring and analyzing effects in long causal chains.

Splitting bias is when the way criteria are grouped affects their weights, for instance, in a value tree. Splitting bias was addressed by making sure all criteria were placed at the same level of hierarchy (the categories of direct, indirect, and environmental impacts were used only as communication aids, not as grouping mechanisms for the analysis).

Results

A Set of Assessment Criteria for Sustainable Transport Appraisal

The primary result of this research is the list of assessment criteria itself. Incorporating both conventional transport objectives and longer-term sustainable development goals, this set of 28 criteria comprises three categories: direct project impacts, indirect societal impacts, and environmental impacts (Table 3). Full descriptions are provided in Table 4.

Operationalization and Compatibility

A further result of this research is the operationalization of these criteria in transport appraisal. This set of assessment criteria was implemented in practice in a MAMCA appraisal of HS2. Thus both the process of developing the list and the list itself have been used and tested. This gives confidence in the criteria's usability. For example, the Methods section noted the necessity of defining assessment criteria in terms that could be operationalized in MCA steps 5 (criteria prioritization) and 6 (performance assessment).

The criteria prioritization process (MCA Step 5) involved two steps. First, from the full list of 28 criteria, respondents were asked to select at minimum 6 criteria they deemed contextually relevant (Question: "Which criteria do you think should be used for assessing the pros and cons of HS2 Phase I and its alternatives? Please select at minimum 6 criteria from the list below"). Second, they were asked to "rate the relative importance of each criterion" they had selected by applying direct weights using a sliding scale from 0 to 10.

Defining the criteria in terms of impacts facilitated the process of performance assessment (MCA Step 6). One of the advantages of MCA over BCA is that performance can be assessed without distinguishing between costs and benefits or separating mixed impacts into costs and benefits. The assessment question for each criterion is formulated as follows: "Which project performs better in terms of criterion X?" This formulation works equally well for negative impacts like "water contamination" and for neutral/positive impacts such as "journey experience" or "air quality."

Other aspects of practicability and compatibility with the MAMCA approach of incorporating multiple stakeholder perspectives were also verified. Evidence that comprehensiveness in the list was achieved is suggested by the fact that in discussing early drafts of the list with interviewees, many additional criteria were suggested, but with later versions, far fewer. In the final version of the questionnaire, very few respondents commented or added to the list provided, thus reaching a satisfactory level of data saturation (*26*).

Table 2. Considerations for Naming Assessment Criteria

Issue	Considerations	Examples
Absolute vs. Relative	• Is the criterion expressed in absolute terms (can be measured/ understood on its own) or relative terms (is inherently comparative to some assumed base number)?	• "Time savings," commonly used in transport appraisal, is a relative measure requiring additional information (Savings compared with what?); this was changed to *Journey time*, an absolute measure.
Gross (total) vs. Net	• In some cases this is the same as Absolute vs. Relative (e.g., passenger capacity vs. passenger capacity increase; journey time vs. journey time savings), but not always (e.g., operating costs: is this total, or net of revenue from train tickets?). • Is the criterion measured in terms of total impact or in terms of net impact (taking into consideration what it replaces)? • Is the project described in a vacuum, or is it considered in the context of various scenarios (i.e., assumptions about follow-on effects or what developments would occur otherwise)?	• The initial question of whether project costs were gross or net of ticket revenues led to the more important question of "cost to whom?" (Answer: taxpayers at project level; user at journey level; but not contractors, builders, rail industry.) • These considerations led to an important discussion about opportunity cost, which is now explicitly covered (under *Equity & distributional effects*).
Positive vs. Negative Phrasing	• Is the criterion phrased in positive, negative, or neutral terms (e.g., air quality vs. air pollution; energy efficiency vs. carbon footprint)? Note: some terms may not be clearly one or the other (e.g., is "air quality" positive or neutral? is "journey time" negative or neutral?). • Does this matter for the measurement of costs and benefits? (Answer: yes for BCA, not necessarily for MCA.) • Is the juxtaposition of positive and negative impacts confusing for respondents (making it harder to evaluate criteria side-by-side)? • Does the phrasing introduce unwanted bias?	• It is generally best to phrase things in neutral terms, and when that is not possible, to strive for consistency one way or the other; however in the context of MCA, this is not a critical issue, so choosing the best terms based on other considerations is probably more important. • "Time savings," in addition to being a relative quantity, is phrased as an explicit benefit and may introduce bias by causing people to value it more highly. A more neutral term such as "journey time" is preferable.
Brevity vs. Clarity	• Short, concise names are preferable, as long as understanding is not compromised.	*Journey reliability & system resilience*: in this case clarity was preferred over brevity. The original name, *Reliability & resilience*, sounded better and may have been understandable to transport experts, but it was not clear to a more general audience (Reliability of what? Resilience of what? Confusion was compounded by juxtaposing the two terms when the "of what" is different for each).
Expert vs. Lay Terms/ Standard, Established vs. New, More Intuitive, Broadly Comprehensible Terms	• In general, there is a benefit to sticking with standard, established terms and not reinventing the wheel. • Sometimes, however, the established term is problematic or not useful and should be modified. • Established terms are sometimes "expert" terms understandable only to experts in a particular area and would be better replaced by more intuitive terms that are understandable to a broader audience.	One of the criteria is defined as "Land-Use & Transport Interaction (LUTI)," an industry-standard phrase. However, the criterion is named *Land use & urban development* instead. One of the interviewees had stressed the importance of housing in the context of land use and transport planning, and it was not clear that people would associate LUTI with housing. "Urban development" is a well-known term that includes housing and is easily understood by non-transport specialists.
Same vs. Different Terms	• In descriptions, terms can be used in different ways (with explanation), but not in names. • Across the list of names, use the same term to mean the same thing and different terms to mean different things.	• Two different types of affordability are covered in the criteria and detailed in the descriptions. Making the distinction clear in the names was not feasible, so the word "affordability" is included in only one criterion name to avoid confusion. • The repeated use of "journey" across several criteria indicates that these are all journey-level impacts, as contrasted with the project-level impacts (for which the word "project" is repeated).

(continued)

Table 2. (continued)

Issue	Considerations	Examples
Alphabetical Order	• If criteria are to be listed alphabetically, choosing names that put related concepts together is helpful.	*Freight capacity* and *Passenger capacity* would have been more concise than *Rail capacity for freight* and *Rail capacity for passengers*, but the added value of keeping these two related concepts next to each other in the list was deemed greater than the extra time required to read extra words. Although in the context of rail projects, it should be obvious that capacity is rail capacity, it was thought that including the word "rail" might serve as a subtle reminder of transport mode and therefore of the concept of modal shift (for those respondents who might be interested in this impact but do not see it in the list).

Several factors aided in keeping the list tractable despite its comprehensiveness. First, the organization of criteria into impact categories (direct, indirect, environmental), though irrelevant for the assessment and analysis, helped respondents comprehend and consider the criteria more easily. Second, effort was put into choosing criteria names that in relation to each other would help respondents grasp the essence of each criterion when looking across the whole set (even to the extent of considering alphabetical order). Finally, respondents were not asked to carry through assessments for all 28 criteria—just the 6 to 9 they selected (not every respondent has to assess every criterion in order for the analysis to work).

Conclusion

This paper makes three practicable contributions:

1. A viable list of STA criteria that is readily applicable to other rail projects and, with adaptation, to other transport projects.
2. A framework for systematically defining and naming assessment criteria.
3. A replicable criteria development process that emphasizes interaction and the inclusive approach of the MAMCA appraisal.

The set of 28 assessment criteria presented in this paper is the result of substantial effort on the part of several individuals, the input of many interviewees, and a methodical process for putting it all together. Furthermore, this list has been successfully applied in a MAMCA-style appraisal setting, demonstrating the practicability of both the list itself and the framework used to develop it. As such, this list serves—at a minimum—as a credible starting point for transport planners wishing to conduct sustainable transport appraisal, particularly when using an MCA approach.

Adopting this set of assessment criteria does not entirely eliminate the need for transport planners to go through a criteria development process, but the expectation is that this research, by providing a viable starting point along with a framework for adapting the list to fit the specific circumstances of other projects, will significantly speed things up—provided the dimensions and considerations presented in tables 1 and 2 for defining and naming criteria are applied.

Finally, no matter how abbreviated the criteria selection process may become, the solicitation of outside input should not be entirely eliminated. A key take-away from this research is the value of adopting an iterative and interactive process that entails gathering input from people with diverse perspectives and expertise (both external

Table 3. Overview of Assessment Criteria for Sustainable Transport Appraisal

Direct project impacts (internal costs & benefits)	Indirect societal impacts (externalities: people)	Environmental impacts (externalities: planet)
1. Journey cost & affordability	11. Accessibility	20. Agriculture, forestry & soils
2. Journey experience	12. Accidents & safety	21. Air quality
3. Journey reliability & system resilience	13. Community disruption & severance; blight	22. Biodiversity & nature
4. Journey time	14. Equity & distributional effects	23. Carbon footprint
5. Project costs	15. Land use & urban development	24. Material footprint
6. Project delivery risks	16. Landscape/townscape & cultural heritage	25. Noise & vibration
7. Rail capacity for freight	17. Prestige & image	26. Solid waste & disposal
8. Rail capacity for passengers	18. Rail industry growth & innovation	27. Water & land contamination
9. Traffic & transport disruption	19. Regional economic development & regeneration	28. Water resources & flood risk
10. Transport integration & connectivity		

Table 4. Assessment Criteria for Sustainable Transport Appraisal

Criterion	Description
Direct project impacts (internal costs and benefits)	
1. Journey cost & affordability	Journey cost (proposed or expected) is the price passengers pay for a trip. In this context, affordability is defined narrowly in terms of whether this particular means of transport is expected to be affordable to would-be users. If your concern is about what might happen to other means of transport, whether currently in existence or proposed for the future, as a result of this project being realized, please refer to "Equity and distributional effects."
2. Journey experience	Journey experience is a measure of the real and perceived physical and social environment experienced while traveling. It includes the overall quality of facilities and infrastructure (stations and rolling stock), as well as more tangible factors like availability of seats, comfort, provision of relevant information, safety and security, crowdedness and other stress factors. Moreover, experience of time as perceived by users includes entertainment and scenery.
3. Journey reliability & system resilience	Journey reliability refers to variability in journey time that individuals are unable to predict due to recurring events (e.g., congestion) or non-recurring events (e.g., accidents). For public transport this is usually measured as the standard deviation of lateness divided by average lateness. Resilience refers to the ability of a transport system or network to recover from disruptions caused by natural disasters or human factors. Resilience and reliability are related in that when resilience is high it is reflected in reliability as well.
4. Journey time	Journey time is defined as travel time from station of origin to station of destination. High-speed rail is aimed at reducing travel time.
5. Project costs	Total cost of project to taxpayers, comprising both upfront and ongoing costs. Upfront capital investment includes construction costs from main work contracts (stations, tracks, rolling stock, power, and signaling), land and property costs (acquisition and compensation), administrative costs (design, management, consultation). Ongoing costs include train and station operations, maintenance, and renewal costs.
6. Project delivery risks	Uncertainties during planning and construction stages regarding final project outcome. These risks include cost overruns, construction delays, and underperformance of new technology. Risks can be mitigated by consulting key experts and stakeholders early in the process and by conducting pilot studies. Risks can be exacerbated when cost estimates are subject to optimism bias (underestimation, whether intentional or unintentional).
7. Rail capacity for freight	An increase in rail capacity anywhere in the system will generally also increase the capacity available for freight, whether directly through the construction of new tracks available to both freight and passengers, or indirectly through the construction of new passenger lines, thus freeing up tracks that were previously shared with passengers. An increase in rail capacity available for freight transport would be able to meet growing demand for freight transport and might reduce rail costs for freight users. This in turn could result in shifting freight from roads to rail. If your interest in rail capacity is primarily related to modal shift and the resulting possibility of reducing CO_2 emissions, please refer to "Carbon footprint." If your interest is more generally in increasing capacity and reducing costs for freight users, please select this criterion.

(continued)

Table 4. (continued)

Criterion	Description
8. Rail capacity for passengers	Passenger capacity is the total number of people a means of transport is able to transfer in a given period (measured for example in person-km per day). Increasing passenger capacity is particularly relevant when travel demand is expected to grow and policy makers want to satisfy that demand. One way of increasing passenger capacity is to increase frequency of service, which is also made possible by increasing train speed. Other methods of increasing capacity include lengthening trains or building new tracks/routes. In addition to meeting growing travel demand, increasing passenger capacity may also encourage a shift from other modes of passenger transport (air, car) to rail. If your interest in rail capacity is primarily related to modal shift and the resulting possibility of reducing CO_2 emissions, please refer to "Carbon footprint." If your interest in rail capacity is more directly about meeting passenger needs for travel, please select this criterion.
9. Traffic & transport disruption	Welfare impacts on rail users caused by transport disruption during project construction and on non-rail users caused by traffic congestion from construction trucks. This is primarily a construction-phase impact. If your interest is in longer-term impacts on traffic patterns and other infrastructure conflicts, please refer to "Community severance" or "Land use & urban planning." If your interest in traffic patterns is related to CO_2 emissions, please refer to "Carbon footprint."
10. Transport integration & connectivity	The extent to which the proposed project would be integrated with and connected to other transport. This includes intermodal integration, which refers to how well different modes of transport (train, bus, ferry, etc.) are interconnected. Intermodal integration can also refer to provisions for active modes of transport (cycling, walking, etc.). When such connections are smooth and convenient, intermodal integration can play an important role in promoting green and healthy/active travel. Transport integration also includes connectivity of travel segments within the same mode of transport (train-to-train, bus-to-bus, etc.). Connectivity is a closely related concept that considers, alongside total travel time, passenger discomfort associated with waiting, transfer, and access/egress times. Transit connectivity aims to provide attractive and "seamless" transfers along multimodal paths as part of the door-to-door passenger chain.
Indirect societal impacts (externalities: people)	
11. Accessibility	Accessibility reflects the range of opportunities and choices people have in connecting with employment, education, essential services, and social networks. It can be measured as a catchment area and is concerned with travel horizons (journey times and distances). It is more holistic than transport user benefits, as it considers the availability and physical accessibility of a transport service in connection with the location of other services and activities. Accessibility also includes option value (the value of the existence of the service for convenience or unplanned trips) and non-use value (appreciating that a service is available for others).
12. Accidents & safety	Risk of individuals (both transport users and non-users) being killed or injured as a result of accidents, usually measured in number of casualties, fatalities, or injuries in a given period. Refers to both construction and operation phases. Expected casualties will vary with infrastructure design and route, depending for example on road intersections, crossings/bridges, and size of the neighboring population.
13. Community disruption & severance; blight	Community disruption refers to the temporary, disruptive impacts of project construction on residential property, public spaces, and communities as a whole. It includes impacts on the amenity of residents in the remaining parts of the residential development, including pleasantness and visual intrusion. Community severance refers to the more lasting impacts on a community, even after construction is finished and the transport system is operating. Severance is the real or perceived isolation of residential properties or community facilities due to physical or visual barriers caused by transport infrastructure or traffic flows. Severance is usually measured as a physical barrier to pedestrian movement, but also to cyclists, equestrians, children, or other vulnerable groups. Blight is the reduction in property or neighborhood value near proposed project sites. Blight takes place as soon as a potential project or route is proposed and is particularly of concern when the design and consultation period extends over many years. Blight becomes less of a problem once a decision has been reached, as property values in areas of rejected projects may recover and property owners affected by realized projects may be compensated. However, it is very much a problem for anyone who wants to sell affected property before a final decision has been reached.

(continued)

Table 4. (continued)

Criterion	Description
14. Equity & distributional effects	Concerns about the distribution of project costs and benefits across different social groups or geographical locations. Equity is concerned about equality of opportunity, with a particular focus on vulnerable groups. Affordability of transport (usually measured as proportion of income spent on transport) is related to equity when the realization of a particular project results in a change in the options available to others. This could happen if existing transport receives less funding as a result of another project being selected. Closely related is the concept of opportunity cost: what else could the money be spent on? Whereas equity emphasizes *who* receives the benefits (e.g., transportation for whom?), opportunity cost emphasizes *what* the money is spent on (transportation? education? sports? and within transportation, *what kind* of transport?) Both opportunity cost and equity involve prioritizing the spending of public money: on what and for whom?
15. Land use & urban development	Impacts that new transport projects have on land use (residential, industrial, or commercial) and allocation of activities – also referred to as Land-Use & Transport Interaction (LUTI). Changes in the transport system affect accessibility, which reallocates land uses and determines the location of new activities. Transport policies and measures (e.g., transit-oriented development, TOD) can lead to more energy-efficient urban forms. Of particular importance in the context of urban development is whether housing will be actively built along with transport infrastructure.
16. Landscape/townscape & cultural heritage	Landscape/townscape impacts refer to changes in the physical and cultural characteristics of the land and perceptions that make up and contribute to landscape character ("sense of place"). It consists of affected topography, views, tree cover (for landscape) and all aspects of the urban form (for townscape), from construction plans along the route to overhead lines, stations, depots, tunnels and ventilation shafts, fences and barriers, bridges, and so forth. Cultural heritage impacts refer to archaeological and paleo-environmental remains (ancient burials, ancient environments), historic landscapes and buildings, and the built environment (both designated and non-designated assets), known collectively as heritage assets.
17. Prestige & image	Prestige, public recognition, and positive media coverage that may be generated for the nation, a region, or public officials. Public exposure that a project may generate for its proponents or for politicians. This may also include the potential contribution of a project to creating a sense of regional or national identity or pride.
18. Rail industry growth & innovation	Contribution to the strategic goals of encouraging technological innovation, growth of commercial expertise within a specific industry, and development of new skills in the labor force. This may include the creation of dedicated training centers.
19. Regional economic development & regeneration	Regional economic development refers to welfare benefits that are broader than transport user benefits. These wider economic impacts affect the labor market, product market, and land market: 1) agglomeration effects: accessibility of firms to other firms, products, and workers; the increase in labor productivity from increased proximity, knowledge, and technology spillovers; 2) welfare gain for firms whose goods and services require transport; and 3) increased tax revenues from the labor market. Regional development also includes rebalancing the national economy and bridging the North–South divide.. When infrastructure is located in areas designated for economic development under UK or EU regeneration programs, projects can help meet regeneration goals by stimulating economic activity and employment.
Environmental impacts (externalities: planet)	
20. Agriculture, forestry & soils	Soil quality impacts on agricultural and forestry land. The driver is land "taken" from agriculture and forestry for an infrastructure project, but the impact of concern is not an acreage issue, but rather the loss of high-quality farm soil. For infrastructure projects, a hectare is a hectare. But for farms, there are differences in soil quality, and this may not get compensated for in the land price. Protecting high-quality agricultural and forestry land may also be considered a priority for society.
21. Air quality	Impacts on local and regional air quality from dust and emissions during both project construction and operation. Impacts can be both positive (e.g., decreased emissions from road transport if modal shift occurs) and negative (e.g., increased emissions from diesel locomotives or increased road traffic around stations and depots).. Pollutants impacting local and regional air quality include oxides of nitrogen (NOx), volatile organic compounds/hydrocarbons (VOC/HC/PAH), particulate matter, oxides of sulfur (SOx), ozone (O_3), carbon monoxide (CO), and trace metals. Global air pollutants (CO_2 and other greenhouse gases) are covered under "Carbon footprint."

(continued)

Table 4. (continued)

Criterion	Description
22. Biodiversity & nature	Impacts on nature conservation arising from habitat loss and degradation, fragmentation of sites, severance of ecological corridors and networks, noise and visual disturbance, barrier effects to movement of fauna, artificial lighting, changes in water quality and quantity, air pollution, and mortality as a result of collisions with trains. Included here are impacts on protected species and habitats, such as Sites of Special Scientific Interest, Areas of Outstanding Natural Beauty, National Parks, Environmentally Sensitive Areas, and ancient woodlands, as well as general ecological value beyond site boundaries.
23. Carbon footprint	All greenhouse gas (GHG) emissions associated with a project expressed as carbon dioxide equivalent (CO2e). Aside from CO_2, which is the most significant GHG associated with transportation and energy use, GHGs include methane (CH_4) and nitrous oxide (N_2O). A project's carbon footprint comprises both embedded carbon and fuel carbon. Embedded carbon refers to CO_2 emitted during project construction as well as in connection with producing the materials used in infrastructure (cement, steel etc.). Fuel carbon refers to CO_2 emitted during transport operations. Fuel carbon emissions are driven by two factors: (1) efficiency of energy form (electricity vs. liquid fuel); and (2) energy source (renewable vs. fossil fuel). To the extent that rail transport is more fuel and carbon efficient than road transport, modal shift from road or air to rail could lead to a net reduction in carbon footprint.
24. Material footprint	Similar to carbon footprint, this measure looks at a project's use of raw materials (e.g., metal and minerals), both embedded in the infrastructure and related to operations. It measures both "used" extraction (the portion of materials that end up in the infrastructure) and the "unused" extraction (the material waste associated with mining and extraction).
25. Noise & vibration	Nuisance to people caused by noise and vibration from road and rail traffic, during both project construction and transport operation. Impacts can be on individual dwellings (residential) and on communities (non-residential, e.g., open spaces, schools, hospitals, offices, hotels).
26. Solid waste & disposal	Refers to solid waste from construction (e.g., earthworks) and operating activities (e.g., wastes from stations, rolling stock, or track operation and maintenance), as well as the impacts associated with its disposal. Of particular concern in the case of large infrastructure projects is the need to excavate enormous quantities of dirt to make way for the project and the resulting need to dispose of vast quantities of soil, some of which may be contaminated from previous land use. In addition to the quantity of waste generated, this criterion covers impacts associated with where and how (e.g., landfill vs. recycling) waste will be disposed of.
27. Water & land contamination	Water contamination can result either directly from the disposal of liquid waste into surface and ground water or indirectly through soil contamination. Soil contaminants often leach into surface and ground water, which is why land and water contamination are interconnected. Ground contamination can occur either during construction or during operation and includes leaks and spillages from line-side equipment and trains, in depots, or along train tracks.
28. Water resources & flood risk	Impacts (other than contamination) on structure and flow of surface waterways, both natural and artificial (e.g., changing the course of streams or channels, emptying lakes or reservoirs). Impacts on drainage networks (e.g., bridges and embankments obstructing path of floodwaters) and associated implications for flood risk.

stakeholders and internal discussion team members). Although existing rating systems and assessment procedures may have gone through similar iterative and interactive processes, such frameworks do not always make explicit the considerations and stakeholder involvement processes involved in selecting, defining, and naming criteria. This paper describes those processes explicitly and systematically, thus making them replicable.

A remaining question open for future research is whether improvements in the comprehensiveness of assessment criteria in transport appraisal (and the instrumental use that is expected of the results) would suffice in influencing decision making and guiding a genuine transition toward a more sustainable transport system.

Acknowledgments

The authors are grateful to the Strategic Research Council of Denmark (Innovationsfonden) for financially supporting this research under the SUSTAIN project (see http://www.sustain.transport.dtu.dk/). The authors thank Christina Georgouli of SCP Transport Planning & Infrastructure Design consultancy for contributing to in-depth discussions during the development and refining of criteria definitions. The authors also thank four anonymous reviewers for their insightful and helpful comments.

Author Contributions

The authors confirm contribution to the paper as follows: study conception and design: MJB, YC; data collection: MJB, YC;

analysis and interpretation of results: MJB, YC; draft manuscript preparation: MJB, YC. All authors reviewed the results and approved the final version of the manuscript.

References

1. Hickman, R., S. Saxena, D. Banister, and O. Ashiru. Examining Transport Futures with Scenario Analysis and MCA. *Transportation Research Part A: Policy and Practice*, Vol. 46, No. 3, 2012, pp. 560–575. https://doi.org/10.1016/j.tra.2011.11.006.
2. Ward, E. J., H. T. Dimitriou, and M. Dean. Research in Transportation Economics Theory and Background of Multi-Criteria Analysisã£: Toward a Policy-Led Approach to Mega Transport Infrastructure Project Appraisal. *Research in Transportation Economics*, Vol. 58, 2016, pp. 21–45. https://doi.org/10.1016/j.retrec.2016.08.003.
3. Macharis, C., and A. Bernardini. Reviewing the Use of Multi-Criteria Decision Analysis for the Evaluation of Transport Projects: Time for a Multi-Actor Approach. *Transport Policy*, Vol. 37, 2015, pp. 177–186. https://doi.org/10.1016/j.tranpol.2014.11.002.
4. Pryn, M. R., Y. Cornet, and K. B. Salling. Applying Sustainability Theory to Transport Infrastructure Assessment using a Multiplicative Ahp Decision Support Model. *Transport*, Vol. 30, No. 3, 2015, pp. 330–341. https://doi.org/10.3846/16484142.2015.1081281.
5. Cornet, Y., M. B. Barfod, M. J. Barradale, and R. Hickman. Giving Current and Future Generations a Real Voice: A Practical Method for Constructing Sustainability Viewpoints in Transport Appraisal. *European Journal of Transport and Infrastructure Research*, Vol. 18, No. 3, 2018, pp. 316–339.
6. Cornet, Y., M. J. Barradale, H. Gudmundsson, and M. B. Barfod. Engaging Multiple Actors in Large Transport Infrastructure Project Appraisal: An Application of MAMCA to the Case of HS2. Journal of Advanced Transportation, Vol. 2018. https://doi.org/10.1155/2018/9267306.
7. Browne, D., and L. Ryan. Comparative Analysis of Evaluation Techniques for Transport Policies. *Environmental Impact Assessment Review*, Vol. 31, No. 3, 2011, pp. 226–233. https://doi.org/10.1016/j.eiar.2010.11.001.
8. Barfod, M. B., and K. B. Salling. A New Composite Decision Support Framework for Strategic and Sustainable Transport Appraisals. *Transportation Research Part A: Policy and Practice*, Vol. 72, 2015, pp. 1–15. https://doi.org/10.1016/j.tra.2014.12.001.
9. Von Winterfeldt, D., and W. Edwards. *Decision Analysis and Behavioral Research*. Cambridge University Press, Cambridge, UK, 1986.
10. Macharis, C., A. De Witte, and J. Ampe. The Multi-Actor, Multi-Criteria Analysis Methodology (MAMCA) for the Evaluation of Transport Projects: Theory and Practice. *Journal of Advanced Transportation*, Vol. 43, No. 2, 2009, pp. 183–202. https://doi.org/10.1002/atr.5670430206.
11. Macharis, C., L. Turcksin, and K. Lebeau. Multi Actor Multi Criteria Analysis (MAMCA) as a Tool to Support Sustainable Decisions: State of Use. *Decision Support Systems*, Vol. 54, No. 1, 2012, pp. 610–620. https://doi.org/10.1016/j.dss.2012.08.008.
12. Cornet, Y., and H. Gudmundsson. Building a Metaframework for Sustainable Transport Indicators: Review of Selected Contributions. *Transportation Research Record: Journal of the Transportation Research Board*, 2015. 2531: 103–112.
13. Gudmundsson, H., R. P. Hall, G. Marsden, and J. Zietsman. *Sustainable Transportation: Indicators, Frameworks and Performance Management*. Co-publication by Springer-Verlag, Heidelberg, Germany, and Samfundslitteratur, Frederiksberg, Denmark, 2016.
14. Gallopín, G. C. Environmental and Sustainability Indicators and the Concept of Situational Indicators. A Systems Approach. *Environmental Modeling & Assessment*, Vol. 1, No. 3, 1996, pp. 101–117. https://doi.org/10.1007/BF01874899.
15. Joumard, R., and H. Gudmundsson. *Indicators of Environmental Sustainability in Transport*. Institut national de recherche sur les transports et leur sécurité, Lyon, France, 2010.
16. Ramani, T., J. Zietsman, H. Gudmundsson, R. Hall, and G. Marsden. Framework for Sustainability Assessment by Transportation Agencies. *Transportation Research Record: Journal of the Transportation Research Board*, 2011. 2242: 9–18.
17. Marsden, G., C. Kelly, and C. Snell. Selecting Indicators for Strategic Performance Management. *Transportation Research Record: Journal of the Transportation Research Board*, 2006. 1956: 21–29.
18. Zietsman, J., T. Ramani, J. Potter, V. Reeder, and J. DeFlorio. *NCHRP REPORT 708 A Guidebook for Sustainability Performance Measurement for Transportation Agencies*. Transportation Research Board of the National Academies, Washington, D.C., 2011.
19. Department for Transport. *Transport Analysis Guidance (TAG) - The Transport Appraisal Process*. 2014.
20. Department for Transport. *Transport Analysis Guidance: WebTAG*. 2014.
21. Standards for Highways. *Design Manual for Roads and Bridges (DMRB): Volume 11*. http://www.standardsforhighways.co.uk/ha/standards/ghost/dmrb/vol11/index.htm. Accessed September 23, 2018.
22. Booz & Co., and Temple. HS2 London to the West Midlands Appraisal of Sustainability. http://webarchive.nationalarchives.gov.uk/20111005090245/http://highspeedrail.dft.gov.uk/library/documents/appraisalsustainability. Accessed September 23, 2018.
23. HS2 Ltd. *HS2 Phase One Environmental Statement: Documents*. https://www.gov.uk/government/collections/hs2-phase-one-environmental-statement-documents. Accessed September 23, 2018.
24. HS2 Ltd. *HS2 Phase One Environmental Statement Volume 2: Community Forum Area Reports and Map Books*. https://www.gov.uk/government/publications/hs2-phase-one-environmental-statement-volume-2-community-forum-area-reports-and-map-books/hs2-phase-one-environmental-statement-volume-2-community-forum-area-reports-and-map-books. Accessed September 23, 2018.

25. Montibeller, G., and D. von Winterfeldt. Cognitive and Motivational Biases in Decision and Risk Analysis. *Risk Analysis*, Vol. 35, No. 7, 2015, pp. 1230–1251. https://doi.org/10.1111/risa.12360.
26. Guest, G. How Many Interviews Are Enough? An Experiment with Data Saturation and Variability. *Field Methods*, Vol. 18, No. 1, 2006, pp. 59–82. https://doi.org/10.1177/1525822X05279903.

The Standing Committee on Transportation and Sustainability (ADD40) peer-reviewed this paper (18-06528).

Article

TRR
JOURNAL OF THE TRANSPORTATION RESEARCH BOARD

Transportation Research Record
2018, Vol. 2672(3) 118–129

DOI: 10.1177/0361198118797799
journals.sagepub.com/home/trr
SAGE

Planning to Build Resilience into Transportation Assets: Lessons Learned

Beth Rodehorst[1], Brenda Dix[2], Brad Hurley[3], Jake Keller[4], Robert Hyman[5], Brian Beucler[5], Khalid Mohamed[5], and Robert Kafalenos[5]

Abstract

Although many state and local transportation agencies recognize the need to make transportation assets more resilient in the face of a changing climate, there have been few methods and best practices they can draw on to determine which assets may be compromised under future conditions and how to evaluate and select adaptation measures. Federal Highway Administration's (FHWA's) Transportation Engineering Approaches to Climate Resiliency project sought to synthesize lessons learned and innovations from a variety of recent FHWA studies and pilots to help transportation agencies address changing climate conditions and extreme weather events at the asset level. This paper describes considerations for why, where, and how to integrate climate considerations into the project development process. It also discusses the types of climate information that should be considered, and summarizes lessons learned from FHWA's studies and pilots—such as implementing adaptive designs, considering assets in a regional context, and exploring ecosystem-based adaptation solutions—that can be used to guide the process of developing adaptation strategies.

Future sea levels, temperature, and precipitation patterns are expected to deviate from historical norms and extremes because of climate change, potentially placing some of today's transportation infrastructure at risk. As aging infrastructure is rebuilt or upgraded, there are opportunities to plan and design it to be more resilient in the face of future conditions.

Engineers already consider stressors from climate and weather when designing and maintaining infrastructure. Local climate and weather patterns affect everything from how to size a culvert, to which materials to use in pavement, to how to protect assets from erosion. Engineers routinely draw from statistical analyses of long records of observations to make assumptions about the type of climate an asset will be exposed to over its lifetime. There are well-established methods for analyzing historical records, often resulting in a single numeric value to represent potential exposure. For example, historical data are used to generate estimates of flood depths for 4%, 2%, and 1% annual exceedance probability (25-, 50-, or 100-year) events, which may provide the basis for how high to build a bridge or how large to build a culvert (*1*).

Implicit in the use of these data sets is the assumption of *stationarity*, the concept that historical weather conditions are indicative of future ones. However, climate change places this basic assumption into doubt: the climate of the future will likely be nonstationary.

Planning and designing for a future climate that is different from the past is not straightforward. Although the same design standards and engineering methods may still apply, it is more difficult to determine what the future temperature, rainfall, flood levels, and other climate stressors might be. Rather than simply looking at the historical record, engineers now are faced with multiple data sets based on different scenarios of future greenhouse gas levels, climate models, and timeframes. Moreover, climate projections do not always align with the data inputs engineers need, such as 30-minute duration precipitation data.

For these and other reasons, best practices for incorporating novel climate stressors, such as sea level rise, and/or weather extremes outside the bounds of historical records have not been mainstreamed into engineering

[1]ICF, Portland, OR
[2]ICF, New York, NY
[3]ICF, Montréal, QC, Canada
[4]WSP USA, Virginia Beach, VA (retired)
[5]Federal Highway Administration, Washington, D.C.

Corresponding Author:
Address correspondence to Beth Rodehorst: Beth.Rodehorst@icf.com

Table 1. Summary of TEACR Engineering Case Studies

TEACR case study	Purpose
Sea level rise and storm surge impacts on a coastal bridge: I-10 Bayway, Mobile Bay, Alabama (*3*)	Evaluate the engineering implications of wave-induced loads on a typical coastal bridge exposed to surge/wave attack with increases in sea level
Living shoreline along coastal roadways exposed to sea level rise: shore road in Brookhaven, New York (*TEACR Living Shoreline*) (*4*)	Evaluate a living shoreline as a natural and nature-based adaptation for protection of a coastal road threatened by erosion and inundation from tides, surge, and wave action that will be exacerbated by future sea level rise
Temperature and precipitation impacts on cold region pavement: SR-6/SR-15/SR-16 in Maine (*5*)	Evaluate pavement performance in cold regions with future projections of temperature and precipitation using mechanistic-empirical pavement performance prediction models
Comparison of economic analysis methodologies and assumptions: Dyke Bridge in Machias, Maine (*TEACR Economic Assessments*) (*6*)	Assess the extent to which economic analysis results are affected by the methodology (area-under-the-curve or Monte Carlo) when estimating expected cumulative lifecycle damages of an asset owing to extreme weather events, and conduct various input sensitivity tests
Temperature and precipitation impacts to pavements on expansive soils: proposed SR-170 in North Texas (*TEACR Pavement Shrink-Swell*) (*7*)	Evaluate potential impacts of projected changes in temperature and precipitation on expansive soils and the overlying pavement performance
Precipitation and temperature impacts on rock and soil slope stability: I-77 in Carroll County, Virginia (*TEACR Slope Stability*) (*8*)	Evaluate potential impacts of projected changes in precipitation and freeze–thaw cycles on rock and soil slope stability
Barrier island roadway overwashing from sea level rise and storm surge: US-98 on Okaloosa Island, Florida (*TEACR Roadway Surge*) (*9*)	Evaluate an existing countermeasure approach—armoring the shoulder of this coastal highway with sheet pile and rock gabion mattresses—as an engineering adaptation approach to climate change
Incorporating climate change into the design of roadways built on permafrost (*10*)	Provide general background on the topic of climate change and permafrost thaw; explain the implications for roadways, how future thawing can be projected, and what can be done to adapt to the impacts
Wildfire and precipitation impacts on a culvert: US-34 at Canyon Cove Lane, Colorado (*TEACR Culvert*) (*10*)	Investigate the combined impacts of changing precipitation and wildfire risk on the performance of a culvert structure

design processes. Furthermore, different approaches for incorporating climate risks into design may materially impact workforce resource requirements. Little guidance has been available to help engineers and planners align the findings of climate- or weather-related assessments with desired design outcomes and lifetimes. Methods to account for nonstationarity owing to gradual land-use change or rapid change caused by addition/removal of dams exist, but methods to account for climate change-induced nonstationarity need to be developed (*2*).

The Federal Highway Administration (FHWA) acknowledges the evolving state of engineering design practices for climate resilience and has undertaken several efforts to fill knowledge gaps related to incorporating climate and weather vulnerabilities. Among these efforts is the Transportation Engineering Approaches to Climate Resiliency (TEACR) project (2014–2017) to evaluate engineering approaches for assessing climate and extreme weather hazards, and to develop a set of climate-resilient solutions and methodologies that transportation agencies could apply nationally. The project includes:

- An *analysis* of gaps and needs in the current state of practice in incorporating climate change considerations into engineering.
- Nine *case studies* (summarized in Table 1 [*3–10*]) of a diverse set of transportation assets around the country that include facility-level climate change vulnerability assessments and engineering and economic analysis of potential adaptation options.
- A *synthesis* report of specific climate resilience recommendations and approaches for engineers to use in designing transportation projects at both the system and project scales. This report, which builds on the nine case studies in Table 1 (*3–10*) and other FHWA-sponsored studies—including the 2013–2015 Climate Change Resilience Pilots (*Pilot*) (*11*), the Gulf Coast Study Phase 2 (*GC2*) (*12*), and the Post Hurricane Sandy Transportation Resilience Study (*Post Sandy*) (*13*), is titled *Synthesis of Approaches for Addressing Resilience in Project Development* (*14*). It offers lessons learned to help transportation agencies address concerns associated with changing climate conditions and extreme weather events during project-level scoping and to consider ramifications for design.

Using this work as a foundation, this paper provides a basic introduction on where, why, and how to integrate climate considerations into the project development process. It also discusses what types of climate information should be considered, and summarizes key lessons learned relevant to developing adaptation strategies.

Infusing Consideration of Climate Impacts and Adaptive Strategies into the Project Development Process

Before initiating design for a transportation project, planners and engineers go through a series of activities to fully develop the project concept. This project development process helps to identify the issues that need to be addressed in the project design. Several types of studies may be conducted to inform the process, such as a project planning or development study, corridor or feasibility study, or a scoping study. These studies provide a way to assess and consider issues early so practitioners can avoid or minimize potential problems from the start.

The following sections use lessons learned and other findings from FHWA's work to outline why, where, and how to integrate climate change into the project development process, with additional information on accessing and using climate data.

Why to Integrate Climate Change into the Project Development Process

As noted above, changes in average and extreme climate conditions may affect the ability of transportation infrastructure to perform as intended over its lifetime. Infrastructure with a design lifetime of many decades could experience very different conditions from those of the recent and historical past. This is the primary motivation for integrating climate change considerations into project development: past conditions no longer serve as a reliable guide to design requirements for long-lived infrastructure.

Regulatory requirements provide another reason to integrate climate change considerations into the project development process. Engineering-informed adaptation studies can be an important part of addressing recent state and federal guidance and requirements for addressing climate change impacts in transportation project design.

State and federal environmental review requirements and guidance, such as the National Environmental Policy Act (NEPA), establish a framework for transportation agencies to consider the potential environmental consequences of their proposals, document the analysis, and make this information available to the public for comment before implementation. In the transportation project development process, the environmental review phase occurs immediately after the preliminary design and engineering phase. If engineering-informed adaptation studies are developed before undertaking the state or federal environmental review, the studies will help streamline and focus the environmental review on the issues identified as most important or significant.

Some states and regional agencies have requirements that climate change be considered in project scoping and design. The following bullets provide a summary of the requirements to date:

- Washington State Department of Transportation (DOT) *Guidance for NEPA and SEPA Project-Level Climate Change Evaluations* (*15*). Since 2009, Washington State DOT has issued guidance on how to consistently incorporate climate change impacts into the cumulative effects analysis in State Environmental Protection Act (SEPA) and NEPA documents.
- California DOT (Caltrans) *Guidance on Incorporating Sea Level Rise: For use in the planning and development of Project Initiation Documents* (*16*). This guidance is intended for use by Caltrans planning staff and project development teams to determine whether and how to incorporate sea level rise concerns into the programming and design of Caltrans projects.
- Maryland *Climate Change and Coast Smart Construction Infrastructure Siting and Design Guidelines* (*17*). Maryland developed these guidelines to provide "Coast Smart" construction guidance, including recommendations for the siting and design of state structures and infrastructure, institutionalization into state policies and programs, and technical tools and resources.
- Delaware *Avoiding and Minimizing Risk of Flood Damage to State Assets: A Guide for Delaware State Agencies* (*18*). This guidance provides state agencies with step-by-step instructions for avoiding and minimizing current and future flood risk to state assets.
- Port Authority of New York and New Jersey *Design Guidelines: Climate Resilience* (2015) (*19*). These guidelines provide guidance on how project designs should account for changes in temperature, precipitation, and sea level rise.

A common argument *against* considering climate change in project design and development is that projections of future climate are too uncertain to be useful. Using climate change projections may require some adjustments for transportation practitioners who are used to working with historical data, in that there is not one widely accepted data set, model, or set of assumptions. Practitioners may find the uncertainty surrounding the data unsettling, and be cautious about making decisions based on it.

However, uncertainty is not unique to climate change projections. For example, historical data sets are often incomplete and filled in using statistical assumptions. Meanwhile, major investment decisions are routinely made with support of traffic and other types of modeling—all of which are based on uncertain assumptions about future economic and population growth, demographic characteristics, traveler needs and preferences, land-use patterns, and other factors. These assumptions may or may not hold true as future reality unfolds. Thus, although climate data uncertainty may appear different from the types of uncertainty practitioners are familiar with, it will by no means be the first time practitioners make decisions based on uncertain information. The presence of uncertainty is thus not a logical argument for ignoring the potential impacts of climate change in infrastructure design and planning.

There are three main types of uncertainty in climate projections, each of which can be managed through common approaches:

- Scientific (or model) uncertainty: This uncertainty is inherent to scientists' understanding and ability to numerically capture climate processes. To address this uncertainty, it is important to look at data from an ensemble of models, rather than basing an analysis on any single model.
- Scenario (or human) uncertainty: This uncertainty results from our inability to predict human behavior: The scenarios of future greenhouse gas emissions and concentrations that drive climate model projections are built on assumptions about economic activity, energy sources, population growth, and other socioeconomic factors. Different assumptions lead to different levels of greenhouse gas concentrations. To address this uncertainty, it is advisable to include multiple scenarios in an analysis to capture a range of potential outcomes.
- Natural variability in the climate system (natural uncertainty: Climate and weather vary naturally from year to year, and any single year may not be representative of overall trends. This type of uncertainty should be dealt with by averaging climate projections over the course of two to three decades, rather than relying on a projection for a single year (*20*).

The relative importance of each of these types of uncertainty varies over time and across climate stressors. For example, managing the uncertainty in temperature projections requires different approaches from managing uncertainty in projected precipitation changes:

- When developing temperature projections past mid-century, the selection of greenhouse gas emissions scenarios, rather than climate models, will have a greater influence on the projected changes. This is because the further into the future practitioners look, the wider the range of uncertainty in future greenhouse gas emissions production. Despite this uncertainty, the overall trend for temperature should be clear across scenarios.
- When developing precipitation projections, the specific model chosen (and its inherent uncertainties) can have a particularly strong influence on projections, regardless of the emissions scenario chosen. In addition, while most climate projection data sets indicate an increase in springtime rainfall in most locations, the increase is not necessarily larger under higher greenhouse gas emissions scenarios. For example, in the *TEACR Culvert* study, FHWA found that averaging across the climate model precipitation projections created significant dampening of potential change, particularly as some models might suggest a reduction in magnitude while others might suggest an increase (*10*). To deal with this issue and to consider climate model/scenario combinations (i.e., simulations) that span across the range of projections, FHWA isolated and evaluated both the most extreme and the least extreme precipitation simulations.

Where to Integrate Climate Change into the Project Development Process

Although the names may vary by state, the common stages of the transportation project development process are: planning, scoping, preliminary design/engineering, environmental review, final design/engineering, right-of-way acquisition, and construction. As shown in Figure 1, it is important to consider climate change impacts and adaptation in the early stages of this process to ensure that climate resilience is incorporated into the project to the extent possible and appropriate. It is during the first three stages—planning, scoping, and preliminary design/engineering—that engineering-informed adaptation studies can have the greatest impact on the project. In fact, the flexibility available in the final design phase is limited by the decisions made at the earlier stages of planning and project development.

How to Integrate Climate Change into the Project Development Process

Current practices, based on FHWA's case studies, recommend the following steps be included in all engineering-informed adaptation studies:

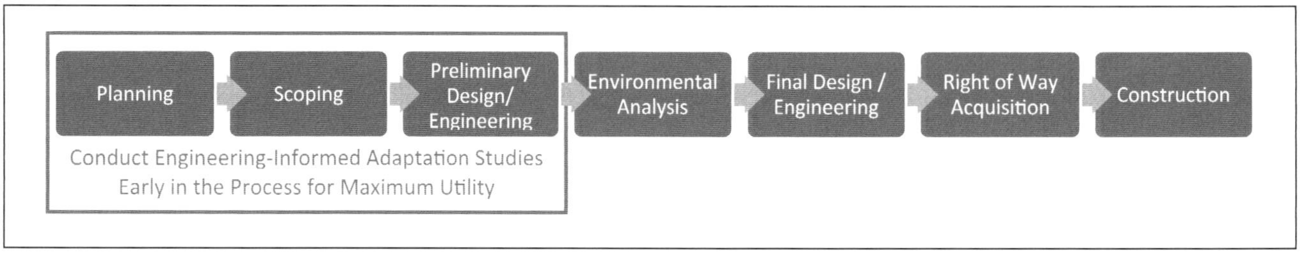

Figure 1. Where to integrate climate change into the project development process.

- Understand the site context and sensitivity to climate hazards: The context of a particular asset determines the scope and scale of a climate impacts and adaptation study. If the asset is vital to the operation of the transportation network, then the analysis may warrant more resources than if the asset provides redundant services. Starting with a basic understanding of how climate hazards (e.g., changes in precipitation and temperature, sea level rise, storm surges) may affect a given location and facility enables the practitioner to drill down, if necessary, and collect downscaled/localized data on the stressors that may affect the asset.
- Test the asset against future climate scenarios: This step involves first identifying or developing regional projections of future climate (for more information see the sections below on *What Climate Information to Consider* and *Sources of Climate Data and Projections*), and then evaluating the asset (or proposed asset) to determine how it would perform under the projected range of future climates. This enables practitioners to better understand whether and how the asset could be affected by future changes in climate.
- Develop, evaluate, and select adaptation measures: If the asset will be negatively affected by climate change, identify plausible adaptation strategies, and then evaluate their efficacy under the projected future climate(s). Selecting the appropriate adaptation measure to implement may depend on both efficacy and cost, so an economic assessment may be warranted. An economic assessment should not only consider the upfront (capital) cost of the various adaptation options, but also the costs and benefits (i.e., economic pros and cons) of adaptation over the lifetime of the asset.
- Review additional considerations: The selection of adaptation measure(s) to carry into environmental review should include consideration of the surrounding environment, the future of the transportation system of which the asset is a part, and socioeconomic considerations. Budgetary (and political) considerations may also come into play.
- Monitor and revisit as needed: The transportation system and surrounding communities the asset supports may change over time. A decision that is preferred today may be outperformed by a different solution in the future. Also, as climate conditions change and model projections are refined, adaptation strategies may need to be adjusted.

What Climate Information to Consider. In addition to climate change stressors with obvious implications for infrastructure, such as changes in extreme floods, sea level rise, and heat waves, climate change will bring variations in smaller, nuisance events that contribute to repeated stress and ultimate failure. Increased frequency of smaller rainfall events, for instance, can increase soil moisture over time and affect pavement reliability. The *TEACR Pavement Shrink-Swell* study discusses the effects of load repetition on pavement with weakened subgrades caused by relatively small changes in temperature and precipitation (*7*).

The potential impacts of these types of changes should be considered when designing and developing infrastructure projects. Another advantage of considering changes in higher frequency events is the lower uncertainty associated with modeling them. Analyzing shorter return interval events can demonstrate a clear impact on future maintenance costs and can therefore help justify investments in adaptation strategies. Information on changes in nuisance events is available from the data sets listed in Table 2 (*21–28*).

It is also important to consider the interactions among diverse climate stressors. For example, in the *TEACR Culvert* study, FHWA found that a wildfire burn of a substantial portion of a watershed may cause a dramatic increase in storm flows and create the potential for debris flows (*10*). Therefore, practitioners should consider the potential for increased rates of wildfire burn in addition to changes in precipitation. Similarly, in the *TEACR Slope Stability* study, FHWA determined that practitioners should consider the relationship between the *timing* and nature of projected precipitation and temperatures when assessing climate change impacts on rock

Table 2. Sources for Localized Climate Data

Resource	Description	Climate stressor
Downscaled CMIP3 and CMIP5 climate and hydrology projections (DCHP) database (*21*)	• Contains publicly available, downloadable, downscaled climate projection data for the contiguous United States • FHWA's preferred data source for temperature and precipitation projections	Temperature; precipitation
U.S. Geological Survey (USGS) geo data portal (*22*)	• Web portal provides access to a suite of climate data sets, including climate projections using different downscaling techniques	Temperature; precipitation
U.S. DOT CMIP climate data processing tool (*23*)	• Excel-based tool to process data from the DCHP database to provide projections for climate variables relevant to transportation planners (e.g., number of days above 95°F, hottest seven-day temperatures, largest three-day precipitation events) • Local-scale (56–224 sq. miles)	Temperature; precipitation
U.S. Army Corps Sea Level change curve calculator (*24*)	• A web-based tool that accepts user input to produce a table and graph of the projected sea level changes at the project site and includes vertical land movement	Sea level rise
NOAA technical report NOS CO-OPS 083: global and regional sea level rise scenarios for the United States (*25*)	• Identifies reasonable ranges for global mean sea level rise through 2200 • Six different scenarios, plus probability that each could be reached under various greenhouse gas scenarios	Sea level rise
NOAA's sea level rise viewer (*26*)	• Web mapping tool to visualize community-level impacts from coastal flooding or sea level rise • Contains downloadable sea level rise data for many locations	Sea level rise
National Research Council's sea-level rise for the Coasts of California, Oregon, and Washington (*27*)	• Contains projection information for California, Oregon, and Washington • May not be the most up-to-date resources for California, but a useful resource for Oregon and Washington	Sea level rise
National Climate Assessment (*28*)	• High-level descriptions of how climate stressors may change • Focused on larger geographic areas (e.g., Pacific Northwest, Southwest, Southeast) • Good starting point to identify what types of climate changes are expected in the region	All

slope weathering, because freeze–thaw events that occur within a day or two of a precipitation event will have a more severe negative impact on slope stability (*8*).

The probabilities associated with coincident or sequential future weather events will likely be unknown and uncertain, but the potential for significant impacts to infrastructure may be appreciable. Practitioners would be well served to use a scenarios approach to consider the impacts of joint events, as was done in the *TEACR Culvert* study to analyze the range of impacts that could occur from a series of wildfire and flooding events (*10*).

Sources of Climate Data and Projections. Climate data and projections relevant to transportation engineers are readily available; there is no need to start from scratch to develop necessary data. Tables 2 (*21–28*) and 3 (*2*, *14*, *29–31*) provide a summary of the reputable resources for data and guidance when developing climate projections.

Recommendations for Developing Adaptation Strategies

The TEACR and other FHWA case studies generated a wealth of lessons drawn from first-hand experience in preparing adaptation strategies and measures. The recommendations below are based on these experiences and can be considered by practitioners as they develop approaches for ensuring resilience and reducing vulnerabilities to the impacts of climate change.

The Need for Adaptation and the Timing of Implementation Will Vary from Location to Location and Asset to Asset

Before developing any adaptation strategies, an analysis of possible impacts over the remaining useful life of the structure must be completed. Existing infrastructure designed to withstand current or past climate conditions

Table 3. Guidance on Accessing and Developing Climate Data

Resource	Description	Climate stressor
FHWA synthesis of approaches for addressing resilience in project development (*14*)	• Includes information on accessing and interpreting projections • Provides technical guidance on processing climate model data to obtain variables relevant to transportation engineering and planning	All
FHWA hydraulic engineering circular (HEC)-17, 2nd edition (2016) (*2*)	• Provides technical guidance and methods for determining vulnerability to extreme events and climate change in riverine environments • Includes information on accessing and interpreting temperature and precipitation projections	Temperature; precipitation
FHWA HEC-25 Vol 2 (2014) (*29*)	• Provides technical guidance and methods for determining vulnerability to sea level rise and storms in coastal environments • Includes a methodology for determining local sea level rise and modeling storm surge	Sea level rise; storm surge
U.S. Army Corps Engineering and Construction Bulletin No. 2016–25 (*30*)	• Guidance for incorporating climate change impacts to inland hydrology in civil works studies, designs, and projects	Precipitation
U.S. Army Corps Engineering and Construction Bulletin No. 2016-05 (*31*)	• Guidance for using non-National Oceanic and Atmospheric Administration (NOAA) tide gage records for computing relative sea level change	Sea level rise

may still have a high level of resilience under future conditions. If the remaining lifetime of the asset is short or the avoided costs of climate impacts (i.e., the benefits of adaptation measures) are low, practitioners should consider the extent to which the current design (e.g., freeboard requirements) may offer inherent protection to changes in climate. For example, the Connecticut DOT *Pilot* research team focused its study on the performance of existing culverts, the condition of the culverts, and the resilience of the current system (*32*). The research team found that culverts meeting current design standards can function during flood events that exceed their design frequency, as a result of the existing allowances for freeboard or overtopping. In this case, age and physical deterioration may be more likely to contribute to failure than flow increases attributable to climate change.

If, on the other hand, the existing or planned structure is found to be ill equipped to manage future changes in climate, adaptation will be required. In some cases, adaptation strategies will be very similar to existing practices. For example, coastal climate adaptation measures for roadways will be similar to existing coastal engineering strategies for resilience (e.g., build "low and strong" to accommodate overwashing and minimize the negative impacts of wave attack or "high and dry" to completely avoid wave attack). In the *TEACR Roadway Surge* study, FHWA concluded that the countermeasure installed to protect the barrier island roadway from today's storms (a buried sheet pile wall and gabions), is a sound adaptation option for managing current weir-flow risks and future sea level rise and storm surge climate impacts (*9*).

Although many adaptation strategies will be similar to existing strategies, it is important to remember that every location and asset is unique and therefore requires its own analysis. For example, the MaineDOT *Pilot* study team found that in one town, the most cost-effective adaptation measure is to replace a bridge in-kind, while in the next town over, the most cost-effective solution is to replace the bridge with a significantly longer span because of differences in climate stressor–damage relationships and structural details regarding the range of candidate alternative designs (*33*).

When selecting adaptation measures, the remaining life of the facility is important to consider. Practitioners managing assets that are scheduled to undergo rehabilitation or full replacement in the near term may be able to delay implementing adaptation strategies until the anticipated rehabilitation and replacement cycle. Expected asset lifetime, relative to anticipated changes in climate, can influence the cost-effectiveness of adaptation options. For example, in a *GC2* case study on pavement, FHWA determined that changes to pavement mix designs to adapt to projected increases in temperature could be implemented during routine rehabilitation cycles (which typically run from 20 to 40 years) (*34*). In another example, when the remaining design life of a bridge is short enough that expected sea level rise would not affect it during its design life, such as in a *GC2* case study on a navigable bridge, the bridge should not be raised until its eventual full replacement (*34*). This approach will be a more cost-effective solution than unnecessarily retrofitting the existing structure now. Likewise, in the *Post Sandy* case study on Loop Parkway Bridge, NYSDOT decided to implement adaptation strategies at a bridge coincident with a planned mid-century rehabilitation (*13*). While the potential for climate-related impacts will remain in the interim, the cost savings of waiting to implement the adaptation measure outweighed the benefits of acting earlier. In these situations of delayed adaptation, it is helpful to include the proposed future strategies in long-range plans.

Regional Solutions Can Be Cost-Effective

Before developing adaptation strategies to address potential damages, practitioners should consider whether the impacts to the area would be widespread. If most of the surrounding roads and community assets are projected to be affected by increased or more frequent flooding (or other climate hazards), then a regional approach to adaptation (such as retreating from the region, or holistic adaptation that goes beyond just individual transportation assets) may be more appropriate. For example, adapting a bridge makes little sense if it provides service to an island that will soon be abandoned because of sea level rise. Similarly, adapting a single culvert may not be helpful if upstream/downstream impacts are ignored, or if nearby culverts are not similarly adapted.

To develop regional strategies, look beyond the individual asset to determine if and how the surrounding area/infrastructure is also at risk to the same stressor. For example, in addition to adaptation strategies at the asset level, the Massachusetts DOT *Pilot* investigated regional adaptation strategies (*35*). The regional strategies identified flood pathways from an adjacent section of the coastline that, if blocked, could protect large inland areas, including many transportation assets. Massachusetts DOT determined that regional solutions along the coastline can be more cost-effective than adapting individual assets, but often require coordination and investment across multiple stakeholders.

In addition, it is equally important to consider the condition of downstream infrastructure to avoid passing increased hazard conditions onto undersized assets. For example, if an undersized culvert in an upper stream reach served to attenuate peak extreme event flows downstream, upsizing that culvert would negate the attenuation and increase flows at a downstream bridge that may not be adequately prepared for the increase.

Looking Beyond the Asset Can Illuminate Potential Unintended Consequences

When evaluating adaptation strategies, it is important to consider potential secondary impacts or cascading consequences of a failed asset. For example, in the *Post Sandy* case study on Bergen Avenue, NY, the study area's roadway has very low traffic volumes and therefore may not be automatically prioritized for adaptation; however, the roadway provides the only access to the nearby wastewater treatment plant (*13*). During a storm event, if the plant does not have access to electricity, it is only able to operate for up to 24 hours using back-up generators, reserve fuel, and other supplies stored on site. If these reserves are exhausted and cannot be replenished via the access road, the plant would have to shut down. In this case, the broader consequences of roadway closure would include the potential inability to refuel the treatment plant, which could result in untreated waste flowing into streets, homes, and bodies of water.

There can be unintended impacts to the local environment if adaptation strategies are not designed with broader consequences in mind. In the *TEACR Economic Assessments*, FHWA evaluated the option of replacing a box culvert, which includes a flap gate on the downstream side, with a bridge (*6*). However, the flap gate prevents saltwater from progressing upstream. If the culvert were replaced by a bridge, saltwater and tidal influence would be reintroduced into the estuary. As another example, in the Minnesota DOT *Pilot*, Minnesota DOT built upstream considerations into the design of its adaptation options by including stream restoration and floodplain enhancement for the specific purpose of backwater reduction (*36*). This strategy would also provide benefits for water quality improvement resulting from stabilization of erosion-prone areas and depositional opportunities for suspended solids and nutrients.

Ecosystem-Based Adaptation Solutions Can Offer Low-Cost, Near-Term Benefits

The benefits of ecosystem-based solutions can be overlooked, especially when they aren't viewed as part of a longer-term adaptive management approach. Ecosystem-based solutions in the coastal environment (e.g., living shorelines, beach nourishment) may include a combination of an engineered structure to attenuate wave energy, appropriate vegetation, and sand to stabilize the shoreline and provide nearshore habitat to allow native species of flora and fauna to flourish.

In the *TEACR Living Shoreline* study, FHWA specifically evaluated the use of a living shoreline to determine its physical ability to protect a roadway from sea level rise and storm surge (*4*). While sea level rise by itself may pose minimal problems, it can exacerbate storm surge and wave impacts and cause more frequent storm-related flooding and damage. These roadways could benefit even today from natural and nature-based adaptation options such as living shorelines, in order to prevent wave damage. At a future date, some combination of natural features and engineered adaptations may be required to prevent flooding over the roadway. FHWA concluded that not only would a living shoreline provide ecological benefits, it could provide protection for the roadway for decades before more costly structural adaptation measures would be required.

The *TEACR Culvert* study notes that ecosystem-based post-fire mitigation measures, such as hydro mulching, seeding, tree plantings, debris basins, and upland erosion control can be used to slow flood flow rates and decrease debris flows in recently burned watersheds (*10*).

Use Adaptive Design to Prepare for the Future without Overspending Today

Adaptive management (or adaptive design) means selecting inherently flexible strategies that can be adjusted over time as needed, especially in light of current uncertainty in future climate change projections. For example, constructing a flood wall that can be raised in the future if sea level increases more rapidly than expected, or increasing maintenance in the short term and upgrading at the end of an asset's useful life, or taking an incremental approach to stabilizing landslides, can help practitioners ensure that current investments provide value under a range of futures.

In the *TEACR Slope Stability* study, the study team reviewed how the Virginia Department of Transportation (VDOT) has used an adaptive management strategy to manage climate uncertainty and the inherent uncertainty of unseen soil conditions and landslides (*8*). VDOT began by monitoring slope movement, installed a "toe wall" anchored by soil nails to stabilize the slope, and developed a plan for additional measures that could be taken if the slide continued to progress.

Adaptive management can be a cost-effective way to ensure resilience given an uncertain future. It reduces upfront capital needs while not excluding the possibility of adding supplemental levels of protection to the asset at a future date. The uncertainty of future impacts combined with the potential high cost of worst-case scenario adaptation strategies dictate that an incremental approach to implementing solutions is warranted. Where possible, practitioners should identify "no regrets" adaptation options that provide value under a range of potential futures, and can be readily adjusted or augmented pending changes in environmental conditions. Each measure should not preclude the installation of subsequent measures or reduce their effectiveness. This approach is feasible assuming a relatively slow rate of climate change. For example, in the *Post Sandy* case study on the Governor's Island ventilation building, the New York Metropolitan Transportation Authority recommended designing an improved flood wall around a tunnel ventilation building so that it could be enhanced (e.g., additional height, additional strength) if warranted in the future (*13*). Monitoring of future changes in flooding would inform the need for and timing of enhancement of the adaptation option.

Sometimes a "trigger" can be predetermined to help the agency decide when it is appropriate to add additional protection. This trigger should be based on the most recent climate change science and the time required to plan, design, finance, and construct a chosen adaptation option. For example, a decision may be triggered by exceedance of a particular threshold per year. In the *TEACR Culvert* study, FHWA analyzed the combined threat of a wildfire followed by a storm event to a culvert (*10*). Even though there are risks, in this situation FHWA recommends that the culvert only be adapted to withstand greater flows and additional debris *after* a wildfire occurs (rather than preemptively) because the probability of a wildfire in the watershed is expected to be low and the cost of adaptation is high. However, when the culvert reaches the end of its useful life and is scheduled for replacement, upsizing it would involve only a minor increase in cost and therefore may be worthwhile. This particular adaptive management example has both a climate stressor trigger (wildfire occurrence) and an inherent condition-based trigger (scheduled culvert replacement).

Use Asset-Management Systems and Routine Monitoring to Implement Adaptive Management

Asset management plans identify a structured sequence of maintenance, preservation, repair, rehabilitation, and replacement actions that will achieve and sustain a desired state of good repair over the lifecycle of the assets at minimum practicable cost. Incorporating adaptation planning into asset management helps ensure that adaptation is considered in a systematic manner alongside other needs for maintenance and repair, and that adaptation strategies are implemented during regular rehabilitation, reconstruction, and replacement activities in order to reduce the costs.

Rather than adhering to a strict schedule for adaptation implementation, the asset management system can be used to monitor the performance of the facility over time. This approach allows the asset owners to periodically revisit climate trends and projections to determine if the asset's design standards are approaching their limits or being exceeded. Such monitoring and periodic assessment can help indicate if it might be necessary to implement additional improvements, change design guidelines, and/or alter operation and maintenance practices. If periodic re-assessment is preferred, there should be clear documented direction as to when and who is responsible for doing so. As mentioned above, in the *TEACR Culvert* study, FHWA acknowledges that the change in wildfire potential under a changing climate is uncertain and therefore recommends continual monitoring to increase the understanding of changing trends in wildfires (*10*). Similarly, in the *GC2* case study on a navigable bridge, bridge clearance issues for the existing asset will only arise if the more extreme long-term sea level rise scenarios are realized (*34*). Therefore, it may make sense to hold off on implementation, incorporate the adaptation consideration into the facility management plan, and continue to monitor the rate of sea level rise.

Conclusions

Significant progress has been made to provide practitioners with resources, examples, and lessons on integrating climate change into asset-level design and the project development process:

- It is clear that uncertainty should not be a barrier to incorporating climate projections into project design and planning. Data are frequently available in formats that can be used in engineering-informed analyses, and methodologies have been established for calculating derived variables more directly applicable to engineering designs (e.g., freeze–thaw cycles).
- Adaptive design is a useful framework for developing climate change adaptation strategies. The phased approach reduces the upfront capital needs while not excluding the possibility of adding additional levels of protection to the asset at a future date. This allows practitioners to track climate change projections and impacts over time, and act accordingly.
- Adaptation solutions cannot be developed in isolation. Climate change will impact broad geographic areas, so a regionally coordinated approach to resilience can minimize expenses and disruption to transportation networks.

Acknowledgments

The *Transportation Engineering Approaches to Climate Resiliency* project was funded by the Federal Highway Administration. Major contributors to the work on which this paper are based include Justin Lennon, Chris Dorney, Ravi Kuchibhotla, Jagannath Mallela, and Suriyanarayanan Sadasivam of WSP; Scott Douglass of South Coast Engineers; and Anne Choate of ICF.

Author Contributions

The authors confirm contribution to the paper as follows: study conception and design: BR, BD, JK, RH, BB, KM, RK; data collection: BR, BD, JK; analysis and interpretation of results: BR, BD, JK; draft manuscript preparation: BD. All authors reviewed the results and approved the final version of the manuscript.

References

1. Shepard, C. C., V. N. Agostini, B. Gilmer, T. Allen, J. Stone, W. Brooks, and M. W. Beck. Assessing Future Risk: Quantifying the Effects of Sea Level Rise on Storm Surge Risk for the Southern Shores of Long Island, New York. *Natural Hazards*, Vol. 60, No. 2, 2012, pp. 727–745. https://doi.org/10.1007/s11069-011-0046-8.
2. Kilgore, R. T., G. R. Hermann, W. O. Thomas, Jr., and D. B. Thompson. *Hydraulic Engineering Circular No. 17, 2nd Edition, Highways in the River Environment: Floodplains, Extreme Events, Risk, and Resilience.* FHWA-HIF-16-018, June 2016. https://www.fhwa.dot.gov/engineering/hydraulics/library_arc.cfm?pub_number=16&id=162. Accessed March 2017.
3. ICF International. *2013–2015 Climate Change Resilience Pilot Program: Outcomes, Lessons Learned, and Recommendations.* FHWA-HEP-16-079, July 2016. https://www.fhwa.dot.gov/environment/sustainability/resilience/pilots/2013-2015_pilots/final_report/index.cfm. Accessed July 2017.
4. USDOT. *The Gulf Coast Study, Phase 2.* FWHA-HEP-12-054. 2015. https://www.fhwa.dot.gov/environment/sustainability/resilience/ongoing_and_current_research/gulf_coast_study/. Accessed July 2017.
5. FHWA. *Post Hurricane Sandy Transportation Resilience Study in New York, New Jersey, and Connecticut.* 2017. https://www.fhwa.dot.gov/environment/sustainability/resilience/publications/hurricane_sandy/. Accessed July 2017.
6. Choate, A., B. Dix, B. Rodehorst, A. Wong, W. Jaglom, J. Keller, J. Lennon, C. Dorney, R. Kuchibhotla, J. Mallela, S. Sadasivam, and S. Douglass. *Synthesis of Approaches for Addressing Resilience in Project Development.* Publication FHWA-HEP-17-082. FHWA, U.S. Department of Transportation, Washington, D.C., 2017.
7. FHWA. *Sea Level Rise and Storm Surge Impacts on a Coastal Bridge: I-10 Bayway, Mobile Bay, Alabama.* FHWA-HEP-17-014. 2017. https://www.fhwa.dot.gov/environment/sustainability/resilience/ongoing_and_current_research/teacr/al_i-10/index.cfm. Accessed July 2017.
8. FHWA. *Living Shoreline along Coastal Roadways Exposed to Sea Level Rise: Shore Road in Brookhaven, New York.* FHWA-HEP-17-016. 2017. https://www.fhwa.dot.gov/environment/sustainability/resilience/ongoing_and_current_research/teacr/ny_shore_road/. Accessed July 2017.

9. FHWA. *Temperature and Precipitation Impacts on Cold Region Pavement: State Route 6/State Route 15/State Route 16 in Maine*. FHWA-HEP-17-019. 2017. https://www.fhwa.dot.gov/environment/sustainability/resilience/ongoing_and_current_research/teacr/me_freeze_thaw/index.cfm. Accessed July 2017.
10. FHWA. *Comparison of Economic Analysis Methodologies and Assumptions: Dyke Bridge in Machias, Maine*. FHWA-HEP-17-022. 2017. https://www.fhwa.dot.gov/environment/sustainability/resilience/ongoing_and_current_research/teacr/dyke_bridge/index.cfm. Accessed July 2017.
11. FHWA. *Temperature and Precipitation Impacts to Pavements on Expansive Soils: Proposed State Highway 170 in North Texas*. FHWA-HEP-17-018. 2017. https://www.fhwa.dot.gov/environment/sustainability/resilience/ongoing_and_current_research/teacr/texas_i-70/index.cfm. Accessed July 2017.
12. FHWA. *Precipitation and Temperature Impacts on Rock and Soil Slope Stability: Interstate I-77 in Carroll County, Virginia*. FHWA-HEP-17-021. 2017. https://www.fhwa.dot.gov/environment/sustainability/resilience/ongoing_and_current_research/teacr/va_slopes/index.cfm. Accessed July 2017.
13. FHWA. *Barrier Island Roadway Overwashing from Sea Level Rise and Storm Surge: US 98 on Okaloosa Island, Florida*. FHWA-HEP-17-015. 2017. https://www.fhwa.dot.gov/environment/sustainability/resilience/ongoing_and_current_research/teacr/fl_us_98/index.cfm. Accessed July 2017.
14. FHWA. *Transportation Engineering Approaches to Climate Resiliency (TEACR) Study*, December 2016. https://www.fhwa.dot.gov/environment/sustainability/resilience/ongoing_and_current_research/teacr/index.cfm#l2. Accessed July 2017.
15. Washington State DOT. *Guidance for NEPA and SEPA Project-Level Climate Change Evaluations*. 2017. www.wsdot.wa.gov/NR/rdonlyres/BDF7C3DA-4F27-4CD5-8D02-6813027A928B/0/WSDOT_Climate_Guidance_2017_FINAL.pdf. Accessed July 2017.
16. California Department of Transportation. *Guidance on Incorporating Sea Level Rise: For Use in the Planning and Development of Project Initiation Documents*. 2011. http://www.dot.ca.gov/hq/tpp/offices/orip/climate_change/documents/guide_incorp_slr.pdf. Accessed July 2017.
17. Johnson, Z. P., ed. *Climate Change and Coast Smart Construction: Infrastructure Siting and Design Guidelines*. 2013. http://climatechange.maryland.gov/wp-content/uploads/sites/16/2014/12/climate_change_and_coast_smart_final_report1.pdf. Accessed July 2017.
18. Delaware Flood Avoidance Workgoup. *Avoiding and Minimizing Risk of Flood Damage to State Assets: A Guide for Delaware State Agencies*. 2016. http://www.dnrec.delaware.gov/energy/Documents/DE%20Flood%20Avoidance%20Guide%20For%20State%20Agencies.pdf. Accessed July 2017.
19. The Port Authority of NY & NJ. *Design Guidelines Climate Resilience*. 2015. https://www.panynj.gov/business-opportunities/pdf/discipline-guidelines/climate-resilience.pdf. Accessed July 2017.
20. Hayhoe, K., and A. Stoner. *Temperature and Precipitation Projections for the Mobile Bay Region. Impacts of Climate Change and Variability on Transportation Systems and Infrastructure: The Gulf Coast Study, Phase II*. FHWA-HEP-12-055. 2012. https://www.fhwa.dot.gov/environment/sustainability/resilience/ongoing_and_current_research/gulf_coast_study/phase2_task2/mobile_infrastructure/mobile_climate_report.pdf. Accessed July 2017.
21. Douglass, S. L., B. M. Webb, and R. Kilgore. *Hydraulic Engineering Circular Vol. 2, No. 25, Highways in the Coastal Environment: Assessing Extreme Events*. FHWA-NHI-14-006, October 2014. https://www.fhwa.dot.gov/engineering/hydraulics/pubs/nhi14006/nhi14006.pdf. Accessed July 2017.
22. US Army Corps of Engineers (USACE). *Guidance for Incorporating Climate Change Impacts to Inland Hydrology in Civil Works Studies, Designs, and Projects*. CECW-CE-2016-25, September 2016. http://www.iwr.usace.army.mil/Portals/70/docs/frmp/eo11988/ECB_2016_25.pdf. Accessed July 2017.
23. USACE. *Using Non-NOAA Tide Gauge Records for Computing Relative Sea Level Change*. CECW-CE-2016-5, January 2016. https://wbdg.org/FFC/ARMYCOE/COEECB/ARCHIVES/ecb_2016_5.pdf. Accessed July 2017.
24. *Downscaled CMIP3 and CMIP5 Climate and Hydrology Projections*, October 2016. https://gdo-dcp.ucllnl.org/downscaled_cmip_projections/dcpInterface.html#Welcome. Accessed July 2017.
25. US Geological Survey (USGS). *USGS Geo Data Portal*. Undated. https://cida.usgs.gov/gdp/. Accessed July 2017.
26. US DOT. *DOT's CMIP Climate Data Processing Tool*, March 2016. http://www.fhwa.dot.gov/environment/sustainability/resilience/adaptation_framework/modules/index.cfm?moduleid=4#tools. Accessed July 2017.
27. USACE. *Sea Level Change Curve Calculator*. Undated. http://www.corpsclimate.us/ccaceslcurves.cfm. Accessed July 2017.
28. Sweet, W. V., R. E. Kopp, C. P. Weaver, J. Obeysekera, R. M. Horton, E. R. Thieler, and C. Zervas. *Global and Regional Sea Level Rise Scenarios for the United States*. NOAA Technical Report NOS CO-OPS 083, January 2017. https://tidesandcurrents.noaa.gov/publications/techrpt83_Global_and_Regional_SLR_Scenarios_for_the_US_final.pdf. Accessed July 2017.
29. NOAA. *Sea Level Rise Viewer*. Undated. https://coast.noaa.gov/digitalcoast/tools/slr. Accessed July 2017.
30. National Research Council. *Sea-Level Rise for the Coasts of California, Oregon, and Washington: Past, Present, and Future*. 2012. https://doi.org/10.17226/13389. Accessed July 2017.
31. US Global Change Research Program [Wuebbles, D. J., D. W. Fahey, K. A. Hibbard, D. J. Dokken, B. C. Stewart, and T. K. Maycock, eds.]. *Climate Science Special Report: Fourth National Climate Assessment, Volume I*. U.S. Global Change Research Program. 2017. http://dx.doi.org/10.7930/J0J964J6. Accessed November 2017.
32. Hogan, M., D. Elder, and S. Molden. *Connecticut Department of Transportation Climate Change and Extreme Weather Vulnerability Pilot Project Final Report*, December

2014. https://www.fhwa.dot.gov/environment/sustainability/resilience/pilots/2013-2015_pilots/connecticut/final_report/index.cfm. Accessed July 2017.

33. Merrill, S., and J. Gates. *Integrating Storm Surge and Sea Level Rise Vulnerability Assessments and Criticality Analyses into Asset Management at MaineDOT*, December 2014. https://www.fhwa.dot.gov/environment/sustainability/resilience/pilots/2013-2015_pilots/maine/final_report/index.cfm. Accessed July 2017.

34. FHWA. *Impacts of Climate Change and Variability on Transportation Systems and Infrastructure, The Gulf Coast Study, Phase 2, Task 3.2*. 2015. https://www.fhwa.dot.gov/environment/sustainability/resilience/ongoing_and_current_research/gulf_coast_study/phase2_task3/task_3.2/page04.cfm. Accessed July 2017.

35. FHWA. *FHWA Climate Resilience Pilot Program: Massachusetts Department of Transportation*. FHWA-HEP-16-073. 2016. https://www.fhwa.dot.gov/environment/sustainability/resilience/pilots/2013-2015_pilots/massdot/index.cfm. Accessed July 2017.

36. Almodovar-Rosario, N., C. Dorney, M. Flood, J. Lennon, and J. T. Lockman. *MnDOT Flash Flood Vulnerability and Adaptation Assessment Pilot Project*, November 2014. https://www.fhwa.dot.gov/environment/sustainability/resilience/pilots/2013-2015_pilots/minnesota/final_report/index.cfm. Accessed July 2017.

The Standing Committee on Transportation and Sustainability (ADD40) peer-reviewed this paper (18-05253).

Article

TRR
JOURNAL OF THE TRANSPORTATION RESEARCH BOARD

Transportation Research Record
2018, Vol. 2672(3) 130–144
© National Academy of Sciences: Transportation Research Board 2018
Article reuse guidelines:
sagepub.com/journals-permissions
DOI: 10.1177/0361198118791382
journals.sagepub.com/home/trr
SAGE

Quantifying the Sustainability, Livability, and Equity Performance of Urban and Suburban Places in California

Alexander Rijiro Frost[1], Bruce Appleyard[2], Joseph Gibbons[2], and Sherry Ryan[2]

Abstract

For years, researchers and practitioners have worked toward measuring urban form, but a gap still remains in the research to quantify how urban and suburban place-types affect economic, social, and environmental outcomes at small geographic scales. To provide such analysis, this paper describes the development of a place typology and sustainability performance measurement framework for all census tracts in California. This study found there were clear trade-offs between urban and suburban living. Compared with suburbs, the households in urban places benefited from a 57.9% reduction in annual vehicle miles traveled, 37.2% lower transport-related greenhouse gas emissions per capita, and saved more than US$2,675 in annual transportation costs, while consuming less electricity (39.9%) and water per capita (63.8%). However, the cost of urban homeownership was 40% higher, despite rents being 18.5% cheaper. And although obesity and cardiovascular disease rates were 10.3% and 8.9% lower in urban places, asthma rates were 7.5% higher. From 1970 to 2015, urban housing decreased from 34% to 21%, whereas statewide it dropped 7.5%. Despite ambitious climate action and smart growth goals, the majority of growth in California continues to be in low-density suburban/rural areas, responsible for 80% of the state's total household carbon emissions. This analysis and place typology could prove useful in identifying areas with the highest potential for lowering vehicle miles traveled and other sustainability, livability, and equity goals. This is made even more significant given California's recent move to abolish level of service analysis for traffic impact studies.

The state of California has set a goal to significantly reduce its greenhouse gas (GHG) emissions 40% and 80% below 1990 levels by 2030 and 2050, while trying to accommodate an additional 10 million residents by 2050 (*1*). Integration of housing, transportation, and land use policies to encourage more location-efficient development is an important component in realizing this goal (*2*), but land use reform has been difficult to achieve due to conflicting priorities between state and local governments (*3, 4*).

Furthermore, the combination of a growing population, scarcity of land, community and municipal barriers, and other external influences are slowing the creation of new homes, especially in urban communities and coastal counties, though demand continues to grow, creating a US$50 billion + annual housing affordability gap (*5, 6*). The location, type, and rate of future housing and land use development patterns will be critical to maintaining the state's long-term economic prosperity, social stability, and meeting environmental sustainability targets (*3*).

Neighborhoods are the basic building blocks of towns, cities, and regions and as zoning and street standards have uniformly shaped them for decades (*7*) there is a demand for quantitative research that connects neighborhood typologies with sustainability outcomes (*8–10*). State agencies are also prioritizing the development of performance metrics to compare different economic, environmental, health, and equity outcomes of existing and future land use policies to achieve climate goals and smart growth (*1*).

Building on previous typology studies (*11–14*), this paper improves the typology classification processes and expands the performance metrics to holistically compare how different urban forms affect economic, social, and

[1]City of San Diego, San Diego, CA
[2]San Diego State University, San Diego, CA

Corresponding Author:
Address correspondence to Alexander Rijiro Frost: afrost@sandiego.gov

environmental outcomes of all 8,043 census tracts in the state of California.

Brief Literature Review on the Quantification of Urban Form

There are many approaches to classifying and defining urban qualities. Lynch and Alexander et al. identify archetypal patterns of natural and built environments at various scales – building, street, community, city – and document potential costs and benefits of various urban designs and city forms (*15, 17*), whereas Jane Jacobs, Allan Jacobs, and Donald Appleyard articulate the necessary physical characteristics and conditions required for urban livability and vitality (*18, 19*). Recent studies provide empirical verification of their theories and confirm neighborhoods with mixed use, diverse building types, shorter blocks, and sufficient density have higher intensity and vitality due to more pedestrian activities compared with other more homogenous urban areas in Seoul, South Korea and Milan, Italy (*20, 21*).

A meta-analysis of literature by Clifton et al. identifies five conceptual dimensions to quantify urban form and finds an overall advantage of compact and mixed development at multiple-spatial scales (*8*). However, there has been limited application of these methods in the realm of urban design due to the difficulty quantifying urban form objectively (*22*). On the other hand, in the United States, many regional planning processes are incorporating a "place-type" approach to help citizens and policy makers evaluate existing conditions and trade-offs of future growth patterns (*9, 23*).

Measuring the Built Environment

The study of built environment and travel behavior outcomes has become one of the most heavily researched topics in urban planning (*24*), but not without considerable debate (*25*). Currently, most travel behavior and built environment research relies on the "D-variables," first developed by Cervero and Kockelman, who originally coined the first three variables: density, diversity, and design (*26*). Most research now uses the five D-variables: density, diversity, design, destination accessibility, and distance to transit, and these measures continue to expand by including other non-built environment variables such as demography and desirability (*24, 27*).

Defining Place Types

Though there is some agreement on built environment measures, there is not yet official government or academic consensus on how to define or classify different neighborhood types (*28–30*). Variables related to population and housing density were used in early studies to distinguish urban and suburban place-types (*31*) and the most common technique to delineate place types is a threshold or rule-based approach (*32*). A. Jacobs and D. Appleyard recommend minimum-density thresholds of 15 net dwelling units per acre (du/ac) for city neighborhoods, whereas Duany and Talen advocate for maximum-density limits based on different transect zones with anything below 6 du/ac to be considered Sub-Urban(*19, 33*). On the other hand, Ewing and Hamidi suggests anything below 8 du/ac is suburban, whereas Litman considers urban communities can start at a minimum of 6 du/ac (*34, 35*).

In addition to density, street network pattern, connectivity and its associated measures – often street intersection density – are used as a proxy to identify place and landscape types (*10, 36, 37*). Other studies spatially classify place types, such as urban centers, inner ring areas, and outer-ring suburbs/exurbs, using distances from city centers (*38*). Spatial classification can work well for pre-automobile, mono-centric urban areas, but it is less useful for post-automobile urban areas, found throughout California.

The D-variables are useful metrics to quantify transit-oriented development (TOD)—a compact, walkable neighborhood with frequent and convenient transit access—a key smart growth and climate action strategy (*39, 40*). Pushkarev and Zupan's seminal study first suggested densities above 7 du/ac for optimum public transit performance, whereas Guerra and Cervero found 17 du/ac for heavy rail and 12 du/ac for light rail would lead them to perform in the top one-quarter of cost-efficient rail investments (*41, 42*).

Renne and Ewing utilize jobs and population density, percentage of land use mix, and average block size to categorize three types of TODs (*43*), whereas Appleyard, Ferrell, Taecker, and Frost refine this methodology by including additional "D" variables, such as distance to bus transit and destination accessibility to capture regional and local access to opportunities, finding integrated TODs have a lower driving rate, more walking and bicycling, lower obesity levels, but higher housing costs (*14*).

In 2007, Song and Knaap presented a quantitative method for classifying neighborhood types using a combination of principal components analysis and cluster analysis (*44*). In 2013, Song et al. then expanded this work toward a national study identifying seven neighborhood classifications—urban center, inner suburban, outer suburban, greenfield suburban, exurban, rural cluster, and rural (*9*). Considering the dominance of sprawling development patterns in the United States

(*36*), it is not surprising to find, six out of seven types are mostly lower density in nature. In 2014, Salon, Boarnet, and Mokhtarian followed similar methods for classifying the census tracts for the entire state of California into seven distinct residential neighborhoods; they found households in urban places have lower rates of annual vehicle miles traveled (VMT) compared with suburban and rural households (*13*).

Lin and Long and Voulgaris et al. classified most of the U.S. census tracts into distinct neighborhood types using factor and cluster analyses to compare travel behavior outcomes (*30, 45*). One of the most prominent national indices to describe the spectrum of urban form is the sprawl index (*34*). Although useful in exploring relationships between sprawl and quality of life outcomes, continuous composite measures do not have the ability to compare the performance of different neighborhood types and describe the nuance of urban qualities. Finally, Ralph, Voulgaris, and Brown conducted a systematic comparison of activity density, sprawl index, and neighborhood types and how well these measurement types captured the influence of the built environment on travel behavior, finding models that incorporated both composite measures (sprawl index and neighborhood types) performed better (*46*).

Methodology

Building on previous studies, specifically by Song et al., Salon et al., Ewing and Hamidi, and Voulgaris et al. (*9, 13, 34, 45*), this study classifies neighborhood types based on a combination of principal components analysis, K-means cluster analysis, threshold analysis, and overlay analysis. Cluster analysis, a method of identifying the local intensity of a variable or process, is an increasingly common way to classify neighborhood types. With K-means cluster analysis one can group communities with similar physical attributes to categorize neighborhood types. As neighborhood-level measures tend to correlate, methods like factor analysis and principle component analysis are useful ways to identify the core components of interest, which can be used to group similar built environment attributes.

Unit of Analysis and Data Sources

Previous studies on measuring urban and neighborhood forms typically use a variety of geographies to identify neighborhoods, such as traffic analysis zone, zip codes, census tracts, census block groups or buffers of different sizes, and self-defined boundaries, such as a 0.25-mi^2-grid as a unit of analysis (*9*).

Although census block groups are a more detailed unit of analysis, there are several advantages to using census tracts for the study of neighborhood forms, namely the availability of datasets for sustainability measures at this unit. Furthermore, utilizing census tracts allows for longitudinal comparison to study trends in population and housing growth for each community type.

For this study, a California statewide database of the built environment characteristics and sustainability performance metrics was created for 8,043 census tracts. The database contains variables from the following sources: U.S. Census Bureau American Community Survey, California Office of Environmental Health Hazard Assessment's CalEnviroScreen 3.0, U.S. Environmental Protection Agency's (EPA) Smart Location Database 2.0, U.S. Housing and Urban Development (HUD) Location Affordability Index 2.0, University of California Berkeley's Cool Climate Network, University of Utah's Metropolitan Research Center, University of California Los Angeles' AskCHIS Neighborhood Edition, and Walk Score. The majority of socioeconomic and environmental variables were utilized directly at the census tract level. Three variables—Walk Score, obesity, and household carbon footprint—were only available at the ZIP-code level, whereas data from HUD and EPA were available at census block group level. These datasets were aggregated and calculated to conform to the 2010 census tract boundaries. Data from HUD, Cool Climate Network, and UCLA were all secondary (modeled) variables; the rest of the variables came from primary sources. For the longitudinal analysis of housing trends, this study used interpolated U.S. decennial census tract data from 1970, 1980, 1990, 2000, and 2010 (*47*), which were combined with variables from the American Community Survey, 2010–2015.

Framework, Methods, and Tools

This study used both quantitative and qualitative techniques to identify and measure physical attributes of place types in order to compare their economic, social, and environment performance. Using SPSS v.24 and ArcGIS 10.3, four iterative statistical and geospatial methods were used to identify major categories of neighborhood form and their respective sustainability performance. The overall methodological framework (Figure 1) is organized into three phases: 1) developing the place typology; 2) describing and validating the place typology; and 3) exploring relationships.

Phase 1: Developing a Place Typology. The principal component analysis establishes a location-efficient neighborhood design (LEND) index based on six dimensions of community characteristics—density, land use mix, building and regional diversity, street design, transit distance, and destination accessibility to jobs—created from a total of 19 built environment variables. The LEND index provides a scale and a single score to distinguish neighborhoods with

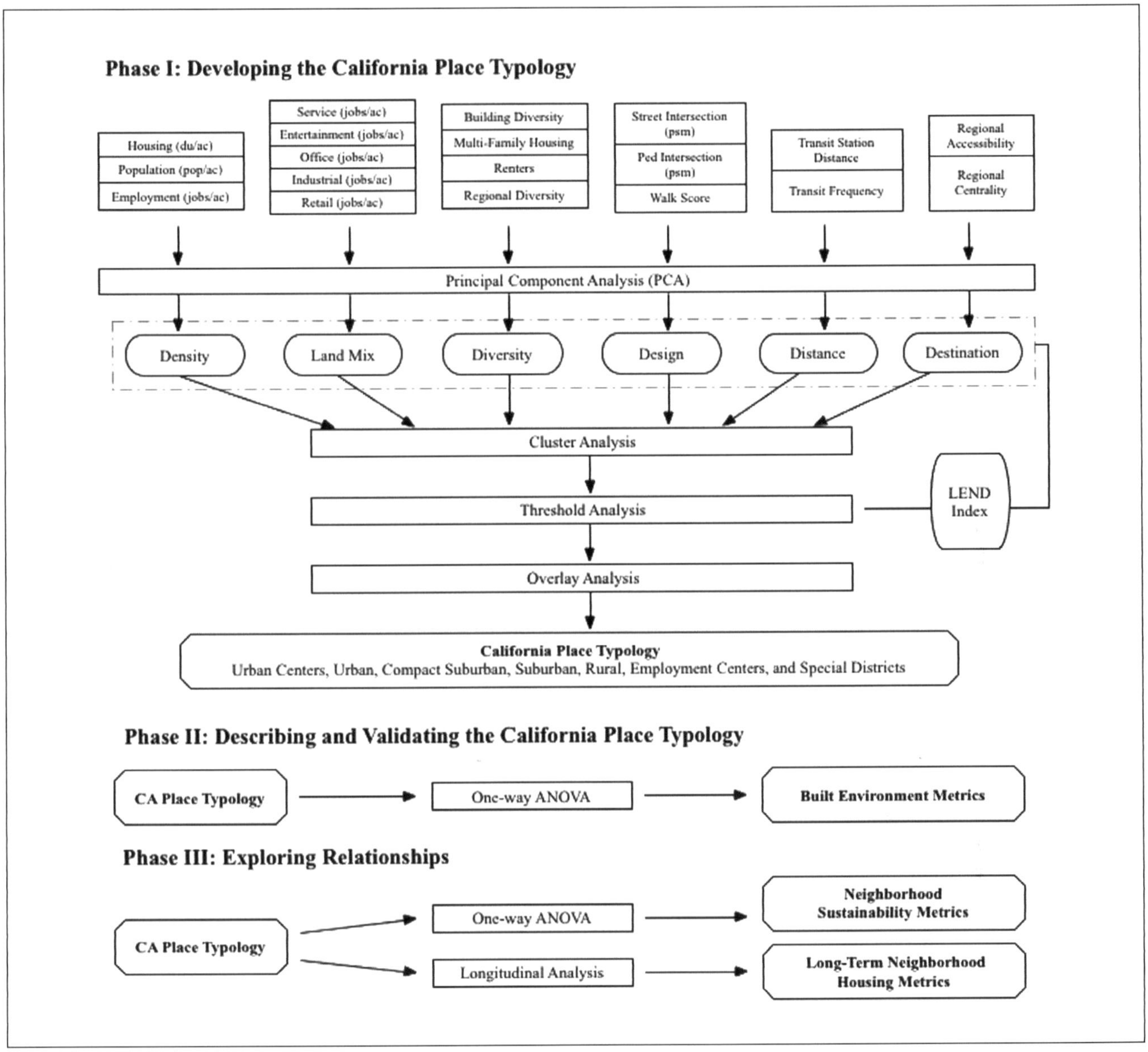

Figure 1. Methodological framework.

high location efficiency, walkability, transit access, and regional accessibility.

The standardized six dimension factors from the principal component analysis were then used in a K-means cluster analysis. The threshold analysis helped identify non-traditional land uses, such as employment centers outside traditional downtowns, and to distinguish clear differences between each place type using both composite measures (LEND and sprawl indices) and key variables, such as dwelling units per acre, transit frequency, and transit job accessibility. The overlay analysis provides a visual assessment via ArcGIS to help confirm the validity of the proposed place types and identification of additional development patterns, such as special districts.

Table 1 describes the results from the principal component analysis, highlighting the selected variables, the correlation of these factors with each variable, and percentages of explained variances. This analysis shows both distinct explanatory powers between the six components, each of which are composed of a specific group of variables. All of the components had robust factor loadings. For example, for the density factor, gross housing density had a loading of 0.958, gross population density had a loading of 0.945, and gross employment density had a loading of 0.762. Together 79.72% of the variance in these variables could be explained by the density factor. Not all of the resulting factors were as robust. The building and regional diversity factor only explained

Table 1. PCA Variable Loadings of Five Factors

Component matrix	Factor loadings	Data sources	Unit
Population and employment density factor			
Gross housing density (du/ac)	0.958	SLD 2010	CBG
Gross population density (pop/ac)	0.945	SLD 2010	CBG
Gross employment density (jobs/ac)	0.762	SLD 2010	CBG
Eigenvalue	2.392		
Explained variance	79.72%		
Land use mix factor			
Gross retail (5-tier) employment density (jobs/ac)	0.857	SLD 2010	CBG
Gross office (5-tier) employment density (jobs/ac)	0.872	SLD 2010	CBG
Gross industrial (5-tier) employment density (jobs/ac)	0.745	SLD 2010	CBG
Gross service (5-tier) employment density (jobs/ac)	0.865	SLD 2010	CBG
Gross entertainment (5-tier) employment density (jobs/ac)	0.839	SLD 2010	CBG
Eigenvalue	3.503		
Explained variance	70.06%		
Building and regional diversity factor			
Building diversity index	0.511	ACS 2014	CT
Renter occupant (percent)	0.92	ACS 2014	CT
Multi-family housing (percent)	0.878	ACS 2014	CT
Regional diversity (deviation of CBG ratio of jobs/pop from regional average ratio of jobs/pop)	0.492	SLD 2010	CBG
Eigenvalue	2.121		
Explained variance	53.00%		
Street design factor			
Street intersection (intersection density per square mile)	0.804	SLD 2010	CBG
Pedestrian intersection (pedestrian-oriented intersections having four or more legs per square mile)	0.67	SLD 2010	CBG
Community walkability	0.825	Walk Score 2016	ZIP
Eigenvalue	1.774		
Explained variance	59.14%		
Transit distance and frequency factor			
Distance to nearest transit stop (meter)	0.907	SLD 2010	CBG
Transit service coverage (aggregate frequency of transit service per square mile)	0.907	SLD 2010	CBG
Eigenvalue	1.645		
Explained variance	82.26%		
Transit destination – regional accessibility factor			
Transit jobs accessibility within 45 min	0.884	SLD 2010	CBG
Regional centrality index (transit employment accessibility expressed as a ratio of total MSA accessibility)	0.884	SLD 2010	CBG
Eigenvalue	1.562		
Explained variance	78.12%		

Note: SLD = Smart Location Database; ACS = American Community Survey; CBG = Census Block Group; CT = Census Tract; ZIP = ZIP Code.

53% of the variance in its variables. However, this still means this factor explains over half of the variance in that given variable, which is the common rubric for an acceptable factor. The cluster analysis in the next step utilizes the standardized six dimension scores to group neighborhood types, whereas the LEND scores are used in concurrence with threshold based analysis to refine the neighborhood place-types.

In determining the neighborhood place-types, key challenges are a) determining how many clusters to create, while b) minimizing subjectivity as much as possible. Considering the various clustering techniques, from Salon et al. using the Ward's method (*9*), and Voulgaris et al.'s use of multiple indices (*45*), we opted for an approach similar to Song et al.'s 2007 article, and Salon et al.'s 2014 article, both of which applied an iterative process using K-means cluster outputs (*9, 13*). In sum, we generated cluster outputs for up to 10 clusters and then culled them down with the aim to a) minimize the variation within clusters, and b) maximize the variation between clusters. A final step, as described in Phase II and shown in Table 2 below, was to examine the average values for each cluster to make sure multiple clusters are clearly distinct from each other, while also relevantly reflecting places in the real world. In the initial steps, we identified four major groupings: urban, suburban, rural

Table 2. Built Environment Characteristics of California Place Types: One-Way Analysis of Variance (ANOVA) Comparison of Means

	Urban centers	Urban	Compact suburban	Suburban	Rural	Employment centers	Special districts	Sig.
Multivariate urban form indices								
LEND index	149.1	135.2	104.7	88.6	69.4	128.1	103.3	0.00
Compact/sprawl index	142.1	127.6	120.4	97.3	56.7	110.2	98.3	0.00
Density								
Housing density (du/ac)	22.8	13.1	5.2	3.4	0.5	6.4	2.8	0.00
Population density (pop/ac)	35.9	32.4	15.1	10.0	1.3	12.8	12.1	0.00
Jobs density (emp/ac)	72.5	9.7	3.4	2.1	0.4	42.7	7.7	0.00
Land use mix								
Retail (emp/ac)	4.2	0.9	0.4	0.3	0.3	2.6	0.5	0.00
Office (emp/ac)	26.0	1.4	0.5	0.2	0.1	16.4	1.3	0.00
Industrial (emp/ac)	9.5	1.0	0.5	0.4	0.1	3.2	1.7	0.00
Service (emp/ac)	17.5	2.4	0.7	0.4	0.0	7.9	1.0	0.00
Entertainment (emp/ac)	8.6	1.5	0.4	0.2	0.0	4.3	0.6	0.00
Diversity (buildings/regional)								
Buildings built before 1959 (%)	42.2	51.8	43.4	23.1	20.1	23.6	21.3	0.00
Building diversity index	2.2	2.5	2.4	2.2	2.6	2.4	2.3	0.00
Multi-family housing (%)	93.7	66.9	40.7	25.5	9.0	68.8	44.0	0.00
Renter occupant (%)	83.5	70.3	50.0	35.5	29.4	59.0	52.4	0.00
Regional diversity (CT ratio of jobs/pop from regional average)	0.4	0.2	0.2	0.2	0.2	0.5	0.4	0.00
Design (street)								
Street intersection (intersection density psm)	178.5	128.0	102.3	73.8	15.9	105.1	89.8	0.00
Pedestrian intersection	42.7	42.7	20.6	15.9	11.1	35.9	27.5	0.00
Walk score	86.7	71.2	45.7	28.7	5.2	56.9	42.1	0.00
Distance (transit accessibility)								
Distance to nearest transit stop (meter)	200.9	320.2	524.9	572.1	559.6	416.8	489.2	0.00
Transit service coverage (aggregate frequency of transit service psm)	12,216.7	2,438.1	283.1	170.4	7.9	784.7	283.6	0.00
Destination (regional accessibility to jobs)								
Transit jobs accessibility (45 min)	63,351.9	28,947.2	8,609.9	4,860.0	718.3	23,478.2	13,419.1	0.00
Regional centrality index	0.28	0.19	0.05	0.51	0.02	0.09	0.08	0.00

Note: LEND = Location Efficient Neighborhood Development; du = Dwelling Unit; ac = Acre; pop = Population; emp = employment; CT = Census Tract; psm = Per Square Mile.

areas, and undefined places. These four major groups could be further described as follows:

1. Urban: Neighborhoods that are dense, regionally accessible, and walkable with high land use mix and diversity;
2. Suburban: Places that have low density and some regional accessibility, poor street connectivity and design;
3. Rural areas: Places that have extremely low density and limited regional accessibility; and
4. Undefined places: Places that could not be easily categorized.

Urban, suburban, rural, and unclassified census tracts were refined using density variables—dwelling units, population, and employment—and compactness/sprawl index scores to further differentiate places types (*34*).

Recognizing the polycentric nature of cities and regions in California, this study also classifies large commercial area as employment centers, if they meet two basic thresholds: 1) the ratio of jobs to residents must be higher than 1, which indicates there are more jobs than bedrooms, and 2) the area must contain roughly 9,000 jobs/mi^2 or around 14 jobs/acre (*48*).

As this preliminary approach using thresholds was not able to automatically distinguish places such as college campuses, airports, large parks, military, ports and industrial areas, and traditional central business districts (CBDs), the final step required a hands-on overlay analysis to provide a visual assessment via mapping (ArcGIS, as shown in Figure 2 below) and Google Earth to confirm the validity of proposed place types and delineate additional categories. Although this step is subjective, it is an important check on whether the place-typology cluster method is working.

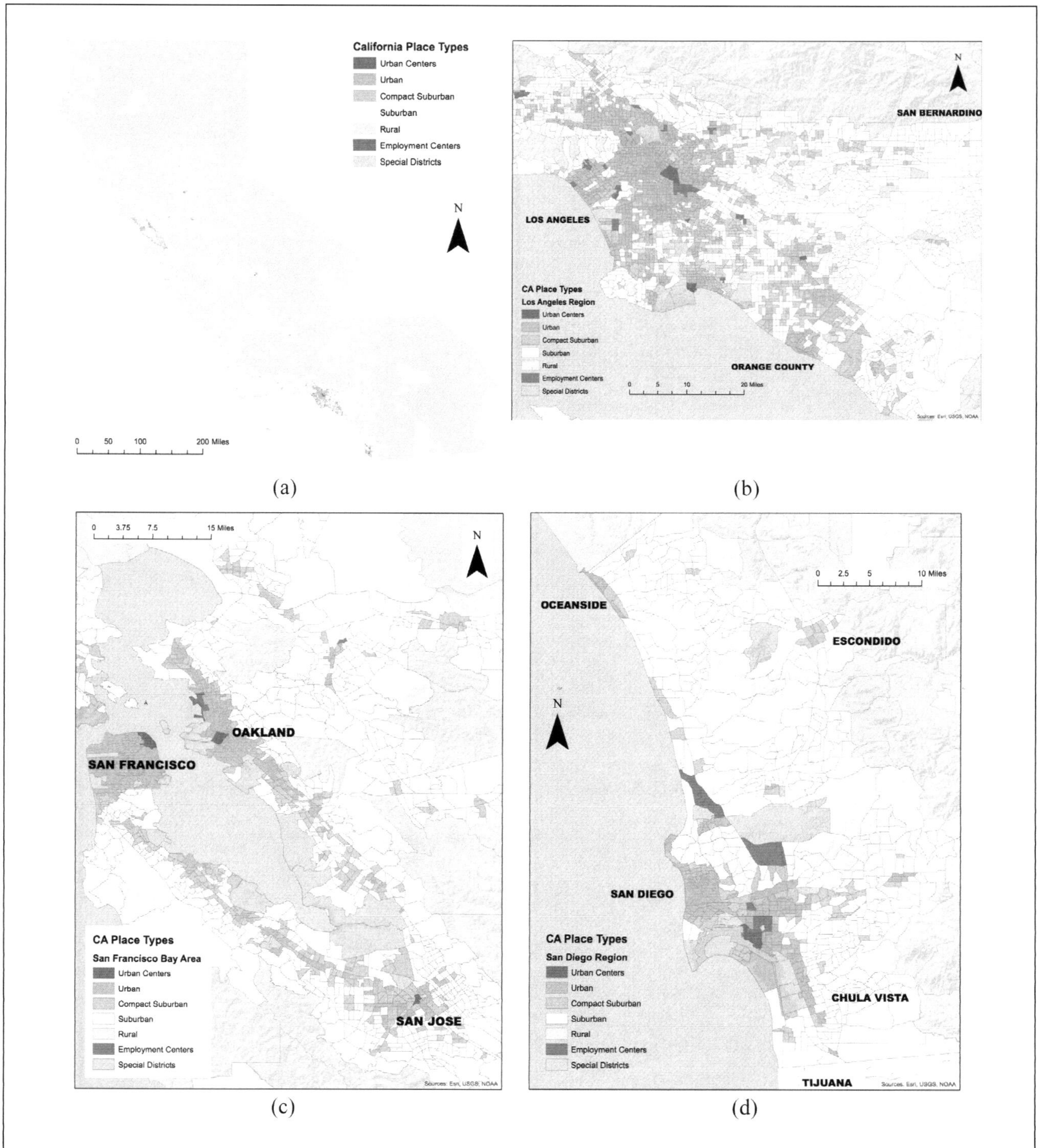

Figure 2. Statewide and regional map of place typology: (*a*) California, (*b*) Los Angeles Region, (*c*) San Francisco Bay Area, and (*d*) San Diego Region.

After this iterative process, seven major archetypes of California neighborhoods were identified as follows: urban centers, urban places, compact suburban places, suburban places, rural places, employment centers, and special districts.

Phase 2: Describing and Validating the Place Typology. One question that should be explored is whether or not these place types are meaningfully different, and statistically distinguishable from one another. This is another key step in the threshold analysis mentioned earlier in the beginning of this section. To address this, a one-way ANOVA difference of means test was used to compare the average scores of composite urban form indices and built environment measures, as shown in Table 2. This helps provide descriptions of the various place types through their various built environment measures, as well as provide an assessment confirming the place types are grounded in their underlying measures, and statistically distinct from each other. This provides a form of validation that the place typology is meaningfully descriptive, and able to provide nuanced information toward guiding policy. All results were statistically significant.

Phase 3: Exploring Relationships. To test the sustainability performance of the proposed neighborhood types, several analytic methods were used, including ANOVA, to explore the short-term and long-term trends. These metrics were organized into the three main sustainability elements of the economy, society, and the environment. The economic elements included indicators related to income, housing, and transportation. The social component included indicators related to demography, equity, and health outcomes. The environmental element illustrated the neighborhood energy and resource use, and the level of pollution exposure and effects.

Results

After an extensive, iterative process, seven major archetypes of California neighborhoods were identified as follows: urban centers, urban places, compact suburban places, suburban places, rural places, employment centers, and special districts. Figure 2 shows the statewide and regional maps of place typology. As expected, the majority of land in California is rural and protected, followed by suburban land use patterns.

Describing and Validating the Place Types

To help readers conceptualize the various neighborhood place-types, and to provide a further check on their relevance in reflecting real-world conditions, Table 2 provides the results from the one-way ANOVA difference of means test comparing the average scores of both the composite urban form measures, as well as the individual built environment measures. This is another key step in the threshold analysis mentioned earlier in the beginning of the Methodology section. This also helps provide a description of place types, illustrating the broad patterns and hierarchies in density, intensity, history, connectivity, and accessibility between different place-typology categories. This analysis helps verify that there are indeed clear and statistically significant differences between the categories of place typology, and providing further confirmation and grounding of the use of these underlying measures.

Urban Centers (UCs)

Urban centers (UCs) are the CBDs or downtowns of the 10 largest cities in California with the highest land use intensity. These neighborhoods are the commercial, civic, and cultural heart of the city containing a full range of horizontally and vertically mixed land uses and extensive pedestrian accessibility. UCs have over 70 jobs per acre and an overall housing density of around 22.8 du/ac with 93.7% multi-family housing units, and 83.5% renter occupants. About 42% of the building stock was built before 1959.

Urban Places (UPs)

Urban places are typically the cities' oldest residential neighborhoods with high regional accessibility, density (13.1 du/ac), and mixed land uses. These communities have a well-connected transit service and pedestrian supportive environment, typically located adjacent to UCs. The majority of housing is multi-family (66.9%), and renters (70.3%) dominate these neighborhoods with 52% of its buildings constructed before 1959. Urban places can also include downtowns of medium-sized cities and towns with a mix of land use, high residential and employment densities, coupled with good street connectivity and transit accessibility.

Compact Suburban Places

Compact suburban places have a hybrid characteristic of both urban and suburban spatial patterns. These neighborhoods are older and more compact than conventional suburbs and are typically located adjacent to urban or suburban places with a mean of 5.2 du/ac. The residential element is a mix of single-family (56.2%), and multi-family (40.7%) with diverse commercial centers. This designation can also include downtowns of small cities, towns, and villages, as well as newer master planned communities characterized by a stronger presence of key community design elements.

Suburban Places

Suburban places are the newest and most dominant neighborhood type in California. The majority of housing (70.2%) is single-family homes with more than three-quarters of buildings constructed after 1959, with a mean of 3.4 du/ac, and an average compact/sprawl index score of 97. Any communities with compact/sprawl index scores below 100 are considered to be sprawling with a dispersed land use that requires the use of an automobile for mobility. These communities have large lot sizes and a low integration of housing with jobs, retail, and services with poorly connected street networks that may be unappealing to walkers or bikers. There is a minimal level of transit service and a large amount of surface parking.

Rural Places (RPs)

Rural places (RPs) are settlements with widely spaced communities separated by farms, vineyards, orchard, and grazing lands. RPs must have less than 2 du/ac and score less than 74 on the compact/sprawl index, which indicates a very spread-out development pattern typical of rural California. This classification includes tourist and recreation destinations, which can significantly affect land use, character, and mobility needs. The majority of RPs is largely undeveloped and uninhabited by people and industries. Protected areas such as forests, lakes, deserts, mountains, wetlands, and habitats are prominent in this place type.

Employment Centers

Employment centers meet two basic thresholds laid out in the methodology section. The average employment density is 42.7 jobs per acre. These job centers are typically located outside the traditional CBDs or adjacent to UCs. There is a concentration of office, service, entertainment, and industrials sectors.

Special Districts

Special districts are places that specialize in a particular activity, with a unique land use pattern that does not conform to the above six typologies. Special districts can include college campuses, airports, prisons, large parks, marinas and ports, industrial zones, military, and other transportation facilities with a fourth highest employment density of 7.7 jobs per acre.

Discussion

Comparing the Sustainability Performance of Urban and Suburban Neighborhoods

Table 3 illustrates the community indicators for four California place typology categories—UCs, urban, compact suburban, and suburban—and highlights the differences between urban and suburban places, the two most common California neighborhood place-types.

Economy: Income, Housing, and Transportation. In 2015, an average urban household made 47.6% less in median income, compared with a suburban household. Median rent in urban places was 18.5% cheaper than suburbs, most likely due to a diversity of housing types. On the other hand, the median housing value in city neighborhood outside downtowns was 16.7% higher, and the overall cost of home ownership was 40.3% more expensive in urban places compared with suburbs, based on a median price to income ratio.

For households in urban places compared with suburbs, the average household transportation cost was 39.8% lower—with annual savings of more than US$2,675. The benefit of location efficiency is evident, with urban residents driving 6.1 fewer miles for their work commutes, but spending 1 min more in mean travel time, likely due to a higher level of congestion and a slower travel speed in city neighborhoods. Overall, the urban households had a 57.9% reduction in annual household VMT compared with suburban households, which is higher than findings from other California studies (*3, 13, 49*).

In 2014, urban commuters on average drove alone to work 14 percentage points less, and rode public transit 11-points more, whereas walking was 2.98-points more, and biking 1.23-points more compared with suburban commuters.

Social: Ethnicity, Education, and Health. For ethnic composition, urban places are, on average, more diverse with 71% of the population identifying as "not-White" compared with 55% "not-White" in the suburbs. On every level, the urban population contained larger shares of disadvantaged people in 2010 compared with suburban places. There were 45% more people over the age of 25 with less than a high school education, and 52% more people where no one aged 14 or over speaks English "very well" or speaks only English.

Differences in public health outcomes are also remarkably different between urban and suburban place-types. There was a more than 10.3% lower obesity rate, and 8.9% fewer incidents of cardiovascular disease for those who were living in urban places compared with suburban neighborhoods. On the contrary, the low birth weight rate was 14.9% higher, and asthma attack rates requiring a hospital visits were 7.52% higher in urban places compared with suburban communities.

Environment: Resource Use, Pollution Exposure and Effects. Table 4 compares the environmental indicators of UCs, urban, compact suburban, and suburban place-types in California. The most consistent difference between urban and suburban places is environmental performance. The household consumption of electricity and natural gas was 39% and 23% lower in the urban area, whereas per capita water use was 63% lower than

Table 3. Economic and Social Indicators of Urban and Suburban Place-Types in California

	Urban centers	Urban	Compact suburban	Suburban	Difference between urban and suburban	%	Sig.
Economy – income, housing and transportation							
Median income (2015)	US$44,027.46	US$52,879.59	US$63,373.12	US$73,853.41	−20,973.82	−47.64%	0.00
Median housing value (2015)	US$396,251.92	US$476,680.11	US$429,247.20	US$410,286.28	66,393.83	16.76%	0.00
Housing value to income ratio (2015)	9.78	9.36	6.63	5.42	3.94	40.29%	0.00
Median rent (2015)	US$1,139.21	US$1,269.54	US$1,387.20	US$1,480.38	−210.84	−18.51%	0.00
Annual household transportation cost (2010)	US$6,719.13	US$8,452.29	US$10,129.07	US$11,127.77	−2,675.48	−39.82%	0.00
Median commute distance: miles (2010)	9.29	8.42	12.08	14.60	−6.18	−66.52%	0.00
Mean travel time: minutes (2014)	24.57	28.49	25.87	27.49	1.00	4.07%	0.00
Annual household VMT (2010)	10,966.15	14,122.42	17,872.51	20,473.78	−6,351.36	−57.92%	0.00
JTW drove alone % (2014)	42.73	62.45	73.99	76.53	−14.08	−32.95%	0.00
JTW public transportation % (2014)	16.66	13.90	4.44	2.75	11.15	66.93%	0.00
JTW walked % (2014)	19.89	4.65	2.65	1.67	2.98	14.98%	0.00
JTW biked % (2014)	2.98	1.95	1.29	0.72	1.23	41.28%	0.00
Social – ethnicity, education, safety, and health							
Not-White % (2010)	64.07	71.04	64.71	55.07	15.97	24.93%	0.00
Education: % of population over 25 years of age with less than a high school education (2015)	20.71	26.05	21.75	16.53	9.52	45.97%	0.00
Linguistic Isolation: % households in which no one aged 14 or over speaks English "very well" or speaks only English (2015)	18.51	17.55	11.56	7.88	9.67	52.24%	0.00
Cardiovascular disease (age-adjusted rate per 10,000) (Ave 2011–2013)	7.13	7.79	8.31	8.43	−0.64	−8.98%	0.00
Asthma (age-adjusted rate of emergency room visit per 10,000) (Ave 2011–2013)	78.72	56.09	54.92	50.17	5.92	7.52%	0.00
Low birth weight (%) (Ave 2006–2012)	3.61	5.21	4.98	4.67	0.54	14.96%	0.00
Obesity: percent of population (aged 18 or over) with BMI >= 30 (2014)	18.54	24.07	25.18	25.99	−1.92	−10.36%	0.00

Note: VMT = Vehicle Miles Traveled; JTW = Journey to Work; Ave = Average; BMI = Body Mass Index.

in the suburbs. Finally, the household carbon footprint (HCF)—a calculation of consumption-based carbon emission—was 28% lower and per capita HCF was 15% lower in urban places compared with suburbs. This difference is primarily due to 37% higher transportation emissions in suburban areas.

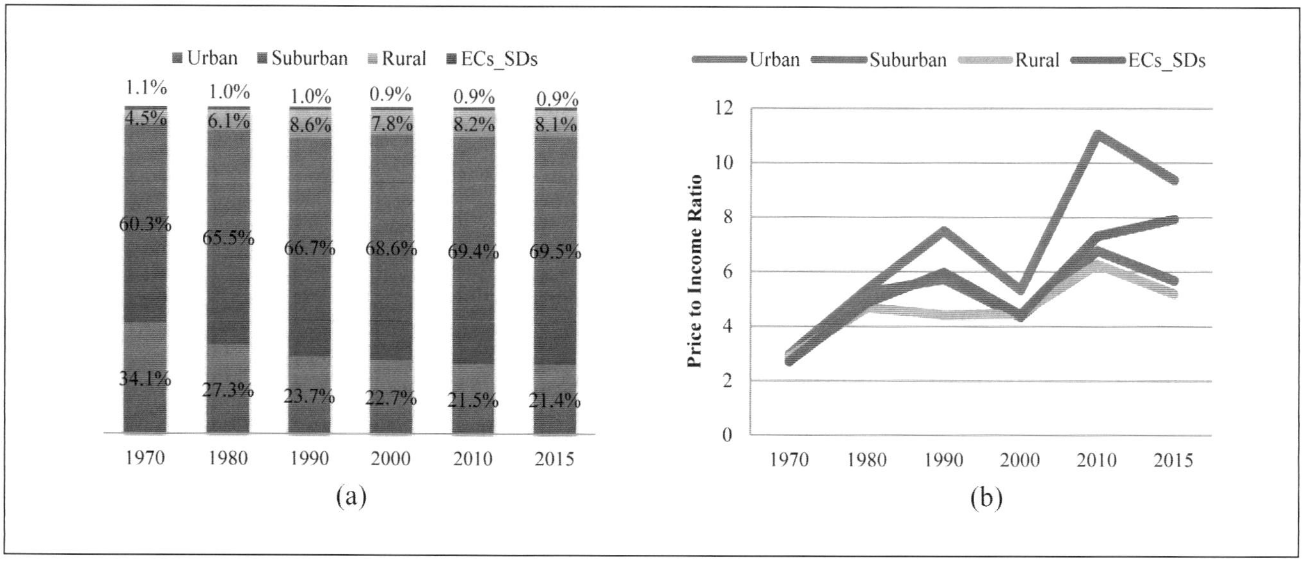

Figure 3. California long-term housing trends by place types 1970–2015: (*a*) California share of housing units by place types, and (*b*) ratio of median housing value to median income by place type.

There are many sources of air pollution in California from industrial facilities, power plants, and automobiles, and there are clear differences in the quality of urban environments by place types. For example, urban areas had 7.2% higher concentration of fine particulate matter (PM)2.5, 23% more exposure to diesel PM, 46% more concentration of toxic releases from facilities, and 43% higher traffic density compared with suburban places. On the other hand in the suburbs, ozone concentration was 15% higher, the risk of drinking water contamination was 20% greater, and potential exposure to pesticides was 80 times greater compared with urban places. Furthermore, the number of solid waste sites nearby suburban neighborhoods was 16% higher than urban places.

Finally, some brownfield site designations, such a cleanup sites (13%), leaking underground storage tanks (9%), hazardous waste generators facilities (6%), and impaired bodies of water (8%) were all higher in urban places compared with suburban places.

Total California Household Carbon Footprint by Place Types. Overall, this study found suburban households were responsible for 73% of consumption-based carbon emissions, followed by urban places with 19% and rural areas contributing 7% of total emissions. The smallest footprint, at 1%, came from employment centers and special districts. In all, 80% of all household carbon emission came from people living in suburban and rural communities in the state of California, which is higher than previous estimates (*50*), however, the majority of current climate policies and strategies are urban in nature.

Exploring Long-Term Trends in Housing Growth by Place Types

Declining Share of Housing in Most Location-Efficient Places. Through an examination of 1970–2015 housing data from the U.S. census (*47*), we found that statewide housing units per capita peaked in 1980 at 389 housing units per 1,000 residents, and declined to 360 housing units per 1,000 residents in 2015 or a 7.5% decrease in the overall availability of housing stock from the peak.

To simplify the urban vs. suburban comparison, the two urban place-types—urban centers and urban places—and two suburban neighborhood types—compact suburban and suburban places—were combined into *urban* and *suburban* categories, respectively. Figure 3 shows the shares of housing units in urban places, employment centers, and special districts all had a smaller share in 2015 compared with 1970, whereas suburban and rural places saw a sharp increase. In 1970, 34.1% of housing units were in urban locations, but by 2015 that share had dropped to 21.4%, whereas the share of suburban housing units increased from 60.3% in 1970 to 69.5% in 2015. Furthermore, the share of homes in rural places nearly doubled from 4.5% in 1970 to 8.1% in 2015.

Declining Household Size, Except in Urban Places. During the same period, on average, the people per unit of housing by place type declined for all neighborhoods, except for urban places in which the number of people per housing unit increased from 2.37 to 2.56 people, an increase of 8.2%—bucking the national and statewide demographic trend toward smaller household size.

Table 4. Environmental Indicators of Urban and Suburban Place-Types in California

	Urban centers	Urban	Compact suburban	Suburban	Difference between urban and suburban	%	Sig.
Household energy (2005), water (2014), and carbon footprint (2007)							
Electricity (kWh)	4,387.77	5,935.66	7,068.38	7,687.40	1,751.74	39.92%	0.00
Natural gas (ft^3)	29,640.81	33,841.29	38,141.25	40,789.07	6,947.78	23.44%	0.00
Daily per capita water use (gal)	65.54	94.70	120.99	136.57	41.87	63.88%	0.00
Household carbon footprint (MTCO2e)	28.89	39.31	45.92	47.68	8.37	28.97%	0.00
Carbon footprint per capita (MTCO2e)	13.73	14.25	15.88	16.37	2.12	15.44%	0.00
Transport carbon footprint per capita (MTCO2e)	3.84	4.57	5.68	6.00	1.43	37.24%	0.00
Neighborhood pollution exposure							
Ozone (daily max 8-h concentration 2012–2014 Ave)	0.0400	0.0430	0.0450	0.0490	0.01	15.00%	0.00
Particulate matter 2.5 (annual mean concentrations 2012–2014 Ave)	10.82	11.07	10.51	10.29	−0.78	−7.21%	0.00
Diesel particulate matter (on-road and non-road sources 2012)	78.56	33.32	20.24	14.77	−18.55	−23.61%	0.00
Drinking water contaminants index (2005–2013)	315.87	429.46	400.15	493.67	64.21	20.33%	0.00
Pesticide use (total pounds per agricultural use per mi^2 2012–2014)	2.72	5.52	134.51	443.61	438.09	16,106.25%	0.00
Toxic releases from facilities (toxicity-weighted concentrations of modeled chemical releases to air from facility emissions and off-site incineration 2011– 2013)	3,394.83	4,348.61	4,505.85	2,772.60	−1,576.01	−46.42%	0.00
Traffic density (vehicle-kilometers per hour per road length of the census tract boundary 2013)	887.61	1,256.28	1,025.30	868.38	−387.90	−43.70%	0.00
Neighborhood pollution effects							
Cleanup sites (sum of weighted cleanup sites within buffered distances to populated blocks of census tracts 2016)	30.75	13.07	13.27	8.83	−4.24	−13.79%	0.00
Groundwater threats (sum of weighted GeoTracker leaking underground storage tank sites within buffered distances to populated blocks of CTs 2016)	52.70	17.15	20.41	11.99	−5.16	−9.79%	0.00
Hazardous waste generators and facilities (sum of weighted hazardous waste facilities and large quantity generators within buffered distances to populated blocks of CTs 2012–2014)	2.69	0.53	0.43	0.36	−0.17	−6.32%	0.00
Impaired waters (sum of number of pollutants across all impaired water bodies within buffered distances to populated blocks of CTs 2012)	8.75	3.56	3.31	2.85	−0.71	−8.11%	0.00
Solid waste sites (sum of weighted solid waste sites and facilities within buffered distances to populated blocks of CTs 2016)	1.30	0.85	0.78	1.07	0.22	16.92%	0.00

Note: kWh = Kilowatt hour; gal = Gallon; MTCO2e = Metric Tons of Carbon Dioxide Equivalent; Ave = Average; CT = Census Tract

This analysis confirms findings from recent studies showing the lack of sufficient housing supply in urban neighborhoods and high-cost coastal cities in California (5). Despite the shift in government emphasis on smart growth and climate-friendly policies encouraging more infill development in California – especially concentrated around transit areas—the majority of housing developments continue to be in suburban and RPs.

Housing Price to Income Ratio Accelerating Faster in Urban Areas. The housing supply gap and higher demand for urban lifestyles are starting to influence the price of housing. In 1970, the housing value was about three times the median income for all neighborhood types in California, making home ownership accessible to a large segment of society, but by 2000, the ratio of median housing value to median income began to diverge between urban and suburban places.

By 2015 the median housing value was 9.4 times the median income in urban neighborhoods, whereas suburban areas were still relatively affordable with housing value at 5.7 times the median income. Overall, purchasing a home in urban neighborhoods today is 40% more expensive than suburban areas in California, which reflects the consequences of a supply gap and the restrictive government policies and regulations that discourage infill development (*3, 5, 6*).

Conclusion and Next Steps

This paper described the development of a place typology and sustainability performance measures for places throughout the state of California. The study approach had several advantages over other classification methods of urban and neighborhood forms. First, the study method applied both threshold and overlay analyses to distinguish areas such as traditional CBDs, college campuses, airports, large parks, military, ports, industrial, and transportation facilities. Second, this study compared more than 65 sustainability, livability, and equity outcome measures at smaller geographic scales. Third, the longitudinal analyses helped to examine 45 years of housing data to explore long-term population, income, and housing trends.

Key Findings

This study found that there were clear trade-offs between urban and suburban living. Compared with suburbs, the households in urban places benefited from a 57.9% reduction in annual VMT, 37.2% lower transport-related GHG emissions per capita, and saved more than US$2,675 in annual transportation expenses, while consuming less electricity (39.9%) and water per capita (63.8%). However, the cost of homeownership was 40% higher in urban places, despite rents being 18.5% cheaper. And whereas obesity and cardiovascular disease rates were 10.3% and 8.9% lower in urban places, there was a 7.5% higher rate of asthma incidents whereas, on average, urban areas experienced higher concentrations of PM2.5 (7.2%), diesel PM (23%), and toxic releases from facilities (46%) compared with suburban communities.

From 1970 to 2015, the total percentage of residential units in urban neighborhoods decreased from 34% to 21% and the statewide housing units per capita dropped 7.5% between 1980 and 2015. A shortage of new housing supply and an increase in demand caused the values of homes to appreciate faster in urban neighborhoods, with the majority of urban in-migration made up of white, affluent, and college educated (*38*). The high demand for an urban lifestyle has disproportionally affected a large population of minority and low-income residents who have been displaced out of their urban communities due to rising rent and housing prices from redevelopment pressures, especially along the new transit corridors.

Limitation of Study

The main limitation of the study design was the use of secondary data for several variables. Furthermore, though the median rental price data from the census provided a useful snapshot, they were based on a 5-year rolling average, did not provide the cost of renting different units, and the data tended to under-report the actual costs of rent (*51*). Finally, the statewide aggregation of neighborhood types and binary comparison can conceal local patterns and contexts that provide more nuanced and finer-grain findings.

Research Application and Conclusion

There are many potential applications for the study results. For example, a place typology could provide sufficient land use context to identify areas with the highest potential for a shared-mobility network, multi-modal transportation, and public transit investment. This is made even more significant given California's recent move to abolish the use of level of service congestion measures for traffic impact studies, moving instead toward valuing reductions in VMT.

For housing, the biggest opportunities for a new residential development are *UCs, employment centers* and *special districts*. Housing in these areas shifts the current jobs-housing imbalance to reduce household VMT while potentially encountering less community resistance, compared with other neighborhood types. Development policies must integrate the goal of economic equality as a core housing strategy or future growth will continue to exasperate the widening economic divide that leads to social instability.

In conclusion, despite the ambitious climate policies and efforts of smart growth advocates, the majority of new growth in California continues to be in low-density suburban and rural areas, which are responsible for about 80% of total household carbon emission in California. This research finding demonstrates the need for innovative strategies to reduce household carbon emissions from urban, suburban, and rural areas, reforming tax policies to encourage more infill development, a need for better community education, and aligning state and local expenditures with transparent sustainability outcomes to meet the state's 2030 climate goals and realize a more inclusive urbanism.

These findings contribute to policy makers', planners', and citizens' ability to understand, measure, and evaluate existing urban form (*23*). Having a clearer picture of economic, social, and environmental trade-offs in land use patterns is critical for informed discussions and intelligent decision-making to accelerate the future urban development and infrastructure investment in areas that provide the greatest benefits and least harm to people and the planet, equitably.

Author Contributions

The authors confirm contribution to the paper as follows: study conception and design: ARF, BA; data collection: ARF, BA; analysis and interpretation of results: ARF, BA, JG, SR; draft manuscript preparation: ARF, BA, JG, SR. All authors reviewed the results and approved the final version of the manuscript.

References

1. *Vibrant Communities and Landscapes: A Vision for California in 2050*. California Air Resource Board, 2016.
2. Barbour, E., and E. A. Deakin. Smart Growth Planning for Climate Protection: Evaluating California's Senate Bill 375. *Journal of the American Planning Association*, Vol. 78, No. 1, 2012, pp. 70–86.
3. Decker, N., K. C. Galante, A. Martin, E. Elkind, and M. Hanson. *Right Place, Right Type: Assessing the Environmental and Economic Impacts of Infill Residential Development through 2030*. UC Berkeley and Next 10, 2017.
4. Wheeler, S. Policy Innovation and Structural Hurdles. *Planning for Climate Change: Strategies for Mitigation and Adaptation for Spatial Planners*, Vol. 6, No. 3, 2009, p. 125.
5. Taylor, M. *California's High Housing Costs: Causes and Consequences*. Legislative Analyst's Office, Sacramento, Calif., 2015.
6. Woetzel, J., J. Mischke, S. Peloquin, and D. Weisfield. *A Tool Kit to Close California's Housing Gap: 3.5 Million Homes by 2025*. McKinsey Global Institute, 2016, pp. 307–329.
7. Ben-Joseph, E., and M. Southworth. *Streets and the Shaping of Towns and Cities*. Island Press, Washington, D.C., 2003.
8. Clifton, K., R. Ewing, G. J. Knaap, and Y. Song. Quantitative Analysis of Urban Form: A Multidisciplinary Review. *Journal of Urbanism: International Research on Placemaking and Urban Sustainability*, Vol. 1, No. 1, 2008, pp. 17–45. https://doi.org/10.1080/17549170801903496.
9. Song, Y., B. Popkin, and P. Gordon-Larsen. A National-Level Analysis of Neighborhood Form Metrics. *Landscape and Urban Planning*, Vol. 116, 2013, pp. 73–85. https://doi.org/10.1016/j.landurbplan.2013.04.002.
10. Wheeler, S. Built Landscapes of Metropolitan Regions: An International Typology. *Journal of the American Planning Association*, Vol. 81, No. 3, 2015, pp. 167–190. https://doi.org/10.1080/01944363.2015.1081567.
11. Appleyard, B., C. E. Ferrell, and M. Taecker. Transit Corridor Livability: Realizing the Potential of Transportation and Land Use Integration. Transportation Research Record: Journal of the Transportation Research Board, 2017. 2671: 20–30
12. Appleyard, B., C. Ferrell, M. Taecker, C. Armusewicz, and C. Schroeder. *Livability Calculator for the TCRP H-45 Handbook, Building Livable Transit Corridors: Methods, Metrics and Strategies*. Transportation Research Board of the National Academies, Washington, D.C., 2016.
13. Salon, D., M. G. Boarnet, and P. Mokhtarian. *Quantifying the Effect of Local Government Actions on VMT*. Institute of Transportation Studies, University of California, Davis, 2014.
14. Appleyard, B., C. Ferrell, M. Taecker, and A. Frost. Building Livable Transit Corridors through Transportation Land Use Integration: An Interactive Guide, Calculator and Exercise. Journal of Transport and Health, Vol. 5, p.S60
15. Lynch, K. *The Image of the City*. MIT Press, Cambridge, Mass., 1960.
16. Lynch, K. *Good City Form*. MIT Press, Cambridge, Mass., 1984.
17. Alexander, C., S. Ishikawa, M. Silverstein, M. Jacobson, I. Fiksdahl-King, and A. Shlomo. *A Pattern Language*. Oxford University Press, New York, 1977.
18. Jacobs, J. *The Death and Life of Great American Cities*. Random House, New York, 1961.
19. Jacobs, A., and D. Appleyard. Toward an Urban Design Manifesto. *Journal of the American Planning Association*, Vol. 53, No. 1, 1987, pp. 112–120.
20. Sung, H., S. Lee, and S. Cheon. Operationalizing Jane Jacobs's Urban Design Theory: Empirical Verification from the Great City of Seoul, Korea. *Journal of Planning Education and Research*, Vol. 35, No. 2, 2015, pp. 117–130. https://doi.org/10.1177/0739456X14568021.
21. De Nadai, M., J. Staiano, R. Larcher, N. Sebe, D. Quercia, and B. Lepri. The Death and Life of Great Italian Cities: A Mobile Phone Data Perspective. *Proc., 25th International Conference on World Wide Web*, Montréal, Québec, Canada, 2016, pp. 413–423.
22. Gil, J., J. N. Beirão, N. Montenegro, and J. P. Duarte. On the Discovery of Urban Typologies: Data Mining the Many Dimensions of Urban Form. *Urban Morphology*, Vol. 16, No. 1, 2012, p. 27.
23. Wheeler, S., and C. Beebe. The Rise of the Postmodern Metropolis: Spatial Evolution of the Sacramento

Metropolitan Region. *Journal of Urban Design*, Vol. 16, No. 3, 2011, pp. 307–332. https://doi.org/10.1080/13574809.2011.572253.
24. Ewing, R., and R. Cervero. Travel and the Built Environment: A Meta-Analysis. *Journal of the American Planning Association*, Vol. 76, No. 3, 2010, pp. 265–294. https://doi.org/10.1080/01944361003766766.
25. Stevens, M. R. Does Compact Development Make People Drive Less? *Journal of the American Planning Association*, Vol. 83, No. 1, 2017, pp. 7–18. https://doi.org/10.1080/01944363.2016.1240044.
26. Cervero, R., and K. Kockelman. Travel Demand and the 3Ds: Density, Diversity, and Design. *Transportation Research Part D: Transport and Environment*, Vol. 2, No. 3, 1997, pp. 199–219. https://doi.org/10.1016/S1361-9209(97)00009-6.
27. Knaap, G. J., U. Avin, and L. Fang. Driving and Compact Growth: A Careful Look in the Rearview Mirror. *Journal of the American Planning Association*, Vol. 83, No. 1, 2017, pp. 32–35. https://doi.org/10.1080/01944363.2017.1251276.
28. Ewing, R., S. Hamidi, and J. B. Grace. Urban Sprawl as a Risk Factor in Motor Vehicle Crashes. *Urban Studies*, Vol. 53, No. 2, 2016, pp. 247–266.
29. Forsyth, A. Defining Suburbs. *CPL Bibliography*, Vol. 27, No. 3, 2012, pp. 270–281. https://doi.org/10.1177/0885412212448101.
30. Lin, J., and L. Long. What Neighborhood Are You in? Empirical Findings of Relationships between Household Travel and Neighborhood Characteristics. *Transportation*, Vol. 35, No. 6, 2008, pp. 739–758. https://doi.org/10.1007/s11116-008-9167-7.
31. Ewing, R., S. Hamidi, G. Tian, D. Proffitt, S. Tonin, and L. Fregolent. Testing Newman and Kenworthy's Theory of Density and Automobile Dependence. *Journal of Planning Education and Research*, 2017, Vol. 38, No. 2, 2018, pp. 167–182. https://doi.org/10.1177/0739456X16688767.
32. Mikelbank, B. A.A Typology of U.S. Suburban Places. *Housing Policy Debate*, Vol. 15, No. 4, 2004, pp. 935–964. https://doi.org/10.1080/10511482.2004.9521527.
33. Duany, A., and E. Talen. Transect Planning. *Journal of the American Planning Association*, Vol. 68, No. 3, 2002, pp. 245–266. https://doi.org/10.1080/01944360208976271.
34. Ewing, R., and S. Hamidi. *Measuring Urban Sprawl and Validating Sprawl Measures*. National Institutes of Health and Smart Growth America, 2014.
35. Litman, T. *Land Use Impacts on Transport*. Victoria Transport Policy Institute, 2016.
36. Barrington-Leigh, C., and A. Millard-Ball. A Century of Sprawl in the United States. *Proceedings of the National Academy of Sciences*, Vol. 112, No. 27, 2015, pp. 8244–8249. https://doi.org/10.1073/pnas.1504033112.
37. Marshall, W., and N. Garrick. Effect of Street Network Design on Walking and Biking. *Transportation Research Record: Journal of the Transportation Research Board*, 2010. 2198: 103–115.
38. Juday, L. J. *The Changing Shape of American Cities*. Demographics Research Group, Weldon Cooper Center for Public Service, University of Virginia, 2015.
39. Calthorpe, P. *Urbanism in the Age of Climate Change*. Island Press, Washington, D.C., 2010.
40. Renne, J. L., S. Hamidi, and R. Ewing. Transit Commuting, the Network Accessibility Effect, and the Built Environment in Station Areas across the United States. *Research in Transportation Economics*, Vol. 60, 2016, pp. 35–43. https://doi.org/10.1016/j.retrec.2017.02.003.
41. Pushkarev, B., and J. M. Zupan. *Public Transportation and Land Use Policy*. Indiana University Press, 1977.
42. Guerra, E., and R. Cervero. Cost of a Ride. *Journal of the American Planning Association*, Vol. 77, No. 3, 2011, pp. 267–290. https://doi.org/10.1080/01944363.2011.589767.
43. Renne, J., and R. Ewing. *Transit-Oriented Development: An Examination of America's Transit Precincts in 2000 & 2010*. UNOTI Publications, 2013.
44. Song, Y., and G. J. Knaap. Quantitative Classification of Neighbourhoods: The Neighbourhoods of New Single-Family Homes in the Portland Metropolitan Area. *Journal of Urban Design*, Vol. 12, No. 1, 2007, pp. 1–24. https://doi.org/10.1080/13574800601072640.
45. Voulgaris, C. T., B. D. Taylor, E. Blumenberg, A. Brown, and K. Ralph. Synergistic Neighborhood Relationships with Travel Behavior: An Analysis of Travel in 30,000 US Neighborhoods. *Journal of Transport and Land Use*, Vol. 10, No. 1, 2017, pp. 437–461. https://doi.org/10.5198/jtlu.2016.840.
46. Ralph, K., C. T. Voulgaris, and A. Brown. Travel and the Built Environment. *Transportation Research Record: Journal of the Transportation Research Board*, 2017. 2653: 1–9.
47. Logan, J. R., Z. Xu, and B. Stults.Interpolating U.S. Decennial Census Tract Data from as Early as 1970 to 2010: A Longitudinal Tract Database. *The Professional Geographer: The Journal of the Association of American Geographers*, Vol. 66, No. 3, 2014, pp. 412–420. https://doi.org/10.1080/00330124.2014.905156.
48. Gardner, T. K., and M. Marlay. Whither the Jobs: Identifying Concentrations of Employment in Metropolitan Areas. *Urban Geography*, Vol. 34, No. 6, 2013, pp. 795–814. https://doi.org/10.1080/02723638.2013.790692.
49. Boarnet, M., and X. Wang. *Urban Spatial Structure and the Potential for Vehicle Miles Traveled Reduction*. National Center for Sustainable Transportation, 2016, (No. CA16-2832)
50. Jones, C. M., and D. M. Kammen. Spatial Distribution of U.S. Household Carbon Footprints Reveals Suburbanization Undermines Greenhouse Gas Benefits of Urban Population Density. *Environmental Science & Technology*, Vol. 48, No. 2, 2014, pp. 895–902. https://doi.org/10.1021/es4034364.
51. Boeing, G., and P. Waddell. New Insights into Rental Housing Markets across the United States: Web Scraping and Analyzing Craigslist Rental Listings. *Journal of Planning Education and Research*, Vol. 37, No. 4, 2016, pp. 457–476. https://doi.org/10.1177/0739456X16664789.

The Standing Committee on Transportation and Sustainability (ADD40) peer-reviewed this paper (18-06140).

Article

TRR
JOURNAL OF THE TRANSPORTATION RESEARCH BOARD

Transportation Research Record
2018, Vol. 2672(3) 145–154

DOI: 10.1177/0361198118798968
journals.sagepub.com/home/trr
SAGE

Modeling Cyclist Route Choice using Revealed Preference Data: An Age and Gender Perspective

Aditi Misra[1] and Kari Watkins[2]

Abstract

Perception of safety varies among cyclists depending on their cycling experience and frequency as well as on their socio-demographic attributes, particularly age and gender. Female cyclists are often more likely to feel unsafe than their male counterparts under similar traffic conditions or surrounding environment. In general, female cyclists are known to avoid slopes and high traffic, and prefer shorter trips than their male counterparts. However, there is no research that has yet explicitly modeled the differences in route choice between female and male cyclists using revealed preference data. In this research, we use GPS enabled smartphone based route data to develop segmented route choice models for different age groups and genders. The results indicate that segmented models are significantly better representatives of route choices made by female and male cyclists than pooled models. The findings suggest that traffic characteristics such as speed and annual average daily traffic (AADT) influence the decision to choose a particular route differently for female cyclists and male cyclists. Similar results are seen across different age groups as well. Preference towards routes with cycling facilities are however universal and no significant difference is noted among gender or age segments in this regard.

In recent times, cycling as an alternative mode of transport has been promoted both at the federal and local government levels. However, the adoption of cycling as a mode of commuting is considerably hampered by a perceived lack of safety that is often equated with a lack of infrastructure. Perceptions of safety and, therefore, a preference for infrastructure, intuitively, may differ among cyclists depending on how long they have been cycling and under what circumstances. A cyclist used to congested main roads may not necessarily view sharing the road with vehicular traffic as threatening but may find cycling on rural high-speed undivided roads unnerving.

In addition to such learned adaptations, attitude and confidence can also be assumed to contribute to an individual's perception of safety. Several studies have linked socio-demographic characteristics and attitudinal variability to perceptions of safety and to reasons for adopting cycling or not. For studies based on the cycling population in the U.S., gender proves to be a strong indicator of greater safety concerns or being less likely to adopt cycling as a primary transportation mode (*1–6*). Multiple studies suggest that the general risk-averse nature of females may deter them from cycling or make them feel less safe than male cyclists under similar conditions (*2, 4–6*). Low self-efficacy, coupled with household responsibilities such as grocery shopping, escorting children to their activities and so forth, are thought to limit women's propensity to cycle (*3*).

Most studies on cycling infrastructure preferences are stated–preference-based research where participants were asked to rank or indicate their preferences for infrastructure and cycling (*7*). One major issue with stated preference is that it represents ideal scenarios. For example, a survey by our research team had found that all cyclists prefer separate infrastructure, irrespective of their cycling confidence (*8*). It is, however, difficult to ascertain from such studies whether a female or less confident cyclist will not cycle at all in the absence of infrastructure as compared to male cyclists or their more confident counterparts who, like them, prefer having facilities, but, unlike them, will still cycle in the absence of infrastructure or take the shortest route irrespective of traffic conditions.

In general, data for explicit preference and decision-making criteria can only be obtained from route traces

[1]University of Michigan Transportation Research Institute, Ann Arbor, MI
[2]Georgia Institute of Technology, Atlanta, GA

Corresponding Author:
Address correspondence to Aditi Misra: aditimis@umich.edu

collected during a trip which can be highly data intensive and, therefore, expensive. Recently, with the advent of sensor technologies, route data collection efforts have become more common which has enabled more modelling of route choices using revealed-preference data. A few such cyclist route-choice models included gender as an explanatory variable and found that women were more likely to avoid slopes and heavy traffic and were also more likely to make short distance trips (*9*, *10*). However, most route choice models for cyclists are designed as pooled models for all cyclists within the sample, without specifically taking into account the possibility of significant differences in trip-making patterns among cyclists of different age, gender, or confidence levels.

This research makes a significant contribution in that area: (i) by including attitudinal variables such as cyclist type along with socio-demographic variables such as age and gender in the route choice model to understand how these variables influence route choice decisions; and (ii) by designing segmented models for different age groups, gender, and type of cyclist. Due to the inherent bias on male, younger, and more confident cyclists in the dataset, any model estimated on the entire dataset has a possibility to mask the preferences and choices of the lesser represented groups. Segmented models provide the opportunity to separate out groups that are homogeneous in their preferences and choices from the ones that are significantly different. This research therefore provides a better estimation of the route preferences of female cyclists, of cyclists from different age groups and confidence levels, which can help planners identify appropriate facilities that take such preferences into account.

Research Background

The primary data source for this research was the revealed preference trip data obtained from users of the GPS enabled smartphone application (app) Cycle Atlanta (*7*). The users of the app could voluntarily provide information on their socio-demographic status, their cycling history, cycling frequency and trip purpose, while the app, once switched on, collected the route of the cycling trips they made at one Global Positioning System (GPS) point per second. Once a trip was completed, upon receiving permission from the user, the GPS data were uploaded to a secure datacenter at Georgia Tech. Three separate databases were created from the data received via the Cycle Atlanta app: (i) users, which stored user-related information like socio-demographics, cyclist history and cycling frequency, uniquely tied to user ids; (ii) trips, which stored information on trip purpose, start and end GPS points, duration of the trip and any other input on the trips from the users, uniquely identified by trip id and user id; and (iii) routes, which stored the complete GPS information for every trip including the latitude and longitude of each point, a timestamp, accuracy measures, speed and altitude, uniquely tied to trip id. Between October 2012 and June 2014, Cycle Atlanta recorded about 20,000 trips by 1,495 users, 60% of whom provided socio-demographic and cycling related information.

Methodology

Two different sets of variables were used in these models—infrastructure related variables and socio-demographic variables. Table 1 shows the link attribute variables, their weighting factors and the route-level aggregation methods that were used for this analysis. The link attributes used for the analysis associated with the street network map were obtained from Atlanta Regional Commission and included traffic volume as measured by annual average daily traffic (AADT), posted speed limit, slope, presence of bicycle facilities, presence of sidewalk, and number of lanes. The percentage of truck traffic was initially considered but was found to be negligible along most corridors within the study area.

One of the issues in route choice modelling is a partial or full overlap of chosen and alternative routes. Any overlap in chosen and generated alternatives makes the paths correlated, which violates the assumption of logit models and produces biased estimates. An alternative solution is to use probit or generalized extreme value (GEV) models which allow for explicit consideration of correlation between alternatives. However, given the volume of road network data, it was decided that logit models would be more tractable and efficient than probit or GEV models, but because of possible partial (segment) overlap between chosen and alternative routes, an overlap correction was also needed. From the available model specifications with path size corrections, a path size correction logit model (PSCL) (*11*) was chosen for this analysis which required estimation of the path size correction term in addition to the other explanatory variables. The path size correction term penalized the utility of a route based on its proportion of overlap with other routes, thereby making it a less attractive option to choose.

The principal idea behind using weighted averages of variables was to understand the effect of each attribute as a function of the link length. Except for the path overlap correction factor (*11*), all attributes were weighted by the link length and then averaged at route level by the route length. For the path overlap correction factor, an indicator variable δ_{al} is defined as $\delta_{al} = 1$, *if the link a is traversed by path l*, 0 *otherwise*. The indicator variable was summed over all the alternatives in a choice set and then weighted by the ratio of the overlapping link length

Table 1. Road Attribute Variable Calculation (Adopted from Misra and Watkins [8])

Attributes for link a of path k	Base value	Link level weighted value	Route level aggregated value
Link length	l_a	l_a	$\sum_{a\in\Gamma_k} l_a$
Annual average daily traffic (AADT)	$(aadt)_a$	$(aadt)_a \times \frac{l_a}{L_k}$	$\sum_{a\in\Gamma_k} (aadt)_a \times \frac{l_a}{L_k}$
Speed limit	s_a	$s_a \times \frac{l_a}{L_k}$	$\sum_{a\in\Gamma_k} s_a \times \frac{l_a}{L_k}$
Number of lanes	n_a (*ranked categorical*)	$n_a \times \frac{l_a}{L_k}$	$\sum_{a\in\Gamma_k} n_a \times \frac{l_a}{L_k}$
Slope	$m_a = m_j - m_i$ (m_i = *elevation at start node*, m_j = *elevation at end node*)	$m_a \times \frac{l_a}{L_k}$	$\sum_{a\in\Gamma_k} m_a \times \frac{l_a}{L_k}$
Percent of trucks	p_a	$p_a \times \frac{l_a}{L_k}$	$\sum_{a\in\Gamma_k} p_a \times \frac{l_a}{L_k}$
Presence of bicycle facilities	δ_b *Binary*(0, 1)	$\delta_b \times l_a$	$\frac{1}{L_k} \sum_{a\Gamma_k} \delta_b \times l_a$
Presence of sidewalk	δ_{sw} *Binary*(0, 1)	$\delta_{sw} \times l_a$	$\frac{1}{L_k} \sum_{a\Gamma_k} \delta_b \times l_a$
Traffic stress	$ts_a = (aadt)_a \times s_a$	$ts_a \times \frac{l_a}{L_k}$	$\sum_{a\in\Gamma_k} ts_a \times \frac{l_a}{L_k}$
Path overlap correction (Bovy et al. 2008)	PSC_k	$\frac{l_a}{L_k} \ln \sum_{l\in C} \delta_{al}$	$\sum_{a\in\Gamma_k} \frac{l_a}{L_k} \ln \sum_{l\in C} \delta_{al}$

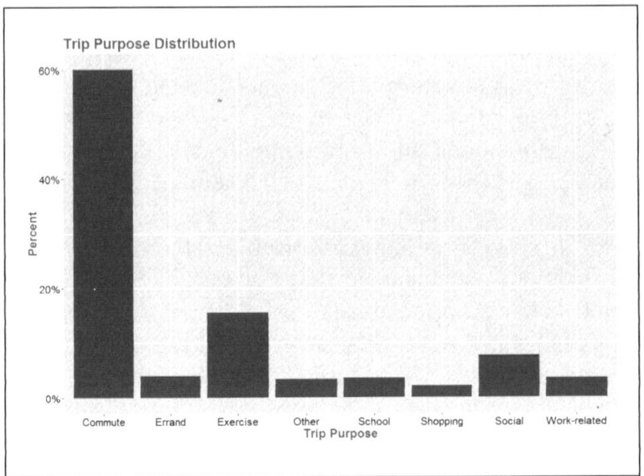

Figure 1. Trip purpose distribution.

to the length of the path. The weighted variable was then aggregated at route level. A detailed definition of the road attribute variables, except for path overlap correction, is provided in Table 1.

Multivariate Analysis

To reduce the size of the study area, trips were clipped to be within a five-mile radius of an intersection in midtown Atlanta, reducing the number of trips to 15,463. Of these 15,463 trips, 9,537 were commute trips (62%), 577 were school trips (4%), 2,461 were exercise trips (16%) and the remainder were errands, shopping, and 'other' trips (18%) (Figure 1). The trip data were then linked to the socio-demographic data via the user id key. Of the 1,495 users, 932 had complete information on age, gender, and cyclist type which were the three socio-demographic variables to be considered for further analysis. Therefore, only trips made by those 932 users were retained in the dataset which reduced the number of trips from 15,463 to 8,323 and the number of commute trips from 10,114 to 5,222. Figure 2 shows the trip purpose across age – cyclists in the age group >45 years use cycling for exercise more than any other group. Figure 3 shows trip purpose by gender, and since the data are heavily dominated by male cyclists, they are the dominating group in all trip purpose categories. However, female cyclists have an almost identical share of shopping trips in spite of being a small fraction of the cyclists. This calls for particular consideration in land use planning to allow women to do trip chaining comfortably and easily.

Figures 4 and 5 show trip frequency distribution across trip length by age and gender respectively considering commute trips. It should be noted though that trips for younger cyclists show their highest frequency at a shorter distance than that of senior cyclists, which is initially counterintuitive. However, one of the reasons may be that senior cyclists are less likely to choose shorter routes if that does not provide sufficient safety and comfort, while younger cyclists may prefer shorter distances to taking a longer detour for a bike facility. The younger cyclists are also dominated by college students, and their commutes may be much shorter in length. Figures 6 and

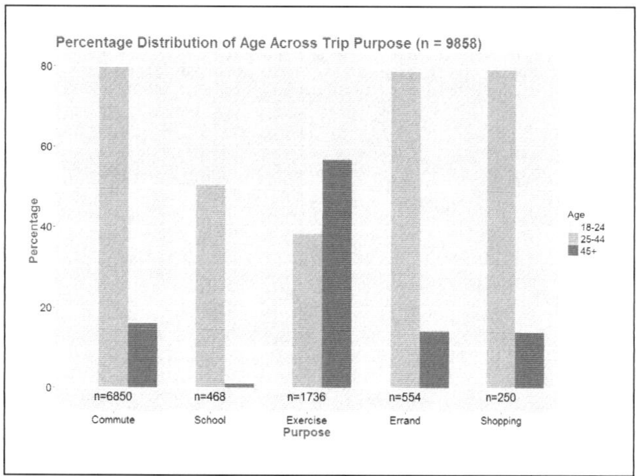

Figure 2. Percent distribution of age groups across trip purpose.

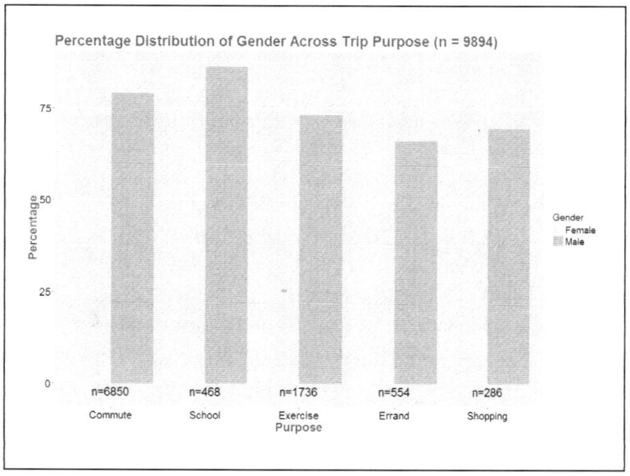

Figure 3. Percentage distribution by gender across trip purpose.

7 show age and gender distribution respectively across trip lengths. The average trip length for female cyclists is shorter than their male counterparts and older cyclists in general make shorter trips.

Although some users recorded multiple trips within the study time period, one trip per user was used for the route choice models resulting in a total of 445 trips to be analyzed. Of these, the choice set generation method failed to generate alternatives for eight trips that were above 13 miles in length, further reducing the dataset to 437 trips. Although the attempt was to have at least 10 alternatives for each chosen route, the number of alternatives that were generated varied between 10 and six, including the chosen trip. Therefore, to maintain a set of the same number of alternatives for all trips, each chosen trip was designed to have five alternatives. For cases where more than five alternatives were generated, only the first five were retained as the choice set generation method used k-shortest path method with utility as the link cost, implying that the further in the generation process a route is generated, the further it may be assumed to be from the optimum choice. Among the sociodemographic variables, 27% of the trips were from female cyclists while the remaining 72% were from male cyclists which is similar to the distribution in the original dataset. Similarly, about 70% of the trips were from cyclists in the age group younger than 45 years while the remaining 30% were from the older population.

Route Choice Models

Given the biased nature of the dataset and previous research suggesting that there may be significant differences in preferences between male and female cyclist as well as between younger and older cyclists, segmented models were run first on the groups and then a pooled model was run. The method followed was: (1) run separate path size logit models on subsets of data which consist only of male cyclists or female cyclists/younger or older cyclists and find the best specifications for those models; (2) run pooled models on the entire data set and find the best specifications for that model; and (3) for variables that are significant in one of the segmented models but not in the pooled model, or if the coefficient estimates are significantly different (tested by t-tests), add them as an interaction term in the final pooled model. For example, if AADT was found to be significant in the female-only model but not in the male-only model or the pooled model then, in the next stage, a pooled model was created with an interaction term female × AADT. If that term was found to be significant in the pooled model only then it was concluded that the variable was significant for that group. Variables were added to the model sequentially and removed stepwise from the model when found to be insignificant. The final segmented models for male and female cyclists and the pooled model are presented in Table 2 while the segmented models for the age groups and their corresponding pooled model results are presented in Table 3. The t- statistic for difference in coefficients between segments was calculated as:

$$t = \frac{\hat{\beta}_1 - \hat{\beta}_2}{\sqrt{var(\hat{\beta}_1) + var(\hat{\beta}_2)}}$$

To test whether the models were significantly different, a $\chi 2$ test was performed as:

$$-2[LL(pooled\ model) - LL(segment\ 1) - LL(segment\ 2)] \sim \chi^2_{df},$$

where the degrees of freedom (df) are equal to $K_1 + K_2 - K$ where K is the number of coefficients in the pooled model and K_s is the number of coefficients in the sth market segment model, s = 1, 2 (*12*).

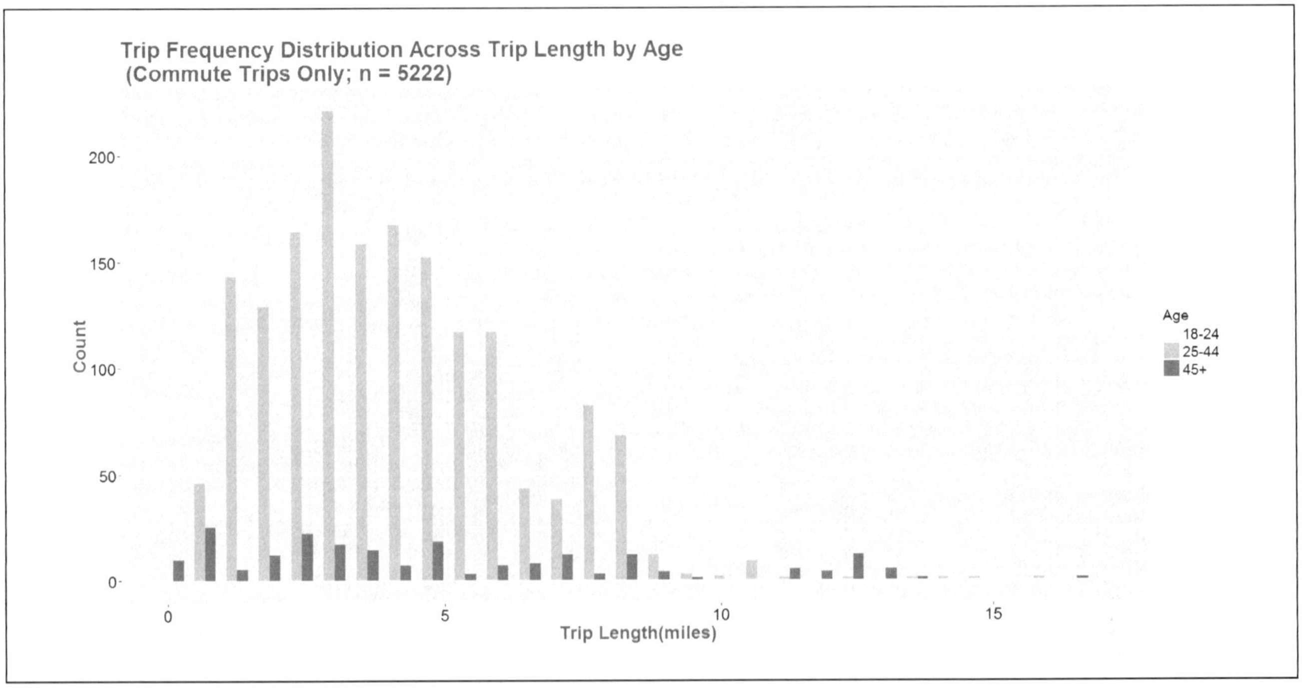

Figure 4. Trip frequency across trip length by age groups for commute trips.

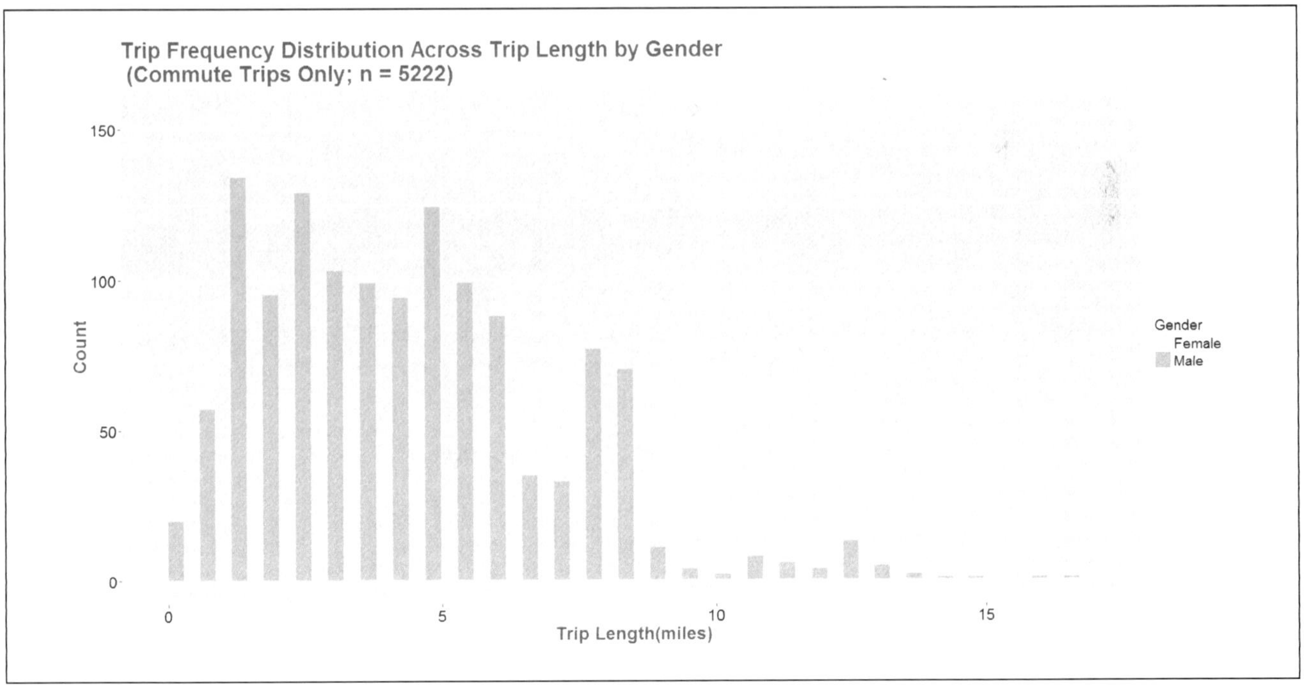

Figure 5. Trip frequency across trip length by gender for commute trips.

Model Results

The model estimated only on female cyclists had a sample size of 119 while that with the male cyclists had 318. The pooled model was estimated on the total sample size of 437. Along with AADT, speed, slope, bicycle facilities, number of lanes, trip distance, and path overlap correction factor, interaction terms of negative road attributes (AADT, speed, slope) with bicycle facilities were experimented with but not found to be significant. AADT was found to be significant across all models except the male-

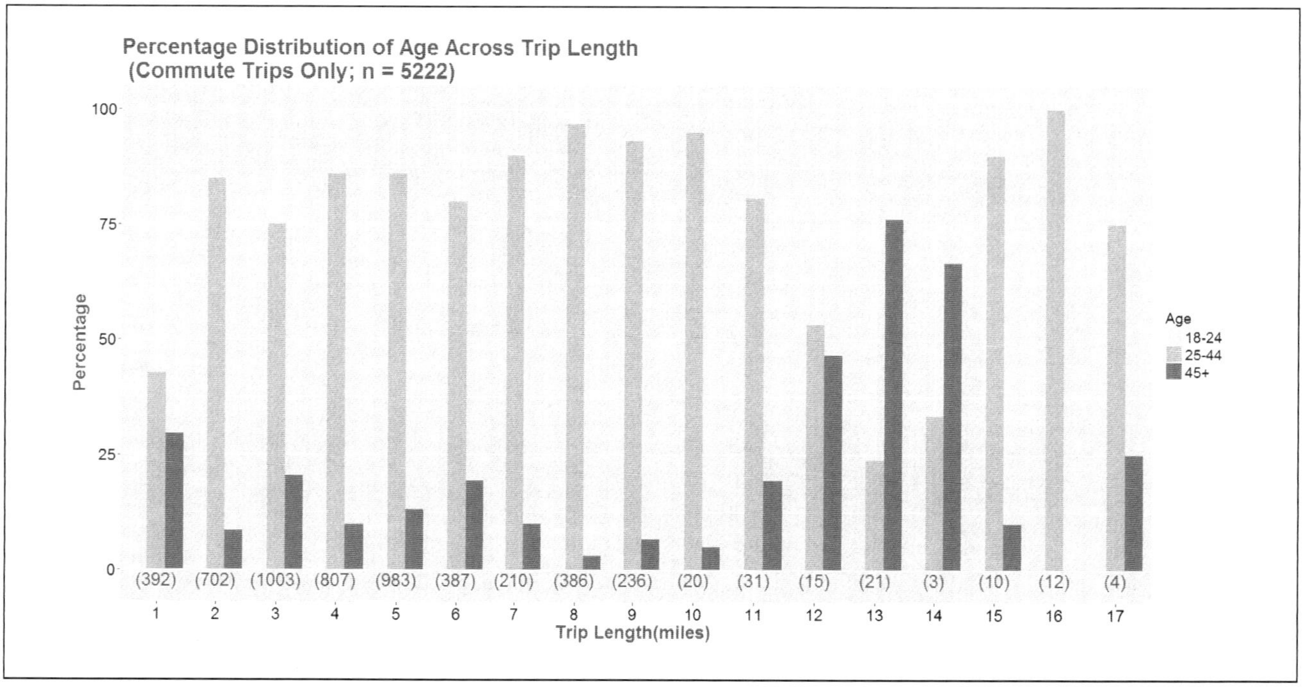

Figure 6. Percentage distribution of age groups across trip length for commute trips.

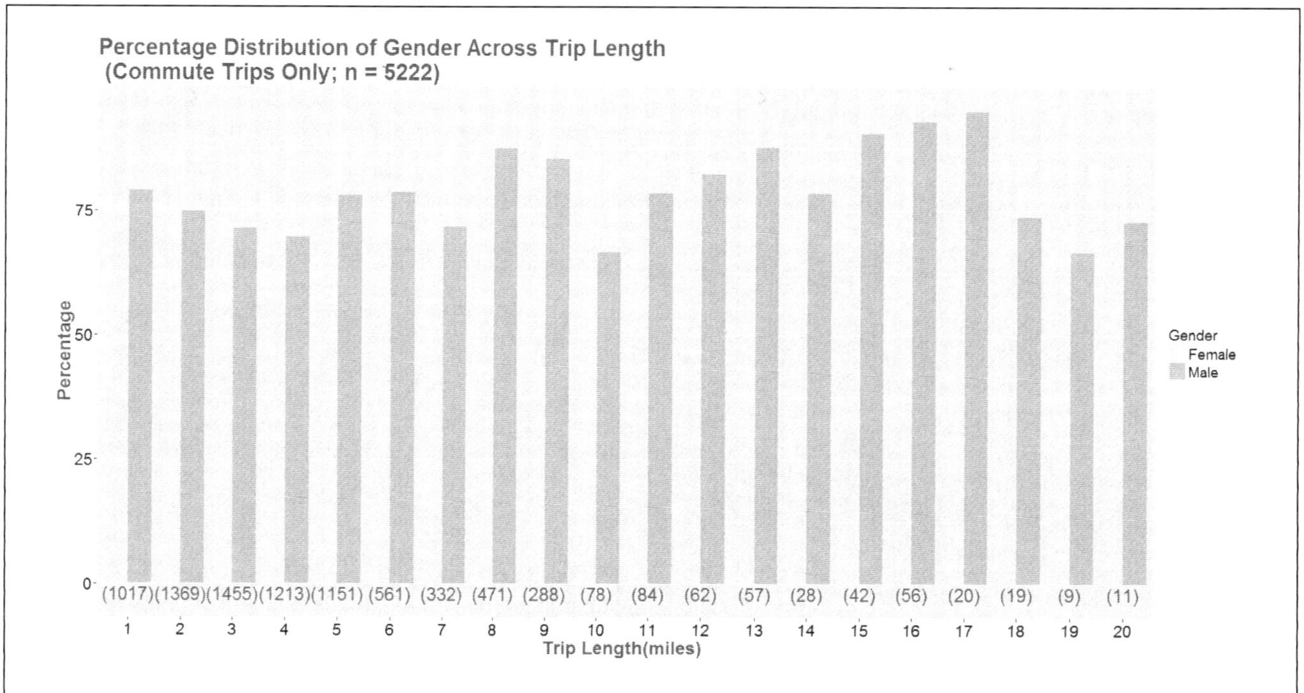

Figure 7. Percentage distribution by gender across trip length for commute trips.

only model and the initial pooled model for gender. Speed, however, was significant across all models as were presence of bicycle facilities, number of lanes and the path size overlap correction factor. Slope was significant in the segmented model for the higher age group and retained its significance in the corresponding pooled model. For gender segmented models, age was included as an explanatory variable, and for age segmented

Table 2. Gender-Segmented and Pooled Path Size Logit Route Choice Models

	Model 1		Model 2		Model 3	Model 4	
Variables	Female only (N = 119)	Odds ratio	Male only (N = 318)	Odds ratio	Initial pooled (N = 437)	Final pooled (N = 437)	Odds ratio
ASC_alternative 2	−0.864*(−1.91)	0.421	−0.934**(−2.89)	0.393	−1.1**(−2.66)	−1.78***(−3.4)	0.169
ASC_alternative 3	−0.968*(−2.35)	0.380	−1.695*(−2.06)	0.184	−1.468†(−1.83)	−1.38**(−2.8)	0.252
ASC_alternative 4	−1.249‡(−1.46)	0.287	−1.668***(−4.15)	0.189	−2.52***(−2.94)	−2.12*(−1.95)	0.120
ASC_alternative 5	−1.545***(−3.19)	0.213	−1.485***(−3.28)	0.227	−2.89***(−3.17)	−3.13***(−2.89)	0.044
ASC_chosen	−3.291***(−3.23)	0.037	−1.36*(−2.56)	0.257	−2.24*(−2.462)	−2.76***(−3.22)	0.063
Distance	−0.72***(−4.37)	0.487	−0.4**(−2.68)	0.670	−0.56**(−2.79)	−0.62***(−3.18)	0.538
AADT	−0.8***(3.42)	0.449					
Speed	−0.5***(−4.02)	0.607	−0.14*(−2.33)	0.869	−0.3*(−2.57)	−0.187* (2.12)	
Percent cycle facilities	1.942***(8.267)	6.973	1.342***(6.27)	3.827	1.116***(5.32)	1.262***(5.72)	3.532
Average number of lanes	−0.401***(−5.21)	0.670	−0.47***(−3.205)	0.625	−0.42*(−2.19)	−0.4*(−2.47)	0.670
Slope	−0.41**(−2.76)	0.664	−0.127†(−1.93)	0.881	−0.29*(−2.21)	−0.32**(−2.85)	0.726
AADT × female						−0.357*(−2.52)	0.811
Speed × female						−0.28*(−2.12)	
Log PS	1.275***(3.27)		2.176***(3.45)		1.973‡(2.83)	1.99**(2.8)	
Model statistics							
LL(0)	−361.22		−967.92		−1328.34	−1328.34	
LL(MS)	−321.7		−859.67		−1181.37	−1181.37	
LL(Model)	−191.55		−436.624		−668.2234	−654.341	
McFadden's ρ2(base MS)	0.4		0.49		0.434	0.45	

***$p < 0.001$. **$p < 0.01$. *$p < 0.05$. $^{\dagger}p < 0.1$. $^{\ddagger}p < 1$.

Table 3. Age-Segmented and Pooled Path Size Logit Route Choice Models

	Model 1		Model 2		Model 3	Model 4	
Variables	Age⩾45 only (N = 131)	Odds ratio	Age <45 only (N = 306)	Odds ratio	Initial pooled (N = 437)	Final pooled (N = 437)	Odds ratio
ASC_alternative 2	0.45†(1.86)	1.569	−1.699‡(−0.961)	0.183	−1.23‡(−1.32)	−1.752*(2.28)	0.173
ASC_alternative 3	−3.693*** (−7.36)	0.025	−0.886**(−2.544)	0.412	−1.526‡(−0.724)	−2.154†(−1.27)	0.116
ASC_alternative 4	−3.654***(−7.301)	0.026	−1.147*(−2.02)	0.318	−1.47***(−2.89)	−1.873*(−2.732)	0.154
ASC_alternative 5	−0.460‡(−0.91)	0.631	−1.353*(−2.08)	0.258	−1.85 **(−2.52)	−1.766*(−2.94)	0.171
ASC_chosen	−1.633***(−2.544)	0.195	−1.765*(−2.25)	0.171	−2.4723*(−2.296)	−2.162*(−2.147)	0.115
Distance	−0.396***(−2.724)	0.673	−0.247*(−1.89)	0.781	−0.278*(−1.986)	−0.275*(−2.654)	0.760
AADT	−0.851***(−6.11)	0.427	−0.291**(−2.89)	0.748	−0.553** (−2.543)	−0.317**(2.37)	
Speed	−0.474**(−2.32)	0.622	−0.164†(1.72)	0.849	−0.175†(−1.69)	−0.18†(1.92)	
Percent bicycle facilities	1.837***(2.763)	6.283	2.061*(2.12)	7.858	2.002**(2.76)	2.098***(2.392)	8.150
Average number of lanes	−0.106**(2.06)	0.899	−0.165**(2.35)	0.848	−0.139*(−2.23)	−0.16‡(−1.74)	0.852
Slope	−0.894***(−5.746)	0.409					
AADT × age ⩾ 45						−0.432**(−2.984)	
Speed × age ⩾ 45						−0.127**(−3.27)	
Slope × age ⩾ 45						−0.881***(−4.317)	
Log PS	1.456***(4.532)		1.222***(3.762)		1.692***(3.85)	1.26***(4.14)	
Model Statistics							
LL(0)	−411.240		−1078.500		−1489.740	−1489.740	
LL(MS)	−343.2		−872.35		−1215.53	−1215.53	
LL(Model)	−223.19		−547.24		−863.223	−801.61	
McFadden's ρ2(base MS)	0.35		0.37		0.29	0.34	

***$p < 0.001$. **$p < 0.01$. *$p < 0.05$. †$p < 0.1$. ‡$p < 1$.

models, gender was included as an explanatory variable but was not found to be significant in either case and was therefore not retained in the final model specification.

Between the two gender segmented models, although speed was significant in both, the coefficients were significantly different ($p = 0.002$) between the two groups. Therefore, an interaction term of speed and female was added in the final pooled model. As mentioned before, AADT was significant for female cyclists but not for the male cyclists, so an interaction term between AADT and female was added to the final pooled model. Bicycle facilities, number of lanes, and the path size correction factor were significant in all three models and were therefore retained in the final pooled model. The model fits as given by McFadden's ρ^2 (base market share model) range between 0.4 and 0.49 for all the models, which is considered relatively good for disaggregate discrete choice models of travel behavior.

The odds ratio calculated shows that a unit (1,000 vehicles) increase in AADT can decrease the odds of choosing a route by 55% for females. Similarly, a unit increase (10 mph increase) in speed can decrease the odds of choosing a route by 40% for females and by about 20% for male users. A unit increase in the percent presence of bicycle facilities (for example, an increase in percent presence of bicycle facilities from 0.7 to 0.8) can increase the odds of choosing that route by almost seven times for female cyclists and by about four times for male cyclists. An increase in path overlap also positively affects the possibility of choosing a route which points to the fact that users generally prefer closely related paths. The significance of speed in the pooled model indicates that female cyclists are significantly more negatively affected by speed compared with their male counterparts, all others being constant. The $\chi 2$ test of significance for model fit is $-2[LL(\textit{final pooled model}) - LL(\textit{male}) - LL(\textit{female})] \sim 54.32_{11}$ which has a corresponding p -value < 0.005. Therefore, the null hypothesis that the pooled model is equivalent to the segmented models can be rejected at a significance level of 0.005. We therefore choose the gender segmented models as better representation of the sample.

For the age segmented models, except for slope, all other variables were significant in all the models. Slope is only significant for cyclists in the older age group. The coefficients of both AADT and speed are significantly different across the two age groups and the significance is retained in the pooled model via the interaction terms. Thus, the final pooled model for age consists of variables AADT, speed, AADT × age group dummy, speed × age group dummy, slope × age group dummy, bicycle facilities, average number of lanes and path size correction term. The ρ^2 (base MS) ranges between 0.32 and 0.39. From the odds ratios, 1,000 vehicles increase in AADT decreases the odds of choosing a route by about 60% for an older cyclist while 10 mph increase in speed limit decreases the odds of choosing the route by 40%. Both AADT and speed affect older cyclists significantly more than younger cyclists for whom the corresponding changes are 25% and 15%. Slope also significantly deters older cyclists from choosing a route and by about 60% for a unit increase (1 Ft) in slope.

To test the hypothesis that the pooled model is the same as the segmented models, the $\chi 2$ test was performed as $-2[LL(\textit{final pooled model}) - LL(\textit{Age}\geq 45) - LL(\textit{Age}<45)] \sim 31.84_{10}$ which is significant at 0.005 level indicating that the hypothesis is rejected with 99.99% confidence and the segmented model describes the sample better than the pooled model.

Discussion

From the segmented models developed and discussed in the earlier section, it is evident that choice of route varies significantly among different types of cyclists – both by age and by gender. Coefficients for variables that can negatively affect route choice (e.g., AADT, speed, slope), are consistently higher for women and older cyclists, whether significant or not. However, since the majority of cyclists in US are male and young, any data on cyclists are likely to be weighted in favor of such cyclists and their preferences. Segmented models help to separate out the preferences and choices of less represented groups.

From a policy point of view, multiple level-of-service measures and types of infrastructure are designed based on the classification of cyclists as experienced or new. Most infrastructure created for new cyclists tends to be off-road and away from the main arterials and therefore involves longer detours for any trip. Experienced cyclists are expected to share the road comfortably with vehicular traffic with minimum or no protection. However, as this research shows, older cyclists and female cyclists are deterred severely by traffic conditions and are also not willing to take longer detours. Any infrastructure created far away from the normal commute or travel road will always fail to attract such cyclists and will remain underused. Therefore, to encourage cyclists, agencies need to plan for infrastructure that is close to the shortest routes and yet provides sufficient buffer between the cyclists and the vehicular traffic.

It should also be noted that although we performed segmentation based on cyclist type, which is a surrogate for cycling experience and confidence, there was no distinct difference between the preferences among different groups. This may be due to the biased nature of the dataset which is heavily dominated by male and younger cyclists. A classification by cyclist type does not preclude the possibility of oversampling those groups particularly because we are also oversampling the more confident cyclists. Therefore, it may not be unexpected that preferences across cyclist types do not vary significantly and are dominated by the preferences of male and younger cyclists. In addition, as mentioned earlier, since this is a revealed preference dataset from a city with few choices for avid cyclists, we are possibly capturing a forced choice rather than a preference which may not differ significantly across the more confident cyclist types (*7*) who, although safety conscious, still identifies themselves as cyclists and would not consider mode shift easily.

Finally, the use of crowdsourced smartphone based data comes with at least two caveats – the sample is generally biased toward participants who are interested and motivated to participate and by participants who have access to smartphones. Misra et al. (*7*) presents a discussion of the data issues and how it may affect the analysis, although the results presented in this paper generally agree with those available in the literature.

Conclusion

Traditionally route choice models have considered only route attributes to be a determinant of a cyclist's decision to choose a route. However, the segmented route choice models developed in this paper indicate distinctly different preferences and effects of road attributes across age and gender. This result is not always evident in cyclist datasets as such data are overwhelmingly dominated by motivated cyclists who, in the US, are male and young in 70% of the cases. Therefore, developing different models for different socio-demographic groups may be necessary to understand the actual effect of any infrastructure related decision.

One caveat of this research is that segmentation was performed deterministically, by separating the data into two or more premeditated groups. Often, this is fraught with preconceived ideas and philosophies on the part of the analyst. A preferred way of approaching segmentation is to let the segmentation happen probabilistically through finite mixture latent class models where class membership is not rigid or predetermined. Future research will conduct the same route choice analysis with latent class analysis and compare the segmentation with the deterministic segmentation presented in this research.

Acknowledgements

This study was funded by the Georgia Department of Transportation and STRIDE. The authors also gratefully acknowledge the advice and suggestions from Dr. Patricia Mokhtarian in the course of this research.

Author Contributions

The authors confirm contribution to the paper as follows: study conception and design: KW; data collection: KW, AM; analysis and interpretation of results: AM; draft manuscript preparation: KW, AM. All authors reviewed the results and approved the final version of the manuscript.

References

1. Krizek, K. J., P. J. Johnson, and N. Tilahun. Gender Differences in Bicycling Behavior and Facility Preferences. *Research on Women's Issues in Transportation*, Vol. 2, 2005, pp. 31–40.
2. Garrard, J., S. Handy, and J. Dill. Women and Cycling. In *City Cycling* (John Pucher, and Ralph Buehler, eds.), MIT Press, Cambridge, MA, 2012, pp. 211–234.
3. Emond, C., W. Tang, and S. Handy. Explaining Gender Difference in Bicycling Behavior. *Transportation Research Record: Journal of the Transportation Research Board*, 2009. 2125: 16–25.
4. Akar, G., N. Fischer, and M. Namgung. Bicycling Choice and Gender Case Study: The Ohio State University. *International Journal of Sustainable Transportation*, Vol. 7, No. 5, 2013, pp. 347–365.
5. Tilahun, N. Y., D. M. Levinson, and K. J. Krizek. Trails, Lanes, or Traffic: Valuing Bicycle Facilities with an Adaptive Stated Preference Survey. *Transportation Research Part A: Policy and Practice*, Vol. 41, No. 4, 2007, pp. 287–301.
6. Segadilha, A. B. P., and S. P. Sanches. Analysis of Bicycle Commuter Routes using GPSs and GIS. *Procedia - Social and Behavioral Sciences*, Vol. 162, 2014, pp. 198–207.
7. Misra, A., K. Watkins, and C. A. Le Dantec. Socio-Demographic Influence on Rider Type Self Classification with Respect to Bicycling. Presented at 94th Annual Meeting of the Transportation Research Board, Washington, D.C., 2015.
8. Misra, A., and K. Watkins. Modelling Cyclists Willingness to Deviate from the Shortest Path using Revealed Preference Data. Presented at 96th Annual Meeting of the Transportation Research Board, Washington, D.C., 2016.
9. Dill, J., and N. McNeil. Four Types of Cyclists? Examining the Typology to Better Understand Bicycling Behavior and Potential. Presented at 92nd Annual Meeting of the Transportation Research Board, Washington, D.C., 2012.
10. Hood, J., E. Sall, and B. Charlton. A GPS-Based Bicycle Route Choice Model for San Francisco, California. *Transportation Letters*, Vol. 3, No. 1, 2013, pp. 63–75.
11. Bovy, P., S. Bekhor, and C. Prato. The Factor of Revisited Path Size: Alternative Derivation. *Transportation Research Record: Journal of the Transportation Research Board*, 2008. 2076: 132–140.
12. Koppelman, F. S., and C. Bhat. *A Self Instructing Course in Mode Choice Modeling: Multinomial and Nested Logit Models*. U.S. Department of Transportation, Federal Transit Administration, Washington, D.C., 2006. http://www.caee.utexas.edu/prof/bhat/courses/lm_draft_060131final-060630.pdf.

The Transportation Policy Section (ABE00) peer-reviewed this paper (18-06265).

Article

TRR
JOURNAL OF THE TRANSPORTATION RESEARCH BOARD

Transportation Research Record
2018, Vol. 2672(3) 155–165

DOI: 10.1177/0361198118793242
journals.sagepub.com/home/trr
SAGE

The Role of Culture and Needs in the Cycling Habits of Female Immigrants from a Driving-Oriented to a Cycling-Oriented Country

Sigal Kaplan[1], Dagmara K. Wrzesinska[2], and Carlo G. Prato[3]

Abstract

The enormous potential of cycling to make cities and regions more sustainable is often not realized. Given the evidence that women cycle less than men, this study looks into the cycling habits of female immigrants from driving-oriented to cycling-oriented countries in relation to the fulfillment of their needs and their immersion in their new culture. A survey measured existence, relatedness, and growth needs with Likert-type items and collected socioeconomic characteristics and travel habits of 570 female Polish immigrants who moved to established and developing cycling cultures. A hybrid bivariate ordered model captured the propensity toward habitual cycling in the adopted country and cycling intentions upon return to the home country in relation to the needs and the traits of the female immigrants. Model results suggest that the cycling habits of female immigrants are related to past travel habits, while future intentions of cycling are connected to tangible (e.g., distance) and emotional (e.g., income stigma) barriers. Relevantly, model results indicate a positive relation between cycling culture and cycling habits in terms of both culture strength and exposure, but also a negative relation between relatedness to the home country and future intentions of cycling. Lastly, model results show that cycling habits relate to the fulfillment of existence needs and even more of relatedness and growth needs. Accordingly, the key to promote cycling among female immigrants is not only to satisfy their functional needs, but also cater to their emotional and self-actualization needs.

Cycling has an enormous potential to contribute significantly to the sustainability of cities and regions worldwide, as shifting from motorized transport to cycling would result in less congestion and significantly less emissions (*1*, *2*). For individuals, cycling positively influences maximal oxygen consumption (*3*), significantly reduces BMI-related risk (*4*), dramatically improves the metabolic and cardiovascular system (*5*), and prolongs life expectancy (*6*).

Although its benefits are obvious, cycling does not generally realize its potential. Moreover, considerable literature provides evidence that women cycle less than men, in particular, women from immigrant groups who might experience accessibility problems and transport-related social exclusion (*7*, *8*). Disadvantages for these groups might relate to lower education and income levels, but certainly also to social norms from their home country (*9*). Cycling might help reduce these disadvantages because it is an easy and affordable mode of transport which provides opportunities for women, who tend to travel shorter distances and to perform trip chaining because of their roles inside and outside the household (*10*, *11*). However, these opportunities might be missed because of the lack of dedicated safe infrastructure, concerns about personal safety, limitations in bicycle storage, the unsuitability of bicycles and accessories for women with children, the effect of life stages on the desire to cycle, thoughts about appearance after cycling, and the ability to ride a bicycle (*11–14*).

Given these premises, this study focuses on female immigrants living in cycling-oriented countries in order

[1]Department of Geography, Hebrew University of Jerusalem, Jerusalem, Israel
[2]Institute for Multidisciplinary Research in Quantitative Modelling and Analysis, Catholic University of Louvain, Louvain-la-Neuve, Belgium
[3]School of Civil Engineering, The University of Queensland, Brisbane, Australia

Corresponding Author:
Address correspondence to Sigal Kaplan: sigal.kaplan@mail.huji.ac.il

to understand how their past affects their current cycling habits, and how their exposure to a cycling culture might affect their future cycling habits. This study looks at travel habits as a part of culture and citizenship that influences place attachment in cycling-oriented countries, and considers that high bicycle traffic and its omnipresence might be overwhelming at first glance, but also how identifying oneself with an essential element of the daily life routine relates to the adaptation to new conditions that female immigrants face (*15*, *16*). In particular, this study explores the hypothesis that, for female immigrants, cycling is strongly associated with a sense of attachment to place in an immigration country with strong social and cultural habits of cycling (*15*), and with a sense of identification with the adopted country (*16*). In turn, place attachment and identification are related to the satisfaction of existence, relatedness, and growth needs (*17*). Notably, this study contributes to the literature by providing evidence about whether female immigrants cycle to satisfy their needs and adapt to the new conditions (*18*), and answering a need for research on cycling and social identity (*16*, *19*) as well as cycling and cultural change (*20*, *21*).

The focus of this study is on Polish women who immigrate to cycling-oriented countries, whether the cycling culture is at an established (the Netherlands and Denmark) or a developing (Germany, Sweden, United Kingdom) stage. Poland is a driving-oriented country with a high popularity of cars and a negative stigma of bicycles, although recent statistics show that Poles purchase up to one million new bicycles every year and over 60% of households own at least one bicycle (*22*). Certainly the current conditions of bicycle infrastructure in Poland leave a lot to be desired, but there exists a social acquiescence to dedicating public funds to cycling infrastructure even if it means less money for car infrastructure. This acquiescence exists in spite of past construction of cycling infrastructure that has not brought benefits because of inappropriate design and poor quality, which might lead to the perception of the funding as a waste of public resources (*22*). Overall, the Polish government has ambitious plans for an emerging cycling culture although at the moment Poland is definitely an established driving culture.

For this study, a survey focused on the cycling habits of female immigrants in their adopted countries in relation to the realization of their needs as well as their provenance from a driving-oriented country. According to the Alderfer's model (*17*), three types of needs were considered: (i) existence needs at the most basic level that concern the fulfillment of basic needs related to mobility, money, and safety; (ii) relatedness needs at the intermediate level that comprise both social and self-esteem components; (iii) growth needs at the higher level that involve both self-esteem and self-actualization components. These latent concepts were measured via 5-point Likert items, while observed characteristics of the female immigrants were retrieved in terms of socioeconomic characteristics and travel habits in both home and adopted country. A multivariate joint model of habitual cycling in the adopted country and intention to cycle when returning to the home country allowed all the information to be used to provide answers about whether they are both related to an adaptation process driven by needs. As the survey collected information about observed and unobserved characteristics of the female immigrants, a hybrid bivariate ordered model was used to accommodate the latent variables within the joint models of current cycling habits and future cycling intentions.

The remainder of the paper is structured as follows. In the next section, the rationale of the survey design is explained and the survey administration is described. Then, the model specification is presented and results of the model estimation are introduced. Last, conclusions are discussed and further research is suggested.

Methods

Survey Design and Administration

This study investigated the habit of cycling in the adopted country, and the intention to cycle when returning to the home country, of female immigrants from a driving-oriented country. In line with existing literature, travel behavior was assumed to be more a habitual than a conscious and controlled process (*23*) and to be an essential element of the culture to the extent that the transition to a different culture might translate into a different behavior (*19*). Habitual behavior should relate to the lower-order existence needs, while the transition to different behaviors should relate to the higher-order relatedness and growth needs (*17*).

The survey design consisted of four parts. The first part measured the habitual travel behavior of the respondents in the home country and the adopted country on a 5-point Likert scale from "rarely" to "daily" (*23*, *24*). The second part presented a hypothetical scenario of returning to the home country and measured the intention of using the bicycle there on the same 5-point Likert scale. The third part collected background information on the socioeconomic characteristics of the respondents, the commuting distance, the type of cycling (i.e., recreational, transport, both), the amount of cycling (if at all), and their place of residence and the duration of living in the adopted country. The fourth part adapted items about the three types of needs (*17*) to the study context focusing on cycling in the home and the adopted country.

Existence needs were measured as the perception of time and cost savings, the opportunity for increased mobility and travel independence, the enlargement of activity participation and opportunity space, and the establishment of a routine that includes exercise and focuses on health. Moreover, existence needs were assessed also as coping with difficulties related to perceived safety, adapting to the road rules, dealing with the weather conditions, lacking cycling experience and skills, and improving physical ability. Relatedness needs were measured as social inclusion, place attachment, social network formation, peer acceptance, and conformation to social norms in the adopted country as well as influence of social norms from the origin country. Personal growth needs were appraised as levels of self-esteem, self-confidence, cultural identity, self-identity as a cyclist, and self-actualization on environmental sustainability and fitness condition. Items were presented as 5-point Likert-type statements from "strongly disagree" to "strongly agree." The respondents were also asked about attitudes toward cycling in Poland to provide perspective on their perception of the cycling situation and the cycling policies in Poland.

The survey was administered online in both established and developing cycling cultures, as there exist national differences in the meaning of cycling (*25*). Accordingly, female Polish immigrants were contacted via a snowballing approach via social and professional networks in the Netherlands, Denmark, Sweden, Germany, and the United Kingdom. The survey was open between February and March 2016 and was accompanied by a raffle to incentivize participation and, via the collection of personal information for the eventual prize collection, increase answer trustworthiness. Notably, the survey was initially designed in English but then launched in Polish to increase interpretation homogeneity.

Hybrid Bivariate Model

In this study, the dependent variables were the habit of cycling in the adopted country and the intention to cycle in the origin country upon return. Both variables were expressed on a 5-point Likert scale and accordingly two ordered probit models were specified.

The unit of analysis for both dependent variables was the female immigrant, and hence a bivariate ordered probit model was specified to accommodate the unobserved correlation that might exist between the error terms that consequently followed a bivariate normal distribution. A multivariate probit approach has been used previously in the travel behavior context (e.g., *26, 27*).

Moreover, the model specification included both observable variables (e.g., travel habits in the country of origin, travel habits in the country of adoption, socioeconomic characteristics) and unobservable variables (e.g., existence, relatedness, growth, attitudes). Accordingly, a latent variable model was estimated jointly with the bivariate model to construct a hybrid bivariate ordered model of current habit and future intentions of cycling. The latent variable model consisted of measurement equations that associated the indicators from the survey items to the unobservable variables, and structural equations that linked the unobservable variables to observed characteristics of the female immigrants.

Initially, exploratory factor analysis was performed in order to elicit the latent variables, and measurement equations of the latent variable model were specified accordingly:

$$\boldsymbol{I}_n = \boldsymbol{\alpha} X_n^* + \boldsymbol{v}_n \tag{1}$$

where $\boldsymbol{I}_n$ is a vector of R indicators corresponding to the indicators of survey items for female immigrant n, $\boldsymbol{X}_n^*$ is a vector of L latent variables for female immigrant n, $\boldsymbol{\alpha}$ is a vector of R parameters to be estimated, and $\boldsymbol{v}_n$ is a vector of normally distributed error terms with a mean vector of zeros and covariance matrix $\boldsymbol{\Psi}_v$.

Then, the structural equations of the latent variable model were expressed as:

$$X_n^* = \boldsymbol{\gamma} X_n + \boldsymbol{\omega}_n \tag{2}$$

where $\boldsymbol{X}_n^*$ is a vector of L latent variables for female immigrant n, $\boldsymbol{X}_n$ is a vector of K observed characteristics of female immigrant n, $\boldsymbol{\gamma}$ is a matrix of $(L \times K)$ parameters to be estimated, and $\boldsymbol{\omega}_n$ is a vector of L normally distributed error terms with a mean vector of zeros and covariance matrix $\boldsymbol{\Psi}_\omega$.

The bivariate ordered probit model consisted of two equations, given the total number P and Q of categories for each dependent variable and having defined the indices p and q for the categories of the dependent variables:

$$\begin{aligned} &f_n^* = \boldsymbol{\beta}_f^* X_n^* + \boldsymbol{\beta}_f X_n + \varepsilon_n \textit{ where } f_n = p \textit{ if } \delta_{p-1} < f_n^* < \boldsymbol{\delta}_p, \\ &\boldsymbol{\delta}_0 = -\infty, \boldsymbol{\delta}_P = +\infty \\ &g_n^* = \boldsymbol{\beta}_g^* X_n^* + \boldsymbol{\beta}_g X_n + \eta_n \textit{ where } g_n = q \textit{ if } \chi_{q-1} < g_n^* < \chi_q, \\ &\chi_0 = -\infty, \chi_Q = +\infty \end{aligned} \tag{3}$$

The latent variables f_n^* and g_n^* represent respectively the propensity of a female immigrant n to have a given cycling habit in the adopted country and a level of intention of cycling in the home country. The variables f_n and g_n represent the survey responses about current habitual cycling and future cycling intentions. The parameters to be estimated are the elements of the vectors $\boldsymbol{\beta}_f^*$, $\boldsymbol{\beta}_f$, $\boldsymbol{\beta}_g^*$, and $\boldsymbol{\beta}_g$, and the thresholds δ_p and χ_q.

The error terms ε_n and η_n were assumed to have marginal normal distributions and to be independently and identically distributed (i.i.d.) across female immigrants. For identification purposes, the normal distribution of the error terms was standardized and a joint bivariate standard normal distribution of the error terms ε_n and η_n was assumed to capture their potential correlation. The elements of the vectors $\boldsymbol{\beta}_f^*$, $\boldsymbol{\beta}_f$, $\boldsymbol{\beta}_g^*$, and $\boldsymbol{\beta}_g$, the $(P\text{-}1)$ thresholds δ_p $(-\infty < \delta_1 < \ldots < \delta_p < \ldots < \delta_{P-1} < +\infty)$, and the $(Q\text{-}1)$ thresholds χ_q $(-\infty < \chi_1 < \ldots < \chi_q < \ldots < \chi_{Q-1} < +\infty)$ were estimated by maximizing the log-likelihood function including a binary indicator $I_n(p,q)$ equal to 1 if female immigrant n chose category p for the habit of cycling in the adopted country and category q for the intention of cycling in the home country upon return, and 0 otherwise.

The joint probability of female immigrant n choosing categories p and q was expressed as:

$$\begin{aligned} \boldsymbol{Pr}(\boldsymbol{f}_n = \boldsymbol{p}, \boldsymbol{g}_n = \boldsymbol{q}) &= \boldsymbol{Pr}\left(\boldsymbol{\delta}_{p-1} < \boldsymbol{f}_n^* < \boldsymbol{\delta}_p \ \boldsymbol{and}\ \boldsymbol{\chi}_{q-1} < \boldsymbol{g}_n^* < \boldsymbol{\chi}_q\right) \\ &= \boldsymbol{Pr}\left(\boldsymbol{d}_{n,p-1} < \varepsilon_n < \boldsymbol{d}_{n,p} \ \boldsymbol{and}\ \boldsymbol{c}_{n,q-1} < \eta_n < \boldsymbol{c}_{n,q}\right) \\ &= \boldsymbol{\Phi}_2(\boldsymbol{d}_{n,p}, \boldsymbol{c}_{n,q}; \theta) - \boldsymbol{\Phi}_2(\boldsymbol{d}_{n,p-1}, \boldsymbol{c}_{n,q}; \boldsymbol{\theta}) - \boldsymbol{\Phi}_2(\boldsymbol{d}_{n,p}, \boldsymbol{c}_{n,q-1}; \boldsymbol{\theta}) \\ &+ \boldsymbol{\Phi}_2(\boldsymbol{d}_{n,p-1}, \boldsymbol{c}_{n,q-1}; \boldsymbol{\theta}) \end{aligned} \tag{4}$$

where $d_{n,p} = \delta_p - \boldsymbol{\beta}_f^* \boldsymbol{X}_n^* + \boldsymbol{\beta}_f \boldsymbol{X}_n$, $c_{n,q} = \chi_q - \boldsymbol{\beta}_g^* \boldsymbol{X}_n^* + \boldsymbol{\beta}_g \boldsymbol{X}_n$, Φ_2 is the bivariate cumulative normal distribution function, and θ is the correlation parameter between the two error terms to be estimated. Given that the variables $\boldsymbol{X}_n^*$ in the model specification were latent, the joint probability required integration over their distribution (*28*, *29*):

$$\begin{aligned} &\boldsymbol{Pr}(\boldsymbol{f}_n = \boldsymbol{p}, \boldsymbol{g}_n = \boldsymbol{q}) = \\ &\int_{X_n^*} \boldsymbol{\Phi}_2\left(\boldsymbol{f}_n = \boldsymbol{p}, \boldsymbol{g}_n = \boldsymbol{q} \middle| \boldsymbol{X}_n, \boldsymbol{X}_n^*; \boldsymbol{\beta}_f^*, \boldsymbol{\beta}_f, \boldsymbol{\beta}_g^*, \boldsymbol{\beta}_g, \boldsymbol{\delta}_p, \boldsymbol{\chi}_q, \boldsymbol{\theta}\right) \\ &\boldsymbol{h}_{I_n}(\boldsymbol{I}_n | \boldsymbol{X}_n, \boldsymbol{X}_n^*; \boldsymbol{\alpha}, \boldsymbol{\Psi}_v) \boldsymbol{h}_{X_n^*}(\boldsymbol{X}_n^* | \boldsymbol{X}_n; \boldsymbol{\gamma}, \boldsymbol{\Psi}_\omega) \boldsymbol{dX}_n^* \end{aligned} \tag{5}$$

where $\Phi_2(f_n = p, g_n = q | \boldsymbol{X}_n, \boldsymbol{X}_n^*; \boldsymbol{\beta}_f^*, \boldsymbol{\beta}_f, \boldsymbol{\beta}_g^*, \boldsymbol{\beta}_g, \delta_p, \chi_q, \theta)$ is the bivariate probability in Equation 4, $h_I(\boldsymbol{I}_n | \boldsymbol{X}_n, \boldsymbol{X}_n^*; \boldsymbol{\alpha}, \boldsymbol{\Psi}_v)$ is the distribution of the indicators, and $h_{Xn^*}(\boldsymbol{X}_n^* | \boldsymbol{X}_n, \boldsymbol{X}_n^*; \boldsymbol{\gamma}, \boldsymbol{\Psi}_\omega)$ is the distribution of the latent variables. The log-likelihood function then took the following form:

$$\boldsymbol{logL} = \sum_{n=1}^{N} \sum_{p=1}^{P} \sum_{q=1}^{Q} \boldsymbol{I}_n(\boldsymbol{p}, \boldsymbol{q}) \boldsymbol{Pr}(\boldsymbol{f}_n = \boldsymbol{p}, \boldsymbol{g}_n = \boldsymbol{q}) \tag{6}$$

This function was coded in PythonBiogeme 2.5 (*30*) and its maximization required simulation via 1000 draws from the modified Latin hypercube sampling algorithm (*31*).

Results

Sample Characteristics

The survey collected 570 completed questionnaires from female Polish immigrants in the five countries following its advertisement via professional social networks. Table 1 presents a summary of the sample characteristics. As anticipated, the sample is relatively young and well educated although the income is mostly average if not lower than average as it is known that education is not always recognized in the adopted country and hence immigrants' jobs are not commensurate to their education level. The sample is mainly from Denmark, Germany and the Netherlands, and exhibits the desired variation between established and developing cycling cultures (*25*). Notably, almost two thirds of the sample had immigrated in the last five years.

The transport habits show on aggregate that driving and walking did not vary much from Poland to the adopted country, while a substitution effect is observed between public transport and cycling as the latter is more used than the former in the adopted country. Also, the cycling intentions show on aggregate a higher number of habitual cyclists, although more than half of the sample does not intend to cycle other than "rarely." It should be noted, however, that this is an aggregate measure and there are plenty of states that could provide this measure even with a lot of changes in travel habits at the individual level.

Estimates of the Hybrid Bivariate Ordered Model

The estimation of the hybrid bivariate ordered model implied an extensive specification effort in order to obtain the best model for representing jointly the current cycling habits and the future cycling intentions of the female Polish immigrants.

Initially, an exploratory factor analysis elicited the latent variables for the specification of the measurement equations in the latent variable model. Latent variables relating to the current habit of cycling were analyzed separately from those relating to the future intentions of cycling. The items revealed overall an excellent internal reliability, as the Cronbach's Alpha was 0.886, and an excellent sampling adequacy, as the Kaiser-Meyer-Olkin measure was 0.911. The exploratory analysis revealed the existence of four factors related to the current habit of cycling: (i) existence, (ii) relatedness, (iii) growth, and (iv) safety. The exploratory analysis also showed the existence of four factors related to the future intentions of cycling: (i) existence, (ii) relatedness, (iii) growth, and (iv) cycling difficulty. Consequently, the hybrid bivariate model included 56 measurement equations from the uncovered relations between the items and the latent

Table 1. Sample Characteristics

Questions	Answers	%	Answers	%
Age	<20 years old	11.2	45–54 years old	14.4
	21–28 years old	37.9	55–64 years old	7.2
	28–35 years old	27.0	>65 years old	2.3
Education	Primary school	1.2	High school	32.1
	Middle school	18.9	Higher education	47.8
Income	Much lower than average	16.3	Higher than average	4.7
	Lower than average	49.9	Much higher than average	2.3
	Average	13.0	Not answered	13.8
Commuting distance	< 1 km	8.4	10–20 km	16.1
	1–5 km	27.8	20–30 km	7.5
	5–10 km	32.0	> 30 km	8.2
Country of immigration	The Netherlands	19.4	Germany	24.2
	Denmark	44.5	United Kingdom	8.1
	Sweden	3.9		
Duration of immigration	<1 year	15.4	5–10 years	22.8
	1–5 years	50.8	>10 years	11.0
Current car use in the adopted country	Rarely	30.3	A few times per week	22.9
	A few times per month	11.3	Daily	27.2
	Once a week	8.3		
Current public transport use in the adopted country	Rarely	46.2	A few times per week	16.1
	A few times per month	16.8	Daily	15.8
	Once a week	5.1		
Current bicycle use in the adopted country	Rarely	31.0	A few times per week	22.1
	A few times per month	24.2	Daily	14.5
	Once a week	8.2		
Current walk for transport in the adopted country	Rarely	9.8	A few times per week	28.4
	A few times per month	11.7	Daily	36.6
	Once a week	14.2		
Current car availability	Yes	59.7	No	40.3
Current bicycle availability	Yes	80.9	No	19.1
Current bike sharing use	Yes	23.4	No	76.6
Kilometres cycled per day in the adopted country	0	31.0	10–15 km	6.7
	0–5 km	32.6	15–20 km	3.7
	5–10 km	24.5	>20 km	1.5
Cycling purpose in the adopted country	Utilitarian	17.3	Both	5.3
	Recreational	46.4	Never cycle	31.0
Past car use in Poland	Rarely	26.8	A few times per week	29.1
	A few times per month	13.7	Daily	23.6
	Once a week	6.8		
Past public transport use in Poland	Rarely	28.5	A few times per week	24.0
	A few times per month	14.4	Daily	26.4
	Once a week	6.7		
Past bicycle use in Poland	Rarely	56.9	A few times per week	7.4
	A few times per month	16.5	Daily	6.0
	Once a week	13.3		
Past walk for transport in Poland	Rarely	11.2	A few times per week	28.7
	A few times per month	11.0	Daily	38.5
	Once a week	10.5		
Car availability in Poland	Yes	64.6	No	35.4
Bicycle availability in Poland	Yes	43.9	No	56.1
Bike sharing use in Poland	Yes	13.7	No	86.3
Intention to cycle upon the return to the home country	Rarely	26.0	A few times per week	27.4
	A few times per month	24.0	Daily	13.2
	Once a week	9.3		

variables. The model also included eight structural equations exploring the association between the latent variables and the observed characteristics of the female immigrants.

Table 2. Estimates of the Measurement Equations of the Latent Variables Related to the Current Cycling Habits

	Estimate	t-stat
Existence		
Cycling makes me save time and money	1.000	–
Cycling helps me stay in shape	1.169	15.38
Cycling allows me to explore my neighborhood more	1.516	16.30
I would cycle places where I would not walk (e.g., a park during the night)	1.049	14.57
Cycling allows me to fulfill my basic needs (e.g. do shopping, meet my friends)	1.939	19.39
Cycling is an element of my daily life routine	1.978	19.11
I found some interesting places during cycling (e.g. stores, restaurants, parks)	1.090	9.91
Cycling increases my travel independence and flexibility	2.335	26.16
I cycle regardless of the weather conditions	0.876	11.23
Relatedness		
Cycling allows me to have more activities with my friends and family	1.000	–
I often speak about cycling with my friends	0.917	20.38
Cycling helps me enlarge my social network	1.008	22.91
I join cycling activities with other people	1.041	22.63
I feel that I am part of the community when I am cycling	1.156	23.59
I like to discuss stuff related to cycling (e.g., paths, equipment)	1.026	21.83
Cycling is a part of the social life in (given country)	0.936	19.50
Cycling increases my social potential	1.149	24.98
I share pictures of my bike trips on social network media	0.860	18.30
Growth		t-stat
Cycling makes me feel better	1.000	–
Cycling makes me consider myself as a sportive person	0.896	29.87
Cycling makes me feel that I do something good for the environment	1.014	30.73
Cycling makes me feel like I am part of something bigger and something good	1.119	32.91
Cycling makes me contribute to making the city a wonderful place to live	1.036	31.39
I have a lot of good memories related to cycling	0.999	26.64
Cycling has become my new hobby	1.182	32.83
I give advice on where to cycle in order to explore new areas	1.113	32.26
Safety		
I have adapted relatively fast to the traffic rules in (given country)	1.000	–
Even though I felt unsafe at first, I feel safe to cycle in (given country) now	1.059	24.63
I am willing to cycle in (given country) because of the well-developed infrastructure	1.281	24.63
I have enough skills and experience to cycle	1.036	22.04
I feel that drivers respect me as a cyclist	1.050	23.33

Note: The estimates of the constants and the standard deviations are omitted for brevity.

Several hybrid bivariate ordered models were then estimated, and observed and latent variables were retained in the specification if the associated parameters were significant at the 0.10 level. Tables 2 and 3 present the estimates for the latent variables related to the part of the model that refers to the current habit of cycling. Tables 4 and 5 show the estimates for the latent variables related to the future intentions of cycling upon return to the home country. Table 6 illustrates the estimates of the hybrid bivariate ordered model. The log-likelihood of the best model specification is −1787.32 and the adjusted McFadden rho-square is 0.548. Likelihood ratio tests (LRT) show that the hybrid bivariate ordered model outperforms the two separate models for the current habit of cycling and the future intentions of cycling upon return to the home country (LRT = 269.60, df = 1, p = 0.000) as well as the bivariate model without latent variables (LRT = 117.90, df = 8, p = 0.000). The correlation between the error terms also shows the importance of estimating joint models as it is equal to 0.407 and it is significantly different from zero at the 0.01 confidence level.

For what concerns the latent variables related to the current cycling habits, the measurement equations in Table 2 show the items measuring needs. Existence relates mainly to the fulfillment of basic needs such as travel independence and flexibility, errand completion, mobility around the neighborhood, and getting by in everyday life. Relatedness concerns mainly the fulfillment of higher needs such as having an active social life, feeling part of the community, and just enjoying cycling activities with friends and families. Growth associates mainly with the fulfillment of higher needs such as having a new hobby, feeling part of something bigger, sensing the contribution to the city and the environment, and in general feeling better about oneself. A role is played also by the perception of safety that comes mainly from the existence of infrastructure, the sense of respect from drivers, and the overall cycling culture in the adopted country.

Table 3. Estimates of the Structural Equations of the Latent Variables Related to the Current Cycling Habits

	Estimate	t-stat
Existence		
Car use a few times per week in Poland	−0.158	−1.75
Car use daily in Poland	−0.259	−2.44
Bicycle use a few times per week or daily in Poland	0.264	1.96
I cycled for recreational purposes in Poland	0.084	3.25
I cycled for commuting purposes in Poland	0.043	1.65
Immigrated between 5 and 10 years ago	0.184	1.92
Immigrated more than 10 years ago	0.270	2.33
Constant	0.355	9.59
Relatedness		
Car use a few times per week or daily	−0.211	−2.51
Walking daily in Poland	0.163	1.77
Owned a bicycle in Poland	0.133	2.04
I cycled for recreational purposes in Poland	0.064	2.56
Immigrated to Denmark	0.367	3.60
Immigrated to the Netherlands	0.514	4.85
Immigrated more than 10 years ago	0.188	1.82
Constant	0.386	14.30
Growth		
Car use daily in Poland	−0.243	−1.96
I cycle 5–15 km per day	0.246	1.78
I cycle more than 15 km per day	0.385	1.94
I cycle for recreational purposes	0.221	2.87
I cycled for recreational purposes in Poland	0.065	2.10
Constant	0.587	18.34
Safety		
Car use a few times per week or daily in Poland	−0.299	−2.26
Public transport use daily in Poland	0.202	1.94
Bicycle use a few times per week or daily in Poland	0.283	2.04
Owned a bicycle in Poland	0.160	2.00
I cycle for commuting purposes	0.225	2.42
I cycled for recreational purposes in Poland	0.087	2.90
I use an electric bicycle	−0.042	−2.00
Commuting distance between 1 and 5 km	−0.230	−1.71
Commuting distance between 5 and 10 km	−0.406	−2.96
Commuting distance over 10 km	−0.566	−3.83
Immigrated more than 10 years ago	0.270	2.06
Constant	0.494	15.44

The structural equations in Table 3 reveal that previous travel habits in the home country relate to the needs, as frequent car use is negatively associated with all three needs while previous experience with cycling is positively linked to existence and relatedness needs, especially when cycling was for recreational purposes. Interestingly, immigrants to the two more established cycling cultures, the Netherlands and Denmark, have stronger relatedness needs, while immigrants with a longer time in the adopted country have higher existence and relatedness needs. The habit of using a car and not a bicycle also relates to the perception of safety that is higher for female immigrants who used sustainable transport in their home country. The feeling of safety reduces for longer commuting distances, while the use of the bicycle for both purposes makes the female immigrants feel safer unless they have an electric bicycle.

For what concerns the latent variables related to the future cycling intentions, the measurement equations in Table 4 illustrate the items measuring needs. Existence is associated mainly with the economic and practical advantages of using the bicycle as well as the feeling of making a difference in the home country by cycling. Relatedness relates mainly to the societal contribution and in particular the encouragement of peers and policy-makers to have a cycling friendly environment. Growth links mainly with feeling better and avoiding all the negative stigma that cycling suffers from in Poland. A role is played also by the perception of cycling difficulty in the home country that derives mainly from the lack of infrastructure, the perception of large distances to be covered by bicycle, and the thought of being a nuisance in traffic to the other road users.

The structural equations in Table 5 illustrate that current travel habits in the adopted country affect the needs as perceived upon return to the home country. Owning a car in the adopted country relates to lower existence and relatedness needs, while owning a bicycle has the opposite relation to relatedness needs. Moreover, using a car or public transport quite frequently in the adopted country is negatively related to existence needs, while cycling quite frequently shows the opposite correlation not only with existence, but also with relatedness and growth needs. Notably, travel habits from Poland are linked to lower relatedness when the female immigrant used to drive a car and higher relatedness when the female immigrant used to cycle. Female immigrants have stronger relatedness needs if they immigrated to the two more established cycling cultures, and stronger existence and relatedness needs if a longer time has passed since they moved to the adopted country. Growth needs are associated mainly with active travel in the adopted country, even more so for larger distances cycled. Notably, higher income is negatively associated with existence and growth needs. The feeling of difficulties inherent in cycling in Poland is lower for female immigrants who have more experience and more practice with cycling in their adopted country, especially if they are living in the Netherlands or Denmark. Moreover, both very low and very high income induce the perception of more difficulties, while longer time in the adopted country diminishes the feeling of cycling being difficult.

The hybrid bivariate ordered model reveals that all the latent variables are associated with both the current habit and the future intentions of cycling. Higher existence, relatedness, and growth needs correlate positively with habitual cycling in the adopted country, while the perception of safety also plays a positive role in current

Table 4. Estimates of the Measurement Equations of the Latent Variables Related to the Future Cycling Intentions

	Estimate	t-stat
Existence		
I would cycle regardless of the weather conditions	1.000	–
I would cycle to improve my fitness and health	1.202	16.69
I would cycle because it is easier than using public transport	1.416	15.56
I would cycle to save time and money	1.683	17.90
I would make a difference in Poland if I cycled	1.407	14.07
My experience from my adopted country would let me cycle safely in Poland	1.187	13.65
Relatedness		
My friends would like to cycle with me	1.000	–
Cycling is a conversation topic with my friends in Poland	0.910	12.13
I think that if more people cycled, policy-makers in Poland would care more	1.421	14.80
I will encourage my friends in Poland to cycle	1.884	16.10
Living in my adopted country I got so used to cycling that it would be difficult to stop in Poland	1.288	14.64
I can see a future cycling culture in Poland	1.266	14.22
I can see the need for change in cycling policy in Poland	1.182	13.74
If I came back to Poland right now, I would cycle as much as in my adopted country	1.364	14.83
Growth		
Cycling makes me feel better	1.000	–
My family would be concerned about me cycling on the street	−1.171	−16.26
I feel that people in Poland would find me strange if I were to cycle daily	−1.513	−17.80
I would feel ashamed to cycle in Poland (e.g., someone might think that I cannot afford a car)	−1.281	−16.86
People think that cycling is not cool	−1.367	−14.70
Cycling difficulty		
Lack of infrastructure makes me feel that it is too risky to cycle	1.000	–
I would be concerned about leaving my bike (e.g., outside a store)	0.828	19.71
I think that I would disturb traffic by cycling and cause danger	0.816	19.43
I feel that I would have difficulties with repairing my bike	1.025	22.78
The cities in Poland are too spread-out so the cycling distance is too much	1.406	25.59
Lack of supporting facilities (e.g. showers at work, cycling racks) would be a barrier to cycling	0.711	18.23
I think it would annoy pedestrians if I were to cycle on the sidewalk	0.962	23.46
Drivers would be unpleasant to me if I were to cycle on the street	1.031	25.15

Note: The estimates of the constants and the standard deviations are omitted for brevity.

cycling habits. Similarly, higher needs across the various order levels associate positively with the intentions of cycling upon return to Poland, while the perception that cycling is difficult limits these intentions.

Other direct effects on the respondents' current cycling habits are past travel habits, in that owning and frequently using a car in the past does not relate positively with current cycling while using a car currently shows the same sign of relation with future cycling intentions, and income, in that higher income appears to be related to a lower probability of cycling upon return to the home country. Distance is a barrier to current cycling habits for female immigrants, as longer commute relates negatively to cycling habits, and the location and duration of residence in the adopted country shows that more established cycle cultures make it more likely for female immigrants to cycle in their adopted countries. All these variables also have an indirect effect on both current habits and future intentions of cycling, as higher needs are correlated with the location and duration of the immigration as well. One variable that has only a direct effect is age, and in particular younger female immigrants are more likely to cycle in their adopted country and also to continue cycling upon return to Poland.

Discussion and Conclusion

This study investigated how existence, relatedness, and growth needs motivate cycling among female immigrants from a driving-oriented country to a cycling-oriented country, and whether cycling in the adopted country translates into cycling upon a hypothesized return to their home country.

The findings from the model reveal that observed characteristics of female immigrants relate to their current habits and their future intention of cycling, both directly and indirectly. Directly, younger age makes female immigrants more likely to cycle both currently and in the future. These findings might possibly be related to the fact that younger women are at a life stage where commitments to children and family might not yet hinder their willingness to cycle (*13*). Also, past travel habits affect the habit of cycling in the adopted country in that stronger driving habits limit the transition to

Table 5. Estimates of the Structural Equations of the Latent Variables Related to the Future Cycling Intentions

	Estimate	t-stat
Existence		
Car use daily	−0.128	−2.45
Public transport use a few times per week	−0.139	−1.88
Public transport use daily	−0.181	−2.18
Bicycle use a few times per week or daily	0.199	2.07
Bicycle use a few times per week or daily in Poland	0.207	2.03
Own a car now	−0.154	−2.02
I cycle 10–20 km per day	0.216	1.89
I cycle more than 15 km per day	0.338	2.47
Immigrated between 5 and 10 years ago	0.143	1.83
Immigrated more than 10 years ago	0.174	1.86
Income higher or much higher than average	−0.218	−2.15
Constant	0.249	9.22
Relatedness		
Immigrated to Denmark or the Netherlands	0.272	3.44
Bicycle use a few times per week or daily	0.286	3.86
Car use daily in Poland	−0.144	−2.11
Bicycle use a few times per week or daily in Poland	0.147	2.37
Own a car now	−0.142	−2.54
I cycle for recreational purposes	0.091	2.12
I cycled for recreational purposes in Poland	0.059	3.28
I use an electric bicycle	−0.025	−2.08
Immigrated between 5 and 10 years ago	0.125	2.02
Immigrated more than 10 years ago	0.253	3.28
Constant	0.183	8.32
Growth		
Bicycle use a few times per week or daily	0.307	2.79
Walking a few times per week or daily	0.191	1.77
Own a bicycle now	0.167	1.89
I cycle more than 15 km per day	0.213	2.00
I cycle for recreational purposes	0.123	2.20
I cycled for recreational purposes in Poland	0.058	2.52
I use an electric bicycle	−0.041	−2.56
Commuting distance more than 10 km	−0.208	−1.85
Income higher or much higher than average	−0.218	−2.12
Constant	0.283	10.11
Cycling difficulty		
Car use daily	0.318	2.65
Immigrated to Denmark	−0.298	−2.03
Immigrated to the Netherlands	−0.377	−3.56
Bicycle use a few times per week or daily	−0.336	−2.45
I cycle 5–10 km per day	−0.201	−1.75
I cycle 10–15 km per day	−0.473	−2.93
Commuting distance over 20 km	0.330	2.05
Immigrated more than 5 years ago	−0.214	−2.02
Income much lower than average	0.280	3.15
Income much higher than average	0.467	2.88
Constant	0.509	15.91
Car use daily	0.318	2.65

Table 6. Estimates of the Hybrid Bivariate Ordered Model

	Estimate	t-stat
Current cycling habits		
Car use a few times per week or daily in Poland	−0.356	−2.17
Owned a car in Poland	−0.362	−3.14
Age lower than 35 years old	0.206	1.89
Commuting distance between 20 and 30 km	−0.418	−1.77
Commuting distance over 30 km	−0.466	−2.09
Immigrated more than 10 years ago	0.294	2.26
Immigrated to Denmark	0.335	1.83
Immigrated to the Netherlands	0.446	2.22
Existence	0.228	4.57
Relatedness	0.482	9.27
Growth	0.263	3.56
Safety	0.462	8.78
δ_1	−0.111	−0.50
δ_2	0.556	1.89
δ_3	0.916	2.49
δ_4	1.900	4.29
Future cycling intentions		
Car use a few times per week or daily in the adopted country	−0.386	−2.33
Age between 36 and 55 years old	−0.313	−1.84
Age higher than 55 years old	−0.547	−2.95
Immigrated to Denmark or the Netherlands	0.311	2.14
Income much higher than average	−0.487	−1.78
Existence	0.334	6.23
Relatedness	−0.227	−4.28
Growth	0.314	5.33
Cycling difficulty	−0.137	−2.55
χ_1	0.633	2.89
χ_2	1.183	4.05
χ_3	1.525	4.17
χ_4	2.411	5.47
Correlation of the error terms		
θ	0.407	4.06

cycling, but moving for a longer time to a more established cycling culture increases the propensity to cycle in the adopted country. These findings confirm that travel is a matter of culture, and suggest that stronger cycling cultures are more successful in convincing female immigrants to embrace the new society (*18*). Moreover, the model confirms that distance is a significant barrier to current habitual cycling for women (*32*), while higher income is a barrier to the future intentions of cycling, most likely because of the negative stigma from which the bicycle still suffers in Poland (*22*).

Indirectly, past travel habits negatively affect the needs at the lower existence level in terms of frequent car use, and at the higher relatedness level in terms of bicycle use prior to the immigration. Moreover, immigration to more established cycling cultures translates into higher relatedness needs as female immigrants are likely feeling part of something new and adapt to the new situation (*18*). In particular, the longer the time spent in the adopted country, the stronger the relatedness needs as female immigrants probably need to grow accustomed to the new society they are part of (*18*). The higher-level growth needs correlate with more frequent cycling, confirming that cycling relates to self-realization and self-identity (*16*, *19*), and in turn this increases the perception of safety while cycling. Interestingly, the use of an electric bicycle relates to a lower feeling of safety, as suggested in recent studies that are inconclusive on electric bicycles being safe (*33*). The combination of direct and indirect effects on the current habit and the future intention of cycling is an example of more insightful results obtained from hybrid models that disentangle the contribution of covariates, unlike traditional models (*29*).

Relevantly, both current habits and future intentions of cycling relate to both lower-order and higher-order needs. As theorized in Alderfer's model (*17*), habitual behavior (i.e., cycling in the adopted country) relates more to higher-level needs of relatedness and growth and less to lower-level needs of existence, although arguably the perception of safety is also an existence need and strongly correlates to the habit of cycling in the adopted country. Interestingly, the same does not apply when the female immigrants imagine returning to their home country, as relatedness has a negative correlation with the intention to cycle in the future. Possibly, this indicates that they are thinking about how they relate to Poland while the positive relatedness effect comes from their feelings of belonging to the adopted country. Again, perceiving cycling as difficult is an existence need and strongly relates to the future intention of cycling upon return to the home country.

Limitations should be acknowledged in that the sample is a convenience one and does not intend to represent all female Polish immigrants. However, although the findings cannot be generalized, the model results are interesting because of the rigorous econometric procedure that guarantees unbiased estimates (*34*). Also, limitations should be acknowledged in that the role of infrastructure was not explicitly captured. However, the existence of infrastructure contributed to the feeling of safety, and the immigration to countries with extended bicycle networks contributed to an increase in the relatedness need. Notably, the findings shed light on some interesting concepts that future research could extend to other female populations as well as broader immigrant populations in order to understand how cycling culture and infrastructure shape the travel habits of people in an era where immigration is becoming more and more common. Moreover, future research could extend the role of the culture while examining the effect that life stages have on the intention to cycle both in the adopted country and the home country (*13*).

The findings of this study indicate that female immigrants cycle to satisfy their needs at the higher level, and would return to their country and cycle because they have both existence and growth needs. However, the findings also suggest that relatedness is associated with the cycling culture of the adoption country, but the barriers to cycling not only directly limit the intention to cycle when returning to the home country, but also foster a negative relation between cycling and relatedness to the home country that still does not have the cycling culture found in the new place. Accordingly, the answer about cycling upon return is conditional on the creation of a culture that will be supported by interventions on infrastructure and facilities that are very relevant in women's choices to cycle. This finding is in line with previous findings that cycling relates to the establishment of an identity within the adopted country (*16*, *19*), apparently more strongly than the one within the home country.

If there was one lesson to be learned from this study, it would be that the key to promoting cycling is to create a marketing plan that focuses not only on the construction of infrastructure and facilities that cater to functional needs at the existence level, but also on initiatives that increase the feeling of being part of a community (e.g., bike to work campaigns, "every kid bikes" initiatives) and help women to realize their self-worth by catering to their emotional needs at the relatedness and the growth level.

Author Contributions

The authors confirm contribution to the paper as follows: study conception and design: Kaplan, Wrzesinska, Prato; data collection: Kaplan, Wrzesinska, Prato; analysis and interpretation of results: Kaplan, Prato, Wrzesinska; draft manuscript preparation: Prato, Kaplan, Wrzesinska. All authors reviewed the results and approved the final version of the manuscript.

References

1. Tight, M., P. Timms, D. Banister, J. Bowmaker, J. Copas, A. Day, D. Drinkwater, M. Givoni, A. Gühnemann, M. Lawler, J. Macmillen, A. Miles, N. Moore, R. Newton, D. Ngoduy, M. Ormerod, M. O'Sullivan, and D. Watling. Visions for a Walking and Cycling Focussed Urban Transport System. *Journal of Transport Geography*, Vol. 19, No. 6, 2011, pp. 1580–1589.
2. Lindsay, G., A. Macmillan, and A. Woodward. Moving Urban Trips from Cars to Bicycles: Impact on Health and Emissions. *Australian and New Zealand Journal of Public Health*, Vol. 35, No. 1, 2011, pp. 54–60.
3. Hendriksen, I. J. M., B. Zuiderveld, H. C. G. Kemper, and P. D. Bezemer. Effect of Commuter Cycling on Physical Performance of Male and Female Employees. *Medicine and Science in Sports and Exercise*, Vol. 32, No. 2, 2000, pp. 504–510.
4. Hou, L., B. Ji, A. Blair, Q. Dai, Y. Gao, and W. Chow. Commuting Physical Activity and Risk of Colon Cancer in Shanghai, China. *American Journal of Epidemiology*, Vol. 160, No. 9, 2004, pp. 860–867.
5. Oja, P., S. Titze, A. Bauman, B. de Geus, P. Krenn, B. Reger-Nash, and T. Kohlberger. Health Benefits of Cycling: A Systematic Review. *Scandinavian Journal of Medicine & Science in Sports*, Vol. 21, No. 4, 2011, pp. 496–509.

6. de Hartog, J. J., H. Boogaard, H. Nijland, and G. Hoek. Do the Health Benefits of Cycling Outweigh the Risks? *Environmental Health Perspectives*, Vol. 118, No. 8, 2010, pp. 1109–1116.
7. Hine, J., and M. Grieco. Scatters and Clusters in Time and Space: Implications for Delivering Integrated and Inclusive Transport. *Transport Policy*, Vol. 10, No. 4, 2003, pp. 299–306.
8. Sanchez, T., Q. Shen, and Z. Peng. Transit Mobility, Job Access and Low-income Labour Participation in US Metropolitan Areas. *Urban Studies*, Vol. 41, No. 7, 2004, pp. 1313–1331.
9. World Bank Group. *World Development Report – Gender Equality and Development*. The International Bank for Reconstruction and Development/The World Bank, Washington, D.C., 2012.
10. Rosenbloom, S. Trip Chaining Behaviour. In *Gender, Transport and Employment* (Grieco, M., L. Pickup, and R. Whipp, eds.), Gower Press, Aldershot, 1989.
11. Garrard, J., G. Rose, and S. K. Lo. Promoting Transportation Cycling for Women: The Role of Bicycle Infrastructure. *Preventive Medicine*, Vol. 46, No. 1, 2008, pp. 55–59.
12. Emond, C., W. Tang, and S. Handy. Explaining Gender Difference in Bicycling Behavior. *Transportation Research Record: Journal of the Transportation Research Board*, 2009. 2125: 16–25.
13. Bonham, J., and A. Wilson. Bicycling and the Life Course: The Start-Stop-Start Experiences of Women Cycling. *International Journal of Sustainable Transportation*, Vol. 6, No. 4, 2012, pp. 195–213.
14. Van der Kloof, A., J. Bastiaansen, and K. Martens. Bicycle Lessons, Activity Participation and Empowerment. *Case Studies on Transport Policy* 2, 2014, 89–95.
15. Horton, D., P. Rosen, and P. Cox. *Cycling and Society*. Ashgate, Burlington, Vt., 2007.
16. Law, S. F., and W. Karnilowicz. "In Our Country It's Just Poor People Who Ride a Bike": Place, Displacement and Cycling in Australia. *Journal of Community & Applied Social Psychology*, Vol. 25, No. 4, 2015, pp. 296–309.
17. Alderfer, C. P. An Empirical Test of a New Theory of Human Needs. *Organization Behavior and Human Performance*, Vol. 4, No. 2, 1969, pp. 142–175.
18. Chatterjee, K., H. Sherwin, J. Jain, J. Christensen, and S. Marsh. A Conceptual Model to Explain Turning Points in Travel Behavior: Application to Bicycle Use. *Transportation Research Record: Journal of the Transportation Research Board*, 2013. 2322: 82–90.
19. Aldred, R., and K. Jungnickel. Why Culture Matters for Transport Policy: The Case of Cycling in the UK. *Journal of Transport Geography*, Vol. 34, 2014, pp. 78–87.
20. Pucher, J., C. Komanoff, and P. Schimek. Bicycling Renaissance in North America? Recent Trends and Alternative Policies to Promote Bicycling. *Transportation Research Part A: Policy and Practice*, Vol. 33, No. 7–8, 1999, pp. 625–654.
21. Gatersleben, B., and K. M. Appleton. Contemplating Cycling to Work: Attitudes and Perceptions in Different Stages of Change. *Transportation Research Part A: Policy and Practice*, Vol. 41, No. 4, 2007, pp. 302–312.
22. Polish Transport Ministry. *Strategia Rozwoju Transportu do 2020 roku (z perspektywą do 2030 roku)*. http://mib.gov.pl/media/3511/Strategia_Rozwoju_Transportu_do_2020_roku.pdf. Accessed June 6, 2016.
23. Bamberg, S., and P. Schmidt. Incentives, Morality, or Habit? Predicting Students' Car Use for University Routes with the Models of Ajzen, Schwartz, and Triandis. *Environment and Behavior*, Vol. 35, No. 2, 2003, pp. 264–285.
24. Chataway, E. S., S. Kaplan, T. A. S. Nielsen, and C. G. Prato. Safety Perceptions and Reported Behaviour Related to Cycling in Mixed Traffic: A Comparison between Brisbane and Copenhagen. *Transportation Research Part F: Traffic Psychology and Behaviour*, Vol. 23, 2014, pp. 32–43.
25. Pelzer, P. Bicycle as a Way of Life: A Comparative Case Study of Bicycle Culture in Portland, OR and Amsterdam. *Proc., 7th Cycling and Society Symposium*, Oxford, UK, 2010.
26. Ferdous, N., N. Eluru, C. R. Bhat, and I. Meloni. A Multivariate Ordered-Response Model System for Adults' Weekday Activity Episode Generation by Activity Purpose and Social Context. *Transportation Research Part B: Methodological*, Vol. 44, No. 8–9, 2010, pp. 922–943.
27. Dias, F. F., P. S. Lavieri, V. M. Garikapati, S. Astroza, R. M. Pendyala, and C. R. Bhat. A Behavioral Choice Model of the Use of Car-Sharing and Ride-Sourcing Services. *Transportation*, Vol. 44, No. 6, 2017, pp. 1307–1323.
28. Walker, J., and M. Ben-Akiva. Generalized Random Utility Model. *Mathematical Social Sciences*, Vol. 43, No. 3, 2002, pp. 303–343.
29. Vij, A., and J. Walker. How, When and Why Integrated Choice and Latent Variable Models are Latently Useful. *Transportation Research Part B: Methodological*, Vol. 90, 2016, pp. 192–217.
30. Bierlaire, M. *PythonBiogeme: A Short Introduction*. Report TRANSP-OR 160706, Series on Biogeme. Transport and Mobility Laboratory, School of Architecture, Civil and Environmental Engineering, Ecole Polytechnique Fédérale de Lausanne, Switzerland, 2016.
31. Hess, S., K. E. Train, and J. W. Polak. On the Use of a Modified Latin Hypercube Sampling (MLHS) Method in the Estimation of a Mixed Logit Model for Vehicle Choice. *Transportation Research Part B: Methodological*, Vol. 40, No. 2, 2006, pp. 147–163.
32. Dickinson, J. E., S. Kingham, S. Copsey, and D. J. Pearlman Hougie. Employer Travel Plans, Cycling and Gender: Will Travel Plan Measures Improve the Outlook for Cycling to Work in the UK? *Transportation Research Part D: Transport and Environment*, Vol. 8, No. 1, 2003, pp. 53–67.
33. Jones, T., L. Harms, and E. Heinen. Motives, Perceptions and Experiences of Electric Bicycle Owners and Implications for Health, Wellbeing and Mobility. *Journal of Transport Geography*, Vol. 53, 2016, pp. 41–49.
34. Wooldridge, J. M. *Introductory Econometrics*, 4th ed. South-Western Cengage Learning, Mason, Ohio, 2009.

The Standing Committee on Women's Issues in Transportation (ABE70) peer-reviewed this paper (18-02033).

Article

TRR
JOURNAL OF THE TRANSPORTATION RESEARCH BOARD

Transportation Research Record
2018, Vol. 2672(3) 166–174

DOI: 10.1177/0361198118795005
journals.sagepub.com/home/trr
SAGE

Gender Differences in Danish Road Accidents

Rie Hultqvist Visby[1] and Karoline Lundholt[1]

Abstract

A study based on the Danish Road Directorates accident statistics shows that there are different gender patterns in how we behave in traffic and our involvement in traffic accidents. Overall men drive more by car and take longer trips. Women take more trips, but they are shorter, and travel more by bicycle or walking. Although men and women spend the same amount of time in traffic, twice as many men as women are involved in traffic accidents. Men constitute 62% of all persons killed and injured in road accidents, and are involved in 66% of the accidents. Not only gender, but education, income and age are also important factors in the accident patterns of men and women in Denmark. People with primary/lower secondary school as the highest education level are over represented in accident statistics in proportion to other educational groups. Young people aged 18–24 constitute about 20% (the largest group) of persons killed and injured in traffic accidents, although they only constitute approximately 8% of the population. In this group, there are about twice as many men killed or injured as women. The overall pattern is that, the older you become, the higher educated you are and the higher income you have—the smaller is your risk of involvement in accidents in Denmark. This knowledge is of importance when planning communication strategies with different segments of the population and when reaching out to young people in schools, who are at the highest risk of getting killed or injured in accidents.

What are the differences in how men and women travel, and does it influence their involvement in road accidents? The Danish Road Directorate set out to investigate these issues in the publication: Gender differences in road accidents. The focus of the investigation was to see if we could obtain new knowledge on the gender topic by comparing the Danish Road Directorate's accident statistics with other sets of data to challenge myths such as: men are more involved in traffic accidents because they travel more. The aim was to achieve more awareness on how to improve traffic safety by focusing campaigns and other safety efforts on the right target groups in relation to gender or age.

Methodology

To obtain more knowledge on the topic of gender differences in road accidents, the Danish Road Directorate collaborated with one of the leading Consultant Groups, COWI, to study correlations between Danish traffic accidents and kilometers travelled based on gender, age, education and income in the Danish population. The study is based on data from the Danish Road Directorate's accident statistics, police-registered traffic accidents, the Danish National Travel Survey and the registered data on population, income, education and commuting distance from Statistics Denmark.

In this study, moving in traffic is called "kilometres travelled". This means the distance people cover when they walk, ride a bicycle, take the train or drive a car. Only traffic accidents, in which one or more persons have been injured, are included in the study, and the focus of the study is on road users between the ages of 6 and 80 years old. A total of 16,937 persons aged between 6 and 80 years old were injured or killed in Danish road accidents during the period 2009–2012 according to the Danish Road Directorate's accident statistics.

When we look at how men and women are involved in Danish road accidents, it is important to distinguish between three concepts: accidents; injuries; and persons involved (see Figure 1). In the period 2009–2012, the police registered a total of 14,321 traffic accidents with personal injury. In these accidents, 17,695 persons were injured or killed, but a total of 29,524 persons were

[1]Department of Traffic Safety and Cycling, The Danish Road Directorate, Copenhagen K, Denmark

Corresponding Author:
Address correspondence to Rie Hultqvist Visby: rhv@vd.dk

involved in these accidents according to the Danish Road Directorate's accident statistics. This means that when reports state that several men or women have been involved in a road accident, it is not necessarily synonymous with the number of persons injured.

To further explain the correlation between the three concepts, it is often the case that an accident occurs between two or more vehicles, and pedestrians and cyclists may also be involved. Consequently, several persons may be injured in the same accident. Furthermore, drivers, who have not been injured in the accident, are also included in the number of persons involved, which explains why this number is bigger than the number of persons injured. Passengers, who were not injured, however, are not included.

Differences in Danish Gender Travel Patterns

Differences between the genders with regards to the travel time and travel length may be related to differences between destinations as shown in Figure 4. For example, men travel longer when commuting to work and driving for business: for example, tradesmen, sales representatives or drivers. Women, however, have more daily errands than men, which may result in more, but shorter trips. This can include shopping or picking up children.

For both genders, the car is the most popular means of transport in Denmark as shown in Figure 3. However, men generally drive more kilometres by car, by motorcycle and by moped, whereas women have more trips by bicycle, on foot and by public transport. Women are also car passengers more frequently than men.

Generally, Danish men cover larger distances than Danish women as shown in Figure 2. This does not mean, however, that men spend much longer in traffic than women. As men more often choose fast means of transport, it means that they can travel greater distances than women without spending significantly more time on daily transport. Women also divide their transport into more trips than men. Every month, women, on average, walk eight more trips than men, and cycle about two trips more than men.

Twice as Many Men as Women are Involved in Road Accidents

Danish men and women spend almost the same amount of time in traffic. However, men are involved in road accidents far more often than women, even when taking into account that they drive more kilometres. About two out of three persons involved in accidents who suffer a personal injury are men.

As the first speedometer shows in Figure 5, the Danish population is evenly distributed with an equal number of

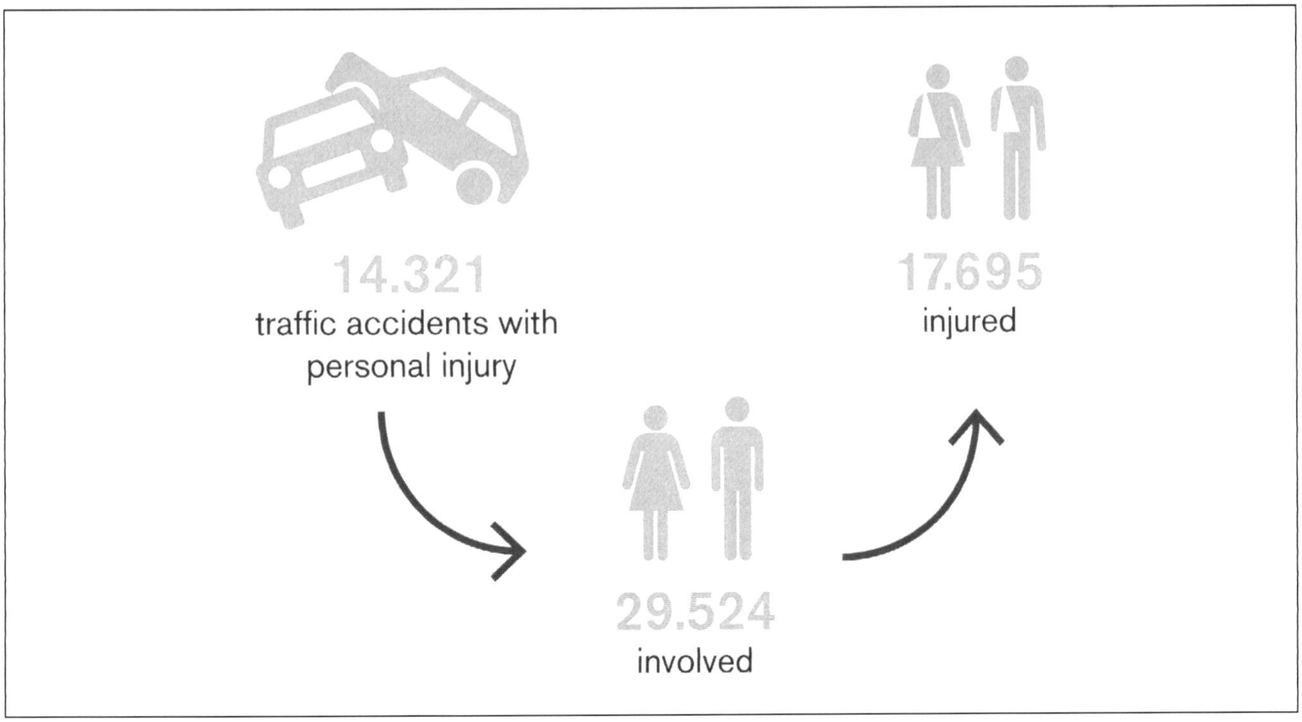

Figure 1. Numbers of accidents, people involved, and people injured in Danish traffic from 2009–2012.

	Travel length, km/day	Travel time, minutes/day	No. of trips per day
Men/boys	45,1	58	2,88
Women/girls	34,7	55	3,01

Figure 2. Kilometres travelled distributed according to purpose of travel and gender, 2008–2012.

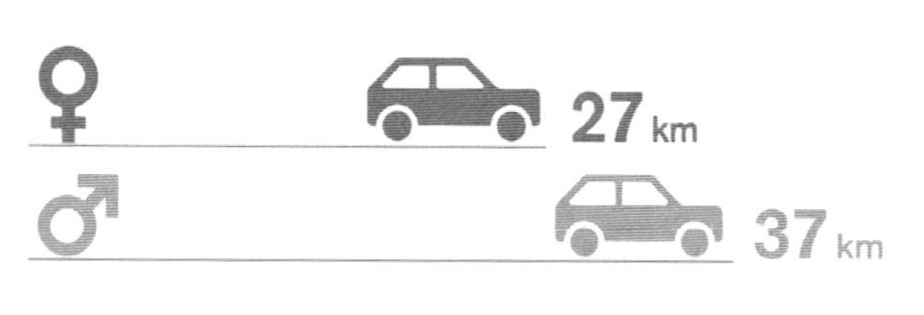

Figure 3. Average travel by car per day by Danish men and women.

men and women. The second speedometer shows the distribution of kilometres travelled; in other words, how much time men and women spend in traffic. Here the number of men is slightly higher than that of women, measured in relation to the number of kilometres driven. The third speedometer shows the gender distribution of persons involved in road accidents. It shows that almost twice as many men as women are involved in road accidents. Compared to the Danish Road Directorate's accident statistics, men constitute 62% of the persons killed and injured, which roughly corresponds to the amount involved in accidents.

Age Differences in Danish Road Accidents

Young people aged 18–24 constitute about 20% of all persons killed and injured in traffic, while they constitute only approximately 8% of the entire Danish population. In this group, there are about twice as many men killed and injured as women as shown in Figure 6.

Men's over representation in the accident statistics becomes gradually lower toward the age of 75, where there is almost a balance between the number of accidents and the amount of kilometres travelled. The oldest age group in the study is the only one in which more women than men are injured or killed in road accidents. This may be because there is an over representation of older women injured or killed in pedestrian accidents. The higher proportion of female pedestrians in accidents may be owing to women constituting a larger proportion of the population in this age group.

Pedestrian Accidents

Walking is a means of transportation that is used by all ages. This is also evident in the Danish safety statistics (see Figure 7). There are significantly more children of both genders, up to the age of 14, who are killed or injured as pedestrians, than is the case for personal injury accidents in general. The same applies for the elderly. More than one out of three of all pedestrian fatalities has a person involved who is 70 years or older.

Kilometres travelled on foot constitute approximately 2% of the total kilometres travelled by all means of transport. If this is calculated on the basis of transport time instead, it is about 16%, and if it is calculated based on the number of trips, it is about 24%.

As pedestrians, women are injured in slightly more accidents than men, but they walk more often than men. On a daily basis, women walk about 30% further than men.

	Travel length, km/day	
	Men/boys	Women/girls
Commuting	13,8	9,2
Business	7,2	1,5
Errands	7,6	7,9
Leisure	16,5	16,1
Total	**45,1**	**34,7**

Figure 4. Kilometres travelled distributed by gender and calculated based on the number of kilometres driven or walked, time spent and number of trips, 2008–2012.

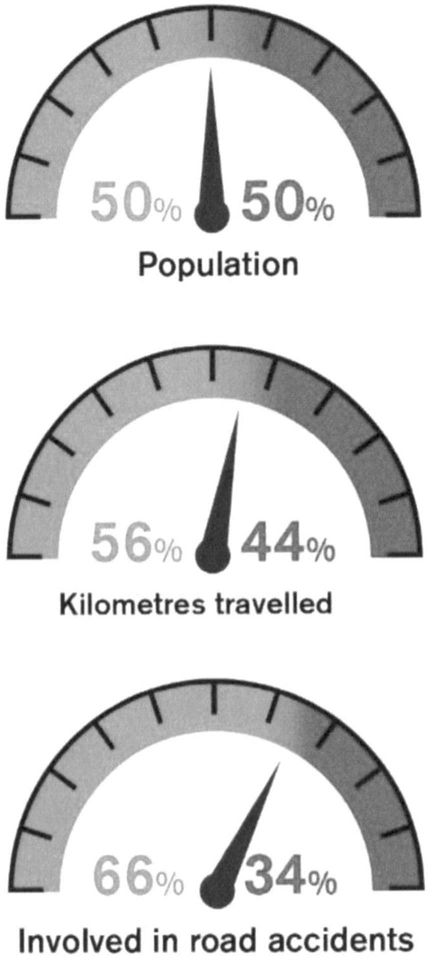

Figure 5. Gender distribution of the population, kilometres travelled and gender distribution of the number of persons involved in road accidents, 2009–2012.

This means that men's risk of injury as pedestrians is higher than the risk of women per kilometre walked. Female pedestrians are involved in more accidents at intersections (34%: women, and 23%: men) and male pedestrians are more often involved in pedestrian accidents on a straight stretch of road (62%: men compared to 50%: women), 87% of women's pedestrian accidents and 78% of men's pedestrian accidents occur in urban zones.

Almost every fourth man involved in a road accident as a pedestrian is under the influence of alcohol, while this only applies to approximately every 12th female.

Influences of Education and Income in People Involved in Danish Road Accidents

Factors such as education and income also affect the accident statistics in Denmark, and certain patterns are visible when we look at the distribution of accidents between men and women. In this case too, there are variations between the genders.

Education seems to influence the relationship between how much people move in traffic, and how often they are involved in road accidents. People with primary/lower secondary school as the highest education level are over represented in accident statistics in proportion to other educational groups.

As shown in Figure 8, women with primary/lower secondary school as the highest completed education constitute 15% of the population, 9% of the kilometres travelled and 13% of the persons involved in road accidents.

Men with primary/lower secondary school as the highest completed education constitute 14% of the population, 12% of the kilometres travelled and 29% of the persons involved in road accidents.

This group is largely young people who have not yet completed education other than primary/lower secondary school. However, 35% of this group are between the ages of 25 and 54, and thereby also consist of older people, who have primary/lower secondary school as their only education. This group is also largely overrepresented in the case of single-vehicle accidents, as will be explained further in the section below.

As previous statistics have shown, men are overrepresented in the accident statistics. However, in relation to income, there is a clear tendency that the higher the income, the smaller the difference between men and women's share of involvement in accidents, with regards to their share of the kilometres travelled.

This means that the over representation of men in road accidents is reduced proportionally with the increase of their income. However, it must be taken into consideration that people with low income are often young people with a low educational level and short driving experience. Therefore, it is uncertain whether the balance between traffic and involvement in accidents for men with higher income is solely because of the income itself, or also because of education or age.

Three Times as Many Men in Single-Vehicle Accidents

Traffic accidents where there is only one party involved (car, bicycle, motorcycle, etc.), are called single-vehicle accidents (although there may be passengers involved also). Every fifth accident involving personal injury is a single-vehicle accident, so single-vehicle accidents constitute a large proportion of the personal injury accidents in Denmark.

In particular young men are injured in single-vehicle accidents as shown in Figure 9. Men between the ages of 18 and 24 constitute approximately 4% of the

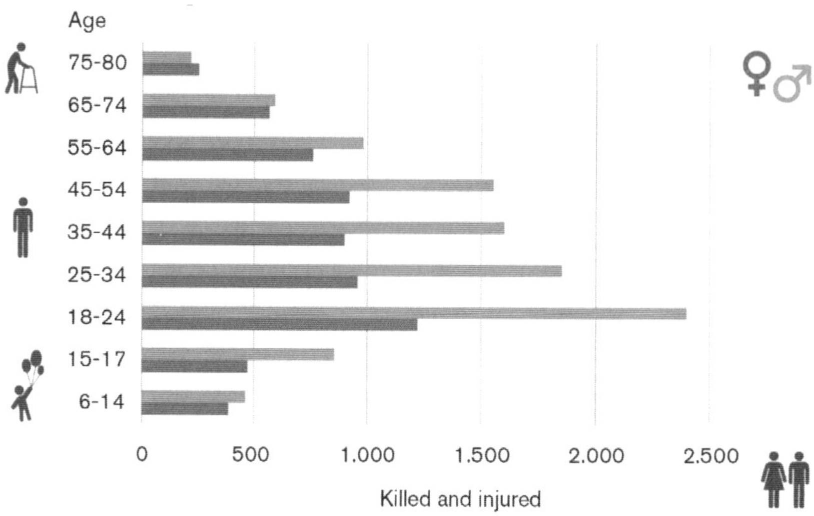

Figure 6. Persons killed and injured broken down by gender and age, 2009–2012.

	Age									
	6-14	15-17	18-24	25-34	35-44	45-54	55-64	65-74	75-80	Total
Women	6 %	4 %	7 %	6 %	4 %	6 %	6 %	7 %	5 %	51 %
Men	6 %	3 %	8 %	7 %	6 %	7 %	6 %	4 %	2 %	49 %
Total	**12 %**	**7 %**	**15 %**	**13 %**	**10 %**	**13 %**	**12 %**	**11 %**	**7 %**	**100 %**

Figure 7. Pedestrians involved in road accidents broken down by gender and age, 2009–2012.

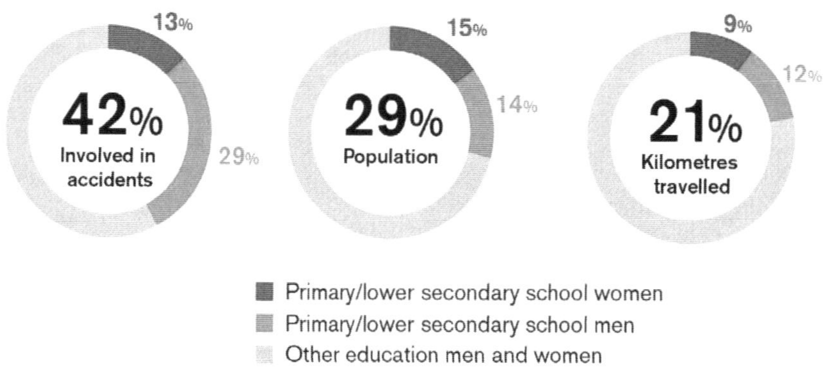

Figure 8. Primary or lower secondary school as the highest education compared to other groups.

population, but 25% of those killed or injured in single-vehicle accidents.

Speeding and lack of awareness are frequent accident factors for young people. This happens when drivers overestimate their own skills for example, are distracted, fatigued or under the influence of alcohol. Failure to use seatbelts also plays a role in aggravating injuries for young people in single-vehicle accidents.

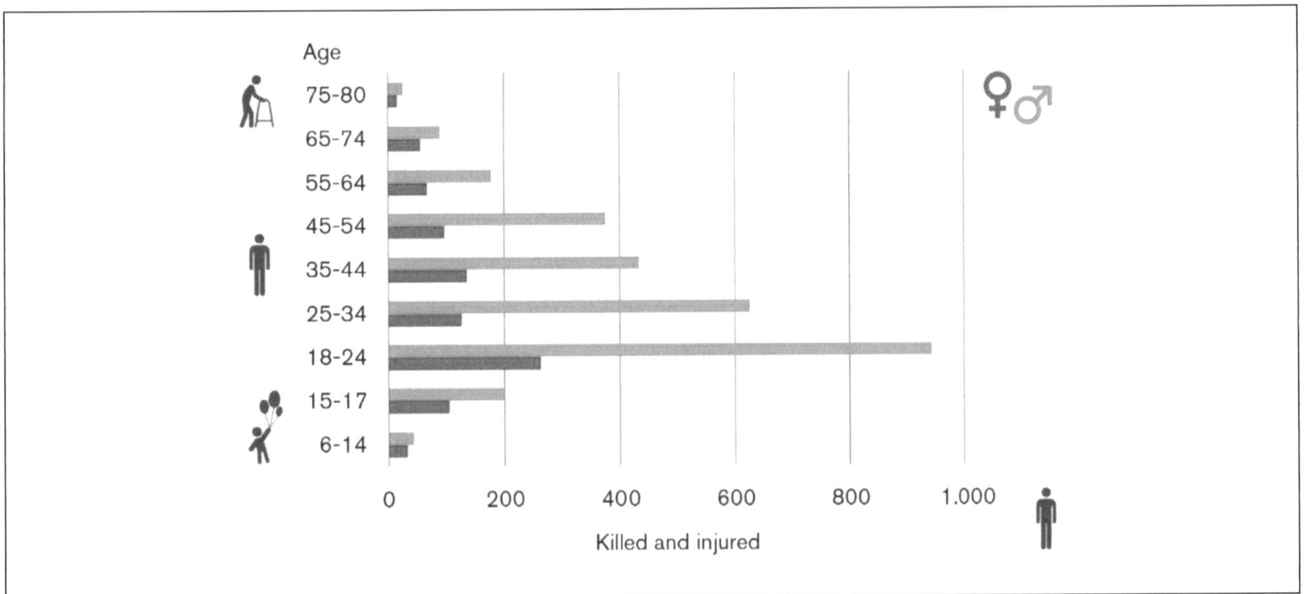

Figure 9. Persons killed and injured in single-vehicle accidents broken down by gender and age.

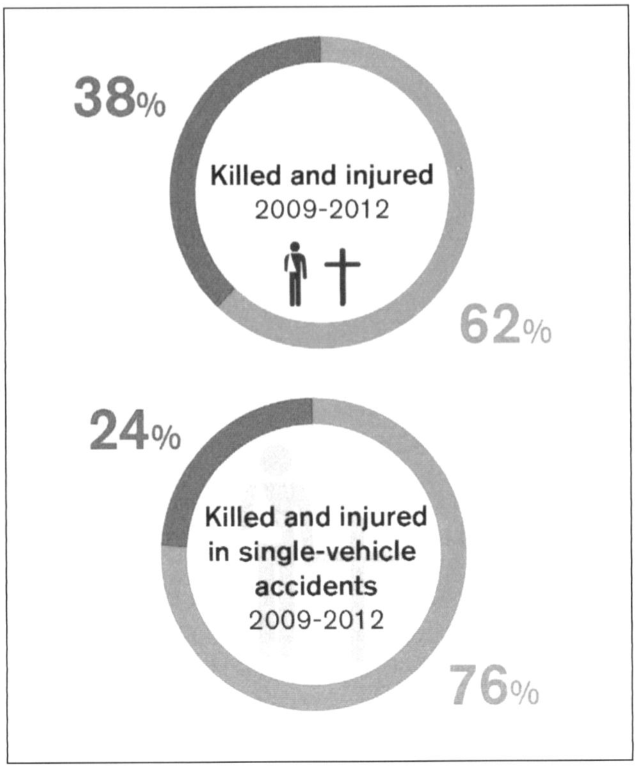

Figure 10. Persons killed and injured broken down by gender, 2009–2012.

More than three times as many men as women are involved in single-vehicle accidents. Making this the accident situation with the largest difference between men and women as shown in Figure 10.

Men between the ages of 18 and 24 constitute almost 26% of all people involved in single-vehicle accidents, while this applies to approximately 7% of the women in the same age group. That means that this age group is most exposed to single-vehicle accidents for both genders, despite this group being relatively small and comprised of fewer years than the subsequent age groups, with which the comparison is made.

Level of Education as a Factor in Single-Vehicle Accidents

In as many as 83% of all single-vehicle accidents a person with primary/lower secondary school or vocational training is involved, even though their share of kilometres travelled is only 51%. Within these groups, men are involved in the vast majority of the single-vehicle accidents, namely 65% of the accidents, whereas women are involved in 18% (see Figure 11).

Driving Experience as a Factor in Single-Vehicle Accidents

Lack of driving experience may also show a correlation between single-vehicle accidents, gender and low educational level. Driving experience is defined as the number of years between the year of the accident and the year of obtaining their first driving license.

For both genders, short driving experience plays a larger role in single-vehicle accidents than in other types of accidents. There is a greater share of drivers in single-vehicle accidents, who have had their driving license for

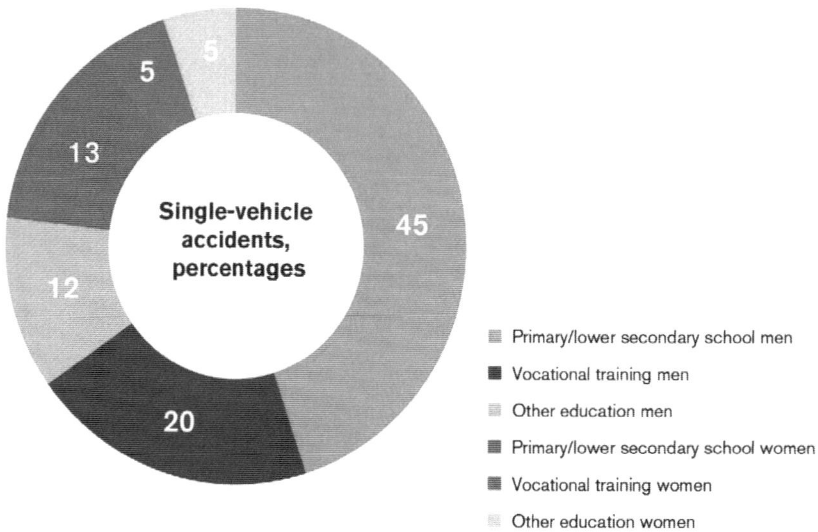

Figure 11. The distribution of single-vehicle accidents by men and women with primary or secondary school as the highest level of education, vocational training and other education.

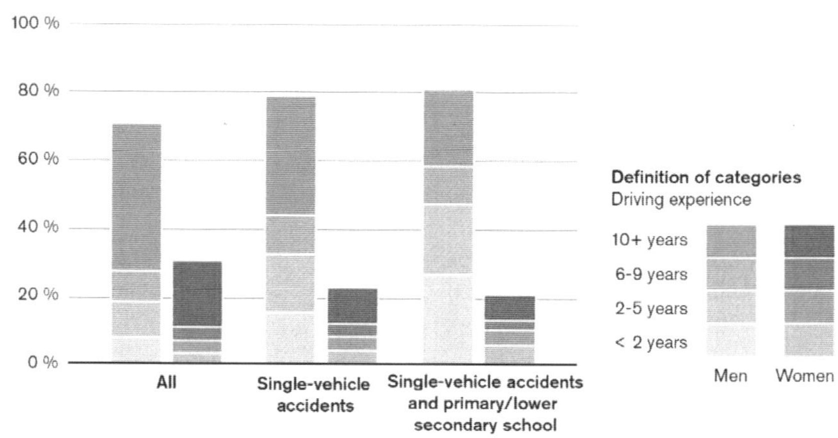

Figure 12. Distribution of driving experience for drivers in different types of accidents and for drivers with primary or lower secondary school as the highest educational level, 2009–2012.

less than two years, compared to all types of personal injury accidents. For people with primary/lower secondary school as the highest level of education, driving experience plays an even bigger role in connection with single-vehicle accidents.

Figure 12 shows how drivers are distributed in relation to driving experience, accident type and primary/lower secondary school as the highest level of education. The blue and red columns are different when it comes to the amount of accidents the genders are involved in, but the correlation between the driving experience of men and women when they are involved in accidents is relatively similar.

Conclusion

Twice as many men as women are involved in Danish road accidents. Only in the age group of 75–80 years are more women killed and injured than men. Education seems to influence the relationship between how much people move in traffic, and how often they are involved in road accidents. People with primary/lower secondary school as the highest education level are over represented in accident statistics in proportion to other educational groups.

Behind the numbers there is a large difference between men and women with regards to the individual type of accident. There are, for example, far more men involved in single-vehicle accidents than women, and slightly more men involved in pedestrian accidents, when the kilometres travelled are taken into account.

A combination of higher age, higher income and higher education can contribute to balancing the gender differences when it comes to how often men and women are involved in traffic accidents. Significantly more children and elderly people are injured in pedestrian accidents than in personal injury accidents in general.

More knowledge on the travel patterns and involvement in road accidents of men and women, children and older people could contribute to better and safer traffic. When we know who is involved in different types of road accidents, we can take better measures at a governmental level to make the necessary precautions and address the right people in our campaigns.

How to Use the Knowledge

The knowledge from this study could contribute to targeting the efforts in the traffic safety area and reducing the number of road accidents.

There are already different campaigns targeting school children in Denmark, as The Danish Road Safety Council provides traffic safety materials for all school grades, which includes tests for pedestrians, cyclists and for moped riders.

When distributing traffic safety information, it is important to consider the best way to reach people who will not continue in the educational system after they have completed primary/lower secondary school, as no educational institution will form a natural platform for this interaction. This group constitutes a large part of those who are involved and injured in traffic accidents—in particular in single-vehicle accidents—and it is therefore important to include them in the traffic safety efforts. It is important to remember, that the group contains immature young drivers as well as older road users, who never completed an education beyond primary or lower secondary school. The design of campaigns should therefore take both groups into account, and examine how best to reach each of them.

A large proportion of the men involved in single-vehicle accidents were driving a car or a moped, and thus driving schools provide a good opportunity to make a focused effort. The effort could, for example, be inspired by a large proportion of the accidents happening in the first year after the drivers have obtained driving licenses. Driving schools could draw attention to this issue and highlight that if the students want to change the statistics, the male students in particular should be more cautious in the first years of driving and be more aware of their traffic behavior in general.

Another option is to continue to focus on vocational training courses and on trading companies, because 20% of the men involved in single-vehicle accidents have a vocational education. It may be important to focus on how young men perceive the traffic behavior of others and themselves, and what understanding they have of risks in different traffic situations or of the vehicles' capacities. This could benefit future campaigns targeting male workmen in particular.

To accommodate vulnerable road users it is important to include safety and accessibility for bicycles and pedestrians at an early stage in the planning of urban areas or new roads and connections. Traffic safety authorities' need, for example, to place pedestrian crossings correctly and safely to ensure that the signal system allows slow pedestrians to cross the road, and to establish speed zones in local areas with many pedestrians and cyclists. Given that children and elderly people have a high involvement in pedestrian accidents it would be favorable to make safe crosswalks and reduce speed limits near senior centers, schools and bus stops as well as in other areas with many pedestrians.

The Danish Road Directorate

The Danish Road Directorate is responsible for the state-owned roads in Denmark. The Danish Road Directorate promotes an integrated road and transport system. With special consideration to the environment, the directorate is working toward a safe, secure and easy journey on the roads for people as well as for goods.

Author Contributions

The authors confirm contribution to the paper as follows: Study conception and design: the Danish Road Directorate (Rie Hultqvist Visby, Lotte Brædder Francke and Marianne Foldberg Steffensen); data collection: COWI (Sven Torp Jespersen and Morten Hørmann) and the Danish Road Directorate (Lotte Brædder Francke); analysis and interpretation of results: COWI (Sven Torp Jespersen) and the Danish Road Directorate (Rie Hultqvist Visby and Lotte Brædder Francke); draft manuscript preparation: the Danish Road Directorate (Rie Hultqvist Visby, Karoline Lundholt and Marianne Foldberg Steffensen). All authors reviewed the results and approved the final version of the manuscript.

References

1. COWI and The Danish Road Directorate Publication. En sammenligning mellem kvinders og mænd trafikadfærd og involvering i trafikulykker. Denmark, August 2015.

2. The Danish Road Directorate's Accident Statistics.
3. COWI Calculations based on Data from the Danish National Travel Survey for the Period 2008–2012.
4. COWI Calculations on the Basis of Register Data from Statistics Denmark.

The Standing Committee on Women's Issues in Transportation (ABE70) peer-reviewed this paper (18-03967)

This paper is based on the publication "Gender Differences in Road Accidents" published by the Danish Road Directorate in January 2017.

Article

TRR
JOURNAL OF THE TRANSPORTATION RESEARCH BOARD

Transportation Research Record
2018, Vol. 2672(3) 175–184

DOI: 10.1177/0361198118796005
journals.sagepub.com/home/trr
SAGE

Traffic Calming and Environmental Justice: New York City's Neighborhood Slow Zones

Jonas Xaver Hagen[1]

Abstract

This paper examines New York City's Neighborhood Slow Zones (NSZ) program in terms of environmental justice. The paper uses both quantitative and qualitative methods. Quantitative analyses show that the areas where the zones are implemented are well represented in regarding environmental justice (low-income and minority) populations, and that risk exposure to traffic injury and traffic casualty counts are similar in NSZ and non-NSZ areas. The qualitative analysis shows that the program was structured in a way that included the participation of environmental justice communities and led to the siting of zones in such neighborhoods. These findings suggest that the NSZ program can address environmental justice's goals of distributing environmental risk more equitably and including low-income and minority communities in planning processes.

Environment-related health burdens are disproportionately borne by minority and low-income populations, a reality that the concept of environmental justice seeks to address (*1*). Further, because minority and low-income populations have been historically excluded from decision-making that led to the inequitable distribution of environmental hazards, another important aspect of environmental justice is the call to include these populations in participatory planning processes related to the distribution of such hazards (*2*).

While much of the research on environmental justice is related to toxic land uses, another environmental hazard that disproportionately affects environmental justice populations is road traffic injury (*3*). One tool that promises to reduce traffic casualties is area-wide traffic calming (*4*). Under this measure, rather than implementing traffic calming devices on individual streets or intersections, they are implemented in entire areas. This paper analyzes an area-wide traffic calming program, New York City's Neighborhood Slow Zones (NSZ), in terms of environmental justice. It examines the zones' spatial distribution (whether they were equitably distributed), as well as the planning process by which they were sited (whether the process included the contributions of minority and low-income residents). The paper uses qualitative and quantitative methods to answer the following research question: *Does the NSZ program further environmental justice in New York City?*

The NSZs consist of 28 areas of approximately five square blocks in which the speed limit is reduced to 20 miles per hour and traffic calming measures, physical alterations to the roadway, are introduced to slow auto speeds. The quantitative analysis examines the NSZs' distribution regarding environmental justice populations, and the qualitative data describe the process that led to the siting of the zones. The analysis shows that the program furthers environmental justice in New York City, both concerning the distribution and the planning processes used to site the zones.

This paper proceeds as follows: after reviewing literature on environmental justice and traffic calming, the methods, then the results of the analyses, are discussed. The paper continues with a discussion of the findings, and finally presents policy implications and conclusions.

Literature Review

Environmental Justice

When the environmental justice movement began in the 1980s, it focused on the siting of toxic land uses, such as hazardous waste disposal sites, in minority and low-

[1]Department of Urban Planning, Columbia University, New York, NY

Corresponding Author:
Address correspondence to Jonas Xaver Hagen: jh3301@columbia.edu

income areas. Pressure from grassroots organizations led to the 1994 Presidential Executive Order 1298, which required Federal Agencies to "make achieving environmental justice part of its mission by identifying and addressing, as appropriate, disproportionately high and adverse human health or environmental effects of its programs, policies, and activities on minority populations and low-income populations" (*5*). Research on environmental justice has examined the siting of Locally Unwanted Land Uses in general in the United States (*6*) and specific uses, including solid waste disposal sites (*1, 7*), and nuclear waste facilities (*8*). Researchers have also documented minority populations' disproportionate exposure to industrial hazards (*9*) and particulate air pollution (*10*).

In addition to focusing on the more equitable distribution of environmental risks, environmental justice activists and researchers have also emphasized community participation in planning processes (*2, 11, 12*). Minority and low-income populations have historically been excluded from mainstream environmental groups and planning processes that locate noxious land uses. Yet, environmental activists and researchers see authentic community participation as a way to help enfranchise such vulnerable populations and lead to more equitable outcomes in which minority and low-income populations do not carry a disproportionate environmental burden.

Environmental Justice and Transportation

The United States Department of Transportation (USDOT) is among the agencies specifically cited in the 1994 Executive Order. The USDOT's directives on environmental justice instruct USDOT-funded agencies to incorporate the following guiding principles into transportation planning: (1) ensure full and fair public participation of minority and low-income populations in transportation decision-making processes, (2) avoid, minimize, or mitigate disproportionate adverse health and human effects on these populations, and (3) ensure that the benefits of transportation systems are delivered equitably and promptly to minority and low-income populations.

Transportation researchers interested in environmental justice have examined issues created by the standard urban transportation planning process, which has "generally ignored" (*13*) the concerns of minority and low-income communities. These issues include disparities in funding for transportation infrastructure, resulting in restricted access to destinations such as jobs, education, health care services, and grocery stores in poor, minority areas (*14, 15*). Discrimination in transportation policy and infrastructure, also called "transportation racism" (*16*), often resulted in the severing or demolition of low-income, minority neighborhoods to make way urban freeways, particularly from the 1950s to the 1970s (*17*). Throughout the United States, minority and low-income populations live in closer proximity to high-volume roads as compared with White and middle- or upper-income populations (*18*). Researchers have also shown that environmental justice populations have higher exposure to air and noise from roadways (*19, 20*). Efforts to improve the environmental justice of the transportation system have led to evaluations of proposed road construction projects (*21, 22*).

Research has also examined environmental justice as related to traffic casualties in the U.S. Nationwide, motor vehicle injuries exact a disproportionate toll on racial and ethnic minorities (*23*) and on low-income neighborhoods (*24, 25*). Further, minorities are overrepresented in pedestrian fatalities in the United States (*13*). Specifically, the incidence of pedestrian–vehicle crashes is relatively elevated in environmental justice areas of the Chicago Metropolitan area (*3*), and pedestrian fatality rates were significantly higher for African-Americans and Latinos in Atlanta (*26*).

Traffic Calming

Standard traffic planning and engineering practices encourage the creation of high-speed roads for autos and increased traffic speeds on road networks, often to the detriment of vulnerable road users (pedestrians and cyclists) (*27, 28*). Traffic calming seeks to slow auto speeds and discourage fast-moving traffic by introducing physical barriers, such as speed humps, curb extensions, and traffic circles, in roadways. Area-wide traffic calming has been shown to reduce traffic casualties (*4, 29*) and emissions of local pollutants and greenhouse gases (*30–32*). Other motivations for implementing traffic calming include improving comfort for vulnerable road users, neighborhood quality of life (*33*), and social inclusion (*34*).

Although much research has been published on the effects of traffic calming, there have been relatively few studies on the equitable distribution and siting of traffic-calmed streets and zones with regard to socioeconomic characteristics. The research for this paper found only two such studies, both from the UK. In those studies researchers also found a higher proportion of traffic-calmed road segments in lower-income areas of England and Wales (*35*), and a study of London's 20 mph zones found that numbers of traffic casualties decreased most in poorer areas of the city (*36*). This paper represents the first attempt at evaluating the distribution of traffic calming through the lens of environmental justice, considering not only the demographic aspect of this distribution in terms low-income and minority areas, but also regarding the inclusion of these populations in participatory planning processes.

Background

Environmental Justice in New York City

New York City has witnessed several important environmental justice initiatives. The "fair share" criteria, adopted in 1990 by the City Planning Commission, ensure that undesirable land uses are borne equitably by every borough and every community in the city (*37*). In addition, in two separate initiatives, community activists and researchers collaborated to document disproportionate exposure to air pollution and asthma rates in minority, low-income areas resulting from heavy bus traffic in Upper Manhattan (*38*, *39*) and truck traffic in Hunts Point, in the Bronx (*40*).

Although previous research has not looked at traffic casualties New York City specifically as an environmental justice issue, researchers have shown that minority and low-income areas are overrepresented in crashes and casualties. The burden of traffic casualties is higher in areas with higher percentages of Black and Hispanic residents (*41*). Further, poor areas of the city have a higher occurrence of pedestrian–vehicle collisions than in middle- or upper-income neighborhoods (*42*).

Researchers have also found that poor areas have a generally less favorable pedestrian environment, including higher crime rates, higher speed limits, and wider streets, and fewer pedestrian amenities (e.g., street trees, landmarked buildings, clean sidewalks) (*42*). Further, high-poverty, minority New Yorkers living in areas with fewer pedestrian amenities and higher homicide rates were less likely to engage in active transportation (walking and cycling) than wealthier, non-hispanic white New Yorkers living in pedestrian-friendly areas (*43*). Finally, studies have shown that poor neighborhoods are subject to elevated levels of air pollution from road traffic, resulting in higher asthma hospitalization rates in low-income minority areas of New York City (*44–46*).

Although environmental justice populations bear the burden of the externalities of motorized traffic, they are less likely to reap its benefits. Workers in minority and low-income areas of New York City were less likely to use a car and more likely to use public transportation than workers in non-environmental justice areas (*47*).

The Origin of the Neighborhood Slow Zones Program

Traffic policy in New York City has increasingly emphasized environmentally sustainable modes (walking, cycling, and public transportation) and safety for pedestrians and cyclists in recent years. Since 2007, the New York City Department of Transportation (the DOT) increased efforts to create safer streets by expanding the cycle network and making pedestrian safety improvements such as Safe Routes to School and Safe Routes for Seniors. The *New York City Pedestrian Safety Study and Action Plan* (*41*) proposed a pilot program that would test the safety performance of neighborhood zones with 20 mph speed limits. This led to the creation of the NSZ program in 2011. The DOT implemented 28 zones from 2011 to 2016. Using data from the 2010 Census, about 540,000 New Yorkers (6.6% of the City's population) reside within the boundaries of the 28 zones.

Methods and Data

This paper uses mixed methods, a research approach that combines quantitative and qualitative methodologies (*48*, *49*). Two analyses, those of the spatial location of NSZs in terms of low-income and minority populations, and of the risk exposure and traffic casualties in those zones, use quantitative data and methods. A third analysis, which evaluates the NSZ program regarding environmental justice and community participation in the planning of the zones, uses qualitative data on the NSZ program.

The paper employed spatial tools to extract data on demographics and crashes in New York City, and statistical software to analyze the data. Population data for race/ethnicity came from the 2010 U.S. Census's Summary File 1 (SF1) at the block group level, and for poverty, from the 2006–2010 American Community Survey at the census tract level. The borders of the zones were included in a shapefile downloaded from a DOT website (*50*). Data on motor vehicle crashes for the years 2005–2010 came from the New York City DOT, acquired via the Freedom of Information Law. This study used data through 2010 because the first NSZ was implemented in 2011.

Qualitative data were gathered from interviews and DOT documents. The data include interviews with two DOT officials and a staff member of a local transportation advocacy group (Transportation Alternatives). The two interviews were semi-structured and about 40 and 20 minutes long, respectively. The documents included the DOT's webpages on the NSZ program, as well as DOT presentations available on the Department's website. The author used "lean coding" (*51*) to organize the content of the interviews, webpages and presentations into themes. In gathering and organizing the qualitative data, the author also relied on previous, informal conversations he had with DOT officials and citizen activists regarding the program, which he had been following as a participant observer since February 2013.

Unit of Analysis and Variables

Following previous research on environmental justice (*3*, *47*, *52*), census tracts were determined to be the most appropriate unit of analysis for the study. Larger than census blocks or census block groups, but smaller than zip codes or community districts, census tracts are spatially disaggregated enough to capture differences in socioeconomic characteristics of areas of New York City. Census tracts in New York City have a mean population of 3,774 (standard deviation: 2,176), and the mean population of NSZs was 19,189 residents living within the borders of the zones (standard deviation: 11,916).

To determine which census tracts were environmental justice or community of concern tracts, thresholds were used for minority and low-income populations, and New York City as the reference area to determine these thresholds. Regarding the threshold for minority census tracts, 66.7% of New Yorkers were minorities in 2010 (decennial census), including Black, Latino, Asian, Native Americans, and Pacific Islanders. The threshold for low-income tracts used the city-wide mean of households living in poverty, which was 18.8% in 2010 (*53*). Using New York City (as opposed to the U.S., New York State, or the New York City region) as a reference area for thresholds ensures that the analysis is relevant to the area of study.

Following Federal guidelines, this paper considers tracts that are above *either* threshold "environmental justice" tracts. Tracts that are above *both* minority and poverty thresholds are considered "communities of concern" (*47*, *54*). Although the threshold method to determine communities of concern has drawbacks as compared with a participatory, self-identification method, it can nonetheless be useful to reach a spatial understanding of the needs and vulnerabilities of a community and to examine a large area (*55*).

Statistical Analyses

This paper includes two statistical analyses. The first analysis compares the proportion of environmental justice populations in the NSZ tracts with the same levels for the entire city, and uses chi-squared to test for significance. The second analysis is a multiple logistic regression that examines the likelihood of NSZ tracts having environmental justice populations, of risk exposure to traffic injury, and of traffic casualties occurring in the NSZ tracts relative to non-NSZ tracts. Of interest was risk exposure and traffic injuries in the NSZ tracts, in order to assess if the zones were implemented in areas that needed them in terms of traffic safety; if these areas were already safe, implementing zones in them would have little consequence for reducing traffic injuries. Both analyses use alpha levels of .05 to determine statistical significance.

This study used population density as a proxy for exposure to risk of traffic injuries and fatalities. Ideally, measures of risk exposure would include volumes of all types of traffic users, including motorized traffic, pedestrians, and cyclists. In practice, researchers often use only motorized traffic volumes, as municipalities rarely, if ever, collect comprehensive data on pedestrian or cyclist volumes (*56*). As data on motorized traffic are not uniformly available for streets in New York City, the regression used population density as a proxy for exposure to the risk of road traffic injury, as density is positively correlated with pedestrian injuries (*57–60*), which disproportionately affect minority and low-income populations.

The multiple logistic regression used the following formula.

$$ln\frac{Pr(NSZ)}{(1-Pr(NSZ))} = a + b_{\text{Minority}} + b_{\text{Low-income}} + b_{\text{CoC}} + b_{\text{Density}} + b_{\text{Vulnerable User}} + b_{\text{MVO}}$$

where:

The dependent variable is the natural logarithm of the odds of the dependent variable being a case, that is, the probability (Pr) of a tract being an NSZ tract or not. *NSZ* is a binary dependent variable (0, 1), where 1 signifies that a census tract was one of the 145 tracts contained within, or intersecting the borders of the 28 NSZs, and 0 was a non-NSZ tract,

a is the intercept,

b are the estimated regression coefficients,

Minority indicates the percentage of minority residents in the tract,

Low-income indicates the percentage families living below the poverty line in the tract,

CoC is a binary variable (0, 1) that indicates whether a tract is a "community of concern," that is, above thresholds in New York City for minority (66.7%) and low-income (18.8% of families living in poverty),

Density is population density per square mile, a proxy for exposure to pedestrian traffic injury, as discussed above,

Vulnerable User indicates the total number of pedestrian and cyclist casualties for each census tract for the years 2005–2010, and

MVO indicates the total number of Motor Vehicle Occupant casualties for each census tract for the years 2005–2010.

Results

As previously stated, the study used New York City as the reference area to determine which census tracts of the city were minority and low-income areas. Figure 1 shows zones that were at or above the threshold for minority residents (66.7%), families living below the poverty line (18.8%), and the NSZs.

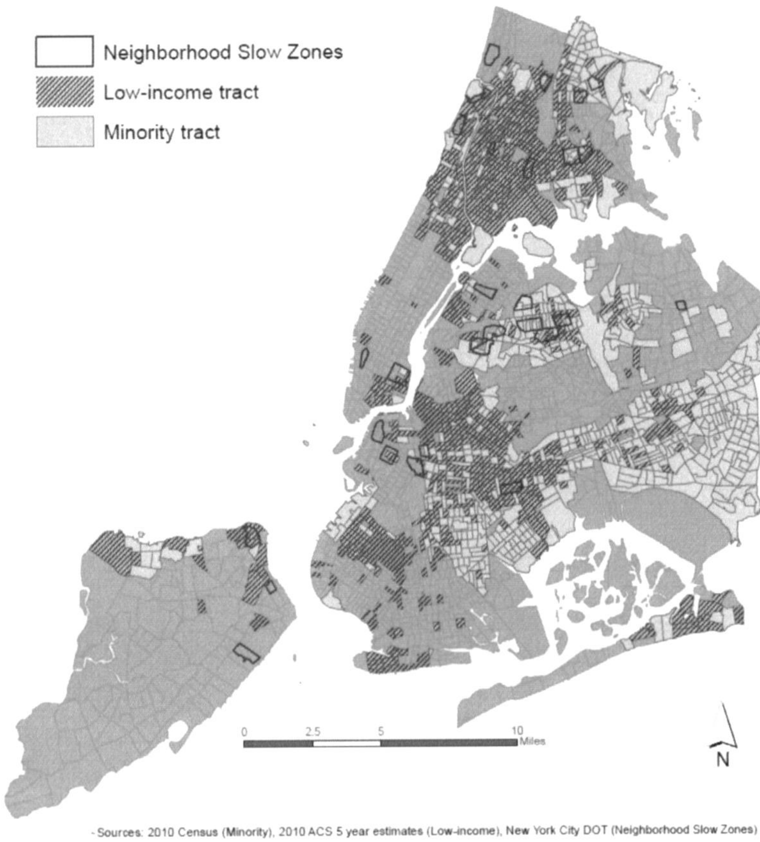

Figure 1. Map of environmental justice census tracts and Neighborhood Slow Zones.

New York City contains 2,166 census tracts. A total of 145 tracts are contained within or intersect the borders of the 28 NSZs. City-wide, 66.7% of the population is minority, and 18.8% is low-income (Table 1). Of all tracts in New York City, 26.5% were communities of concern (minority *and* low-income) and 59.6% were environmental justice (minority *or* low-income). The corresponding proportions in NSZ tracts were 71.8%, 16.7%, 34.5%, and 64.1%, respectively. Although the proportions of minority, communities of concern, and environmental justice populations were slightly higher in NSZ tracts, the low-income population was slightly lower in NSZ tracts than for all of New York City. Chi-squared tests revealed that these differences were statistically significant at the 95% confidence level only for communities of concern. Overall, this analysis shows that environmental justice tracts are represented at similar proportions in the NSZ areas as compared with all of New York City.

The results of the logistic regression, which explores the relationship between NSZ tracts, environmental justice populations, exposure to risk of traffic injury, and counts of traffic casualties, are displayed in Table 2 (below). The table includes the odds ratio (derived from the estimated regression coefficients), the 95% confidence interval, and the p-values of the independent variables.

The results showed two variables to be statistically significant at the 95% confidence level: Density, and Vulnerable User Casualty. However, the odds ratios for these variables were very close to 1 (.99), suggesting that risk exposure and cyclist and pedestrian injuries in NSZ tracts are similar to those of the rest of New York City.

The variable with the largest odds ratio (1.64) is Community of Concern. However, the result for Community of Concern in this model cannot be differentiated from what might occur by chance at the 95% confidence interval.

Minority, Low-income, and MVO Casualty had neither particularly large or small odds ratios, nor significant p-values. These results imply that the likelihood that NSZ tracts are above the minority and low-income thresholds for New York City are the same as for non-NSZ tracts. The same is true for MVO casualties.

Table 1. Environmental Justice Tracts in New York City and Neighborhood Slow Zones

	New York City tracts ($n = 2166$)		NSZ tracts ($n = 145$)				
	Total	Percentage	Total	Percentage	Difference	Chi-squared	p-value
Minority	1187	66.7	90	71.8	5.1%	1.60	0.20
Low-income	681	18.8	47	16.7	−2.1%	0.39	0.53
Community of concern	573	26.5	50	34.5	8%	4.41	0.04*
Environmental justice	1290	59.6	93	64.1	4.5%	1.15	0.28

Note: *Significant at the 95% confidence level.

Table 2. Multiple Logistic Regression of NSZ and Non-NSZ Tracts

Independent variable	Odds ratio	95% CI	p-value
Percent minority	1	.99, 1.01	.47
Percent low-income	1	.98, 1.02	.75
Community of concern	1.64	.90, 3	.10
Density	.99	.98, .99	.00*
Vulnerable user casualty	.99	.98, 1	.01*
Motor vehicle occupant (MVO) casualty	1	1, 1	.13

Note: $n = 2{,}166$. *Significant at the 95% confidence interval.

NSZ Program: Planning and Implementation

What follows are the results of the qualitative research on the NSZ program. The results of the qualitative data are divided into the following categories: (1) applications and review, (2) presentation to the community, and 3) implementation and results.

Application and Review. The DOT created a webpage to enable individuals and organizations to request a zone by submitting an application form. Guidelines on the DOT website instructed applicants to choose an area roughly a quarter-mile in area, or five square blocks, that was primarily residential, avoided wider, heavier volume streets or industrial and major commercial sites, and had strong boundaries, including major streets, parks, and elevated rail lines (*61*). Applicants were instructed to include an inventory of institutions located inside the boundaries of the zones, such as schools or senior centers. The DOT also requested that applicants demonstrate support for the zones by providing letters of support from community organizations such as local civic groups, churches, and Community Boards (a form of local government in New York City), as well as from elected representatives and local police precincts. Successful applications were typically submitted by and included letters of support from a combination of these actors. After the initial pilot NSZ was implemented in 2011, the DOT accepted applications for NSZs for two periods. There were over 100 applications for the first round of zones (13 zones selected 2011–2012). The DOT received 74 applications for the second round, (14 zones selected 2014–2015) (*61*). For the second round of zones, the DOT reached out to community groups, such as the Brownsville Partnership in Brooklyn (from the predominantly African-American neighborhood of Brownsville), to make them aware that the DOT was accepting applications for the zones (personal communication, DOT officials, 03 October 2016).

The DOT reviewed the applications, ranking the proposed zones according to need and feasibility for implementation. The two most important factors in ranking the proposed traffic-calmed areas were support for the zones and the numbers of collisions in the zones. Support for the zones was prioritized for the first round of NSZs with traffic casualty counts weighted second; for the second round of zones, this was reversed, with casualty counts taking first priority and support given second priority. The DOT also considered the presence of facilities such as daycare, schools, senior centers, parks, and proximity to subway stations. After selecting the potential zones for implementation, the DOT adjusted the borders of the zones proposed by the community groups as needed in order to create an area with strong boundaries that could effectively slow traffic.

Presentation to the Community. DOT personnel then presented these proposed zones at Community Board meetings, which are open to the public. These presentations outlined the safety benefits of traffic calming and

20 mph/30 kmph zones, citing the resulting reductions in traffic injuries. They also emphasized the potential gains in neighborhood quality of life of the 20 mph zones, including reductions in traffic noise and through traffic and "more social streets." The presentations gave data on traffic injuries in the zones and evidence of local support for the zones (e.g., applications from community groups and Community Boards, elected representatives, letters of support, and petitions). They then showed the boundaries of the proposed zones and concluded with additional arguments for the implementation of the zones. These arguments emphasized the presence of schools or senior centers and speeding motorists inside the boundaries of the zones.

Implementation and Results. The selected zones that did not encounter considerable opposition at Community Board meetings were implemented. Of 30 selected zones, two were not implemented because of community opposition: Midland Beach, Staten Island, and Crown Heights, Brooklyn. A third zone, Clinton Hill, Brooklyn, was originally planned to encompass the neighboring neighborhood of Bedford-Stuyvesant, but was made smaller because of opposition in the latter area.

DOT officials stated that the feedback regarding the zones from communities has been generally very positive, that "people like the zones." In general, there was "less pushback" from the public toward NSZs than other types of facilities the agency typically implements. The DOT officials also mentioned that several elected representatives, including members of the New York State Assembly and City Council, sought zones for their district and generated positive press from the implementation of the zones. A staff member of an advocacy group for sustainable transportation modes, Transportation Alternatives, praised the program, saying it was at the "intersection of citizen participation and evidence-based policy" (personal communication, Julia Kite, 20 September 2016). She said that her organization was also concerned with the equity and worked to ensure that such safety improvements were distributed fairly throughout the city, not just in those communities that "shout the loudest." Her organization saw NSZs as a cost-effective and "citizen-driven" way to quickly create safer streets.

Discussion

The results of the above analyses of the NSZs are encouraging with regard to environmental justice, both regarding location and community participation. The first analysis shows that NSZs are distributed in environmental justice areas in similar proportions as compared with all of New York City. While the proportion of environmental justice populations was at least the same in NSZ areas as for the entire city, the chi-squared analysis showed that proportions of communities of concern are significantly higher for NSZ tracts than city-wide. The results of the multiple logistic regression were consistent with this result, with communities of concern having the odds ratio with the greatest magnitude, although this result was not statistically significant at the 95% confidence interval. Overall, it is fair to say that the DOT was successful in distributing the zones equitably in minority and low-income areas of the city.

The regression also found that the zones are adequately implemented in New York City when considering risk exposure and counts of traffic casualties. Risk exposure to traffic injury and counts of vulnerable user casualties were almost the same (only very slightly lower) in NSZ tracts and non-NSZ tracts. For MVO casualties, the probability was the same for NSZ tracts. This shows that the DOT located the NSZs in areas of the city that had similar traffic safety needs as compared with the city as a whole.

In addition to the location of the zones, the author was also interested in exploring whether the program furthered environmental justice by incorporating minority and low-income communities in participatory planning processes. The qualitative data gathered for the paper showed that the 28 NSZs were created in a participatory, community-driven, "bottom–up" manner. Further, beyond the participatory nature of the program, the DOT structured the program in ways that made it more likely to reach environmental justice populations. When soliciting NSZ proposals, the DOT reached out to community groups, particularly in low-income and minority areas, such as the predominantly African-American neighborhood of Brownsville in Brooklyn. Next, the DOT gave priority to areas with high counts of traffic casualties, which are more likely to be minority, low-income areas. This was particularly true in the second round of NSZs, which prioritized traffic casualties over levels of support when selecting potential zones for implementation. This finding shows that the NSZ program furthered the environmental justice goal of incorporating minority and low-income communities into the planning process. The equitable siting of the zones reflects the redistributive nature of the planning process the DOT employed.

The location of the zones in environmental justice areas suggests that traffic calming is well received in many environmental justice communities. This is interesting, considering that reportedly, minority, and particularly African-American, community groups have opposed initiatives that support cycling infrastructure, such as bike lanes and bike share systems, in cities including Portland, Austin, Washington D.C., and New York City (*62–66*). This opposition is said to be related to tensions surrounding gentrification, and research has shown an association between bike lanes and gentrification in San Francisco, Chicago, and Portland (*67*, *68*). Thus, compared with bike lanes, area-wide traffic calming may be better received in environmental justice communities.

Policy Implications and Conclusion

The outcomes of this study suggest that the NSZ program has been successful in furthering environmental justice, both regarding the location of the zones and in the participatory process that led to their siting. The healthy representation of minority and low-income areas in the NSZs suggests that traffic calming is well received in such neighborhoods. Further, the large number of applications for zones, which outnumbered the implemented zones by about 6 to 1 (174 applications to 28 implemented zones), suggests that there is a large demand for safer streets and sustainable transportation in New York City.

As such, the NSZ program appears to have important potential to further environmental justice by introducing traffic calming in additional low-income, minority areas of the city. The DOT could build on the program's success by using similar outreach methods and selection criteria to bring the zones to additional environmental justice areas of the city.

However, the participatory approach used by the program may have limits with regard to reaching the most dangerous places for traffic injuries. Of course, the NSZ program is not the only traffic safety program of the DOT. Particularly since adopting a "Vision Zero" policy (the goal of no deaths or serious injuries from road traffic) in 2014, the DOT has increasingly focused on "hot spots"—places where high amounts of injuries occur. The NSZ program's citizen-driven approach may be a good complement to ongoing data-driven traffic safety initiatives adopted under Vision Zero.

The NSZ program has excellent potential to address directly all three objectives of the USDOT's environmental justice directives, that is to: (1) ensure full and fair public participation of minority and low-income populations in transportation decision-making processes, (2) mitigate disproportionate adverse health and human effects on these populations, and (3) deliver the benefits of transportation systems equitably to minority and low-income populations. Accordingly, the USDOT could provide support for the implementation of similar programs in municipalities throughout the country.

Author Contributions

The author confirms sole contribution to the paper as follows: study conception and design; data collection; analysis and interpretation of results; and draft manuscript preparation. The author reviewed the results and approved the final version of the manuscript.

References

1. Bullard, R. D. *Dumping in Dixie: Race, Class, and Environmental Quality*, Vol. 3. Westview Press Boulder, CO, 2000.
2. Schlosberg, D. The Justice of Environmental Justice: Reconciling Equity, Recognition, and Participation in a Political Movement. *Moral Polit Reason Environ Pract*, Vol. 77, 2003, p. 106.
3. Cottrill, C. D., and Thakuriah, P (Vonu). Evaluating Pedestrian Crashes in Areas with High Low-income or Minority Populations. *Accid Anal Prev*, Vol. 42, No. 6, 2010, pp. 1718–1728.
4. Bunn, F., T. Collier, C. Frost, K. Ker, R. Steinbach, I. Roberts, and R. Wentz. Area-wide Traffic Calming for Preventing Traffic Related Injuries. *Cochrane Database of Systematic Reviews*, No. 1, 2003, CD003110.
5. Federal Register. *Federal Actions to Address Environmental Justice in Minority Populations and Low-Income Populations: Executive Order 12898 [Internet]*. 1994, http://www.eahcp.org/documents/1994_Clinton_ExecutiveOrder12898.pdf Accessed January 25, 2017.
6. Been, V. What's Fairness Got to Do with It? Environmental Justice and the Siting of Locally Undesirable Land Uses. *Cornell Law Review*, Vol. 78, 1992, p. 1001.
7. Heiman, M. K. Race, Waste, and Class: New Perspectives on Environmental Justice. *Antipode*, Vol. 28, No. 2, 1996, pp. 111–121.
8. Leonard, L. G. III. Sovereignty, Self-determination, and Environmental Justice in the Mescalero Apache's Decision to Store Nuclear Waste. *Boston College Environmental Affairs Law Review*, Vol. 24, 1996, p. 651.
9. Pais, J., K. Crowder, and L. Downey. Unequal Trajectories: Racial and Class Differences in Residential Exposure to Industrial Hazard. *Soc Forces*, Vol. 92, No. 3, 2014, pp. 1189–1215.
10. Jerrett, M., R. T. Burnett, P. Kanaroglou, J. Eyles, N. Finkelstein, C. Giovis, and J. R. Brook. A GIS–Environmental Justice Analysis of Particulate Air Pollution in Hamilton, Canada. *Environment and Planning A*, Vol. 33, No. 6, 2001, pp. 955–973.
11. Hamilton, C. Coping with Industrial Exploitation. *Confronting Environmental Racism: Voices form the Grassroots*. Vol. 63, 1993, p. 70.
12. Lake, R. W. Volunteers, NIMBYS, and Environmental Justice: Dilemmas of Democratic Practice. *Antipode*, Vol. 28, No. 2, 1996, pp. 160–174.
13. Deka, D. Social and Environmental Justice Issues in Urban Transportation. In The Geography of Urban Transportation (Hansona, S., and G. Giuliano, eds.), Guilford Press, 2004, New York City, NY, pp. 332–355.
14. Karner, A, and D. Niemeier. Civil Rights Guidance and Equity Analysis Methods for Regional Transportation Plans: A Critical Review of Literature and Practice. *Journal of Transport Geography*, Vol. 33, 2013, pp. 126–134.
15. Sanchez, T. W., R. Stolz, and J. S. Ma. Moving to Equity: Addressing Inequitable Effects of Transportation Policies

on Minorities. Civil Rights Project at Harvard University, Cambridge, MA, 2003.
16. Bullard, R. D., G. Johnson, and A. O. Torres. Highway Robbery. *Camb South End Press*, Vol. 56, 2004, pp. 142–150.
17. DiMento, J. F., and C. Ellis. Changing Lanes: Visions and Histories of Urban Freeways. MIT Press, MA, 2013.
18. Rowangould, G. M. A Census of the US Near-Roadway Population: Public Health and Environmental Justice Considerations. *Transportation Research Part D: Transport and Environment*, Vol. 25, 2013, pp. 59–67.
19. Forkenbrock, D. J., and L. A. Schweitzer. Environmental Justice in Transportation Planning. *Journal of the American Planning Association*, Vol 65, No. 1, 1999, pp. 96–112.
20. Jacobson, J. O., N. W. Hengartner, and T. A. Louis. Inequity Measures for Evaluations of Environmental Justice: A Case Study of Close Proximity to Highways in New York City. *Environment and Planning A*, Vol. 37, No. 1, 2005, pp. 21–43.
21. Chakraborty, J. Evaluating the Environmental Justice Impacts of Transportation Improvement Projects in the US. *Transportation Research Part D: Transport and Environment*. Vol. 11, No. 5, 2006, pp. 315–323.
22. Lane, L., S. Hoffeld, and D. Griffin. Environmental Justice Evaluation: Wilmington Bypass, Wilmington, North Carolina. *Transportation Research Record: Journal of the Transportation Research Board*, 1998. 1626: 131–139.
23. McKenzie, B. *Modes Less Traveled—Bicycling and Walking to Work in the United States: 2008–2012*. US Census Bur N Y [Internet]. 2014, https://www.census.gov/content/dam/Census/library/publications/2014/acs/acs-25.pdf. Accessed November 20, 2015.
24. Cubbin, C., F. B. LeClere, and G. S. Smith. Socioeconomic Status and Injury Mortality: Individual and Neighbourhood Determinants. *Journal of Epidemiology & Community Health*, Vol. 54, No. 7, 2000, pp. 517–524.
25. Cubbin, C., and G. S. Smith. Socioeconomic Inequalities in Injury: Critical Issues in Design and Analysis. *Annual Review of Public Health*, Vol. 23, No. 1, 2002, pp. 349–375.
26. Hanzlick, R., D. McGowan, J. Havlak, M. Bishop, H. Bennett, R. Rawlins, B. Raines, K. DeBowles, D. Graves, T. Leet, D. Crites, M. Schmertmann, and K. Powell. Pedestrian fatalities—Cobb, DeKalb, Fulton, and Gwinnett Counties, Georgia, 1994–98. *MMWR. Morbidity and Mortality Weekly Report*, Vol. 48, 1999, pp. 601–605.
27. Southworth, M. Designing the Walkable City. *Journal of Urban Planning and Development*, Vol. 131, No. 4, 2005, pp. 246–257.
28. Noland, R. B. From Theory to Practice in Road Safety Policy: Understanding Risk Versus Mobility. *Research in Transportation Economics*, Vol. 43, No. 1, 2013, pp. 71–84.
29. Cairns, J., J. Warren, K. Garthwaite, G. Greig, and C. Bambra. Go Slow: An Umbrella Review of the Effects of 20 mph Zones and Limits on Health and Health Inequalities. *Journal of Public Health*, Vol. 37, No. 3, 2015, pp. 515–520.
30. Baldasano, J. M., M. Gonçalves, A. Soret, and P. Jiménez-Guerrero. Air Pollution Impacts of Speed Limitation Measures in Large Cities: The Need for Improving Traffic Data in a Metropolitan Area. *Atmospheric Environment*, Vol. 44, No. 25, 2010, pp. 2997–3006.
31. Madireddy, M., B. De Coensel, A. Can, B. Degraeuwe, B. Beusen, I. De Vlieger, and D. Botteldooren. Assessment of the Impact of Speed Limit Reduction and Traffic Signal Coordination on Vehicle Emissions Using an Integrated Approach. *Transportation Research Part D: Transport and Environment*, Vol. 16, No. 7, 2011, pp. 504–508.
32. Hedström, R. Miljöeffekter av 30 km/hi tätort–med avseende p\a a avgasutsläpp och buller. *VTI Medd* [Internet]. 1999, 869p. http://www.vti.se/sv/publikationer/pdf/miljoeffekter-av-30-kmh-i-tatort–med-avseende-pa-avgasutslapp-och-buller-en-forstudie.pdf. Accessed August 18, 2015.
33. Elvik, R., T. Vaa, A. Erke, and M. Sorensen. *The Handbook of Road Safety Measures*. Emerald Group Publishing, 2009.
34. Sauter, D., and M. Huettenmoser. Liveable Streets and Social Inclusion. *Urban Design International*, Vol. 13, No. 2, 2008, pp. 67–79.
35. Rodgers, S. E., S. J. Jones, S. M. Macey, and R. A. Lyons. Using Geographical Information Systems to Assess the Equitable Distribution of Traffic-Calming Measures: Translational Research. *Injury Prevention*, Vol. 16, No. 1, 2010, pp. 7–11.
36. Steinbach, R., J. Cairns, C. Grundy, and P. Edwards. Cost Benefit Analysis of 20mph Zones in London. *Injury Prevention*, Vol. 19, No. 3, 2013, pp. 211–213.
37. New York City Department of City Planning. *Fair Share: an Assessment of New York City's Facility Siting Process* [Internet]. Report No.: DCP# 95-10. 1995, https://www1.nyc.gov/assets/planning/download/pdf/about/publications/fair.pdf, Accessed March 16, 2017.
38. Minkler, M., A. G. Blackwell, M. Thompson, and H. Tamir. Community-based Participatory Research: Implications for Public Health Funding. *American Journal of Public Health*, Vol. 93, No. 8, 2003, pp. 1210–1213.
39. Minkler, M., V. B. Vásquez, and P. Shepard. Promoting Environmental Health Policy Through Community Based Participatory Research: A Case Study from Harlem, New York. *Journal of Urban Health*, Vol. 83, No. 1, 2006, pp. 101–110.
40. Lena, T. S., V. Ochieng, M. Carter, J. Holguín-Veras, and P. L. Kinney. Elemental Carbon and PM (2.5) Levels in an Urban Community Heavily Impacted by Truck Traffic. *Environmental Health Perspectives*. Vol. 110, No. 10, 2002, 1009.
41. Viola, R., M. Roe, and H. Shin. *The New York City Pedestrian Safety Study & Action Plan* [Internet], 2010, http://www.nyc.gov/html/dot/downloads/pdf/nyc_ped_safety_study_action_plan.pdf. Accessed 19 November 2015.
42. Neckerman, K. M., G. S. Lovasi, S. Davies, M. Purciel, J. Quinn, E. Feder, N. Raghunath, B. Wasserman, and A. Rundle. Disparities in Urban Neighborhood Conditions: Evidence from Gis Measures and Field Observation in New York City. *Journal of Public Health Policy*, Vol. 30, No. 1, 2009, S264–S285.
43. Lovasi, G. S., O. Schwartz-Soicher, K. M. Neckerman, K. Konty, B. Kerker, J. Quinn, and A. Rundle. Aesthetic

Amenities and Safety Hazards Associated with Walking and Bicycling for Transportation in New York City. *Annals of Behavioral Medicine*, Vol. 45, 2013, pp. 76–85.
44. Corburn, J., J. Osleeb, and M. Porter. Urban Asthma and the Neighbourhood Environment in New York City. *Health Place*, Vol. 12, No. 2, 2006, pp. 167–179.
45. Gwynn, R. C., and G. D. Thurston. The Burden of Air Pollution: Impacts among Racial Minorities. *Environmental Health Perspectives*, Vol. 109, 2001, p. 501.
46. Claudio, L., L. Tulton, J. Doucette, and P. J. Landrigan. Socioeconomic Factors and Asthma Hospitalization Rates in New York City. *J Asthma*. Vol. 36, No. 4, 1999, pp. 343–350.
47. New York Metropolitan Transportation Council. *Plan 2040: NYMTC Regional Transportation Plan* [Internet], 2013, https://www.nymtc.org/portals/0/pdf/Appendix4.pdf. Accessed December 16, 2016.
48. Creswell, J. W. *Research Design: Qualitative, Quantitative, and Mixed Methods Approaches*, 3rd ed. Thousand Oaks, CA: Sage Publications; 2009. 260 p.
49. Jupp, V., ed. Mixed Methods Research. In: *The SAGE Dictionary of Social Research Methods* [Internet]. SAGE Publications, Ltd., 1 Oliver's Yard, 55 City Road, London England EC1Y 1SP United Kingdom, 2006, http://methods.sagepub.com/reference/the-sage-dictionary-of-social-research-methods. Accessed April 6, 2017.
50. New York City Department of Transportation. *Vision Zero View* [Internet]. https://www.nycvzv.info/. Accessed October 6, 2016.
51. Creswell, J. W. *Qualitative Inquiry and Research Design: Choosing Among Five Approaches*, 2nd ed. Sage Publications, Thousand Oaks, CA, 2013.
52. Sriraj, P.S., G. Fruin, and S. McNeil. Analysis of the Environmental Justice Compliance of the Chicago Transit Authority (CTA). In: *Proc., 2003 Mid-Continent Transportation Research Symposium* [Internet]. 2003, http://origin.wcsu.csu.edu/cerc/documents/AnalysisoftheEnvironmentalJusticeComplianceoftheChicagoTransitAuthorityCTA.pdf. Accessed January 25, 2017.
53. Mayor's Office of Operations. *The CEO Poverty Measure 2005-2016* [Internet]. The City of New York, 2016, http://www.nyc.gov/html/ceo/downloads/pdf/CEO-Poverty-Measure-2016.pdf. Accessed 8 March, 2017.
54. Klancher, W., and S. Ritacco. *Briefing on "Communities of Concern" for the Proposed Enhancements to the Title VI/ Environmental Justice Analysis of the CLRP* [Internet], 2016, https://www.mwcog.org/assets/1/28/12212016_-_Item_10_-_Briefing_-_Communities_of_Concern_Environmental_Justice_Analysis.pdf. Accessed March 17, 2017.
55. Rowangould, D., A. Karner, and J. London. Identifying Environmental Justice Communities for Transportation Analysis. *Transportation Research Part A: Policy and Practice*, Vol. 88, 2016, pp. 151–162.
56. Gemzøe, L. *Are Pedestrians Invisible in the Planning Process? Copenhagen as a Case Study. Proc Aust Walk 21st Century* [Internet]. 2001, http://walk21.com/papers/perth01arepedestriansinvisible.pdf. Accessed August 18, 2015.
57. Wier, M., J. Weintraub, E. H. Humphreys, E. Seto, and R. Bhatia. An Area-level Model of Vehicle-pedestrian Injury Collisions with Implications for Land Use and Transportation Planning. *Accident Analysis & Prevention*, Vol. 41, No. 1, 2009, pp. 137–145.
58. Loukaitou-Sideris, A., R. Liggett, and H.-G. Sung. Death on the Crosswalk: A Study of Pedestrian-Automobile Collisions in Los Angeles. *Journal of Planning Education and Research*, Vol. 26, No. 3, 2007, pp. 338–351.
59. LaScala, E. A., D. Gerber, and P. J. Gruenewald. Demographic and Environmental Correlates of Pedestrian Injury Collisions: A Spatial Analysis. *Accident Analysis & Prevention*, Vol. 32, No. 5, 2000, pp. 651–658.
60. Morency, P., L. Gauvin, C. Plante, M. Fournier, and C. Morency. Neighborhood Social Inequalities in Road Traffic Injuries: The Influence of Traffic Volume and Road Design. *American Journal of Public Health*, Vol. 102, No. 6, 2012, pp. 1112–1119.
61. New York City Department of Transportation. *NYC DOT - Motorists & Parking - Neighborhood Slow Zones* [Internet]. http://www.nyc.gov/html/dot/html/motorist/slowzones.shtml. Accessed December 30, 2015.
62. Alcorn, C. Harlem Activists Fight Citibike "Gateway to Gentrification." *NY Daily News* [Internet]. 2016, http://www.nydailynews.com/new-york/manhattan/harlem-activists-fight-citibike-gateway-gentrification-article-1.2544565. Accessed March 14, 2017.
63. Goodyear, S. Bike Lane Backlash, Even in Portland [Internet]. *CityLab*, 2011, http://www.theatlanticcities.com/neighborhoods/2011/09/portland-bike-lanes-open-racial-wounds/138/. Accessed 29 March 2017.
64. Stein, P. Why are Bike Lanes such Heated Symbols of Gentrification? *The Washington Post*, [Internet], 2015, https://www.washingtonpost.com/news/local/wp/2015/11/12/why-are-bike-lanes-such-heated-symbols-of-gentrification/?utm_term = .a8a1011c5c4c. Accessed March 14, 2017.
65. Stein, S. Bike Lanes and Gentrification: New York City's Shades of Green. *Progressive Planning: The Magazine of Planners Network*, Vol. 188, 2011, pp. 34–37.
66. Walljasper, J. Do Bike Lanes Promote Gentrification? [Internet]. *Huffington Post*, 2013, http://www.huffingtonpost.com/jay-walljasper/do-bike-lanes-promote-gen_b_4178505.html. Accessed March 29, 2017.
67. Stehlin, J. Cycles of Investment: Bicycle Infrastructure, Gentrification, and the Restructuring of the San Francisco Bay Area. Environ Plan A. Vol. 47, No. 1, 2015, pp. 121–137.
68. Flanagan, E., U. Lachapelle, and A. El-Geneidy. Riding Tandem: Does Cycling Infrastructure Investment Mirror Gentrification and Privilege in Portland, OR and Chicago, IL? *Research in Transportation Economics*, [Internet]. 2016, http://linkinghub.elsevier.com/retrieve/pii/S0739885915300287. Accessed March 14, 2017.

The Standing Committee on Environmental Justice in Transportation (ADD50) peer-reviewed this paper (18-00400).

Article

TRR
JOURNAL OF THE TRANSPORTATION RESEARCH BOARD

Transportation Research Record
2018, Vol. 2672(3) 185–193

DOI: 10.1177/0361198118790618
journals.sagepub.com/home/trr
SAGE

The Costs of Compliance: An Environmental Justice Analysis of Automobile Emissions Testing Siting in Metropolitan Chicago

Kate Lowe[1] and Lauren Nolan[2]

Abstract

Ensuring automobile tailpipe emissions meet regulatory standards is an important mechanism for implementing the Clean Air Act Amendments. Research has documented the inequitable distribution of air pollution and its negative impacts, but little is known about how the costs of reducing vehicle emissions are distributed. Spurred by the closure of the sites for mandatory vehicle emissions testing within the city of Chicago, this study examines how implementation details—specifically emissions testing siting—could have unintended negative environmental consequences and how the burdens of traveling to emissions testing sites are distributed by race and income. A conservative estimate projects 1.9 million additional miles traveled from Chicago to suburban emissions testing sites during a two-year testing cycle, following the closures of Chicago's testing sites. Both before and after the closings, higher shares of poor and Black residents, who are protected by federal environmental justice rules, show statistically significant correlations with increased travel time to a testing facility. However, the prior time and distance advantage associated with higher shares of White, non-Latino residents is no longer statistically significant due to worsened access to facilities. From a historic perspective, this does not represent an environmental justice improvement, as rather than improving historically worse access for protected population groups, public facilities became less accessible. The study demonstrates the need for emissions testing programs to conduct environmental justice analyses of travel to testing sites and suggests to researchers the importance of retrospective environmental justice analysis to consider the continued legacy of past inequities.

Since the adoption of the Clean Air Act (CAA) in the United States, there have been significant reductions in the six criteria pollutants (particulate matter, ozone, lead, carbon monoxide, nitrogen dioxide and sulfur dioxide). The CAA and its amendments require that the Environmental Protection Agency (EPA) sets and enforces standards for these pollutants because of their negative effects on public health and welfare (the EPA currently does not set standards for carbon dioxide, the largest contributor to global climate change). Exposure to and public health problems from these and other air pollutants show disparities by race and income (*1–3*), whereas this study considers the burdens of traveling to vehicle emissions testing facilities that are part of CAA implementation.

Implementation of the CAA and its amendments has yielded tremendous benefits. For instance, an EPA study estimated that the 1990 Amendments prevented 13 million lost work days and 130,000 heart attacks in 2010 alone (*4*). Implementation has required that government agencies, from the federal to the local level, adopt measures to reduce air pollution (*5*), and highlighted that strong relationships among agencies and personnel were critical for meeting the new combined requirements of the 1990 CAA Amendments and the 1991 Intermodal Surface Transportation Efficiency Act (*6*).

States play an important role in CAA compliance. States where air quality fails to attain EPA standards must create state implementation plans to move toward compliance. Although states have latitude in determining how to attain air quality standards, the CAA Amendments of 1990 require vehicle inspection and maintenance programs for some areas—like metropolitan Chicago—that fail to meet ozone standards (*7*). Air

[1]University of Illinois at Chicago, Chicago, IL
[2]Woodstock Institute, Chicago, IL

Corresponding Author:
Address correspondence to Kate Lowe: katelowe@uic.edu

quality standards affect transportation planning and investments, as metropolitan planning organizations' transportation plans must conform to these state plans (*5*). Although reducing vehicle miles traveled (VMT) in metropolitan regions can be an important step toward improved air quality (and reduced greenhouse gas emissions, among other benefits), ensuring privately owned automobile tailpipe emissions meet standards is another important mechanism for reducing air emissions. Various types of cost-effective motor vehicle inspection programs are possible (*8*), and this study considers one such program administered in Illinois.

Although research has documented the inequitable distribution of air pollution, little research has examined how the costs of reducing tailpipe emissions are distributed. Spurred by the closure of all sites for mandatory vehicle emissions testing within the city of Chicago, this study examines how the details of implementation—specifically travel to emissions testing sites—could have unintended negative environmental consequences and whether the travel burdens related to emissions testing are inequitably distributed by race and income. The latter could be in conflict with federal environmental justice (EJ) provisions that ensure low-income and minority populations do not face disproportionate transportation burdens. This analysis uses a particular program to consider both aggregated increase in travel time and distance, as well as the relative change and burden across populations. It provides specific recommendations but also contributes to practice and research around historic EJ inequalities and implementation of CAA programs.

Automobile Use and Environmental Justice Populations

Transportation enables access to vital activities like employment, education, health care, and grocery shopping, as well as opportunities for social and civic engagement. One transportation challenge for low-income populations may be holding transportation costs down (*9*). Another challenge could be access to a private automobile. Given the existing transportation system and land use patterns in the United States, easy access to private automotive travel can be transformative for an individual's access to activities. Automobile access to jobs, for example, is significantly higher by private vehicle than by transit (*10, 11*).

Some studies thus assert that access to a car could matter more than relative spatial location between residence and employment for ensuring job accessibility (*10, 12*). Among housing voucher recipients, car ownership shows a positive statistical association with maintaining or *securing* a new job, whereas robust transit access shows a positive association only with *maintaining* employment (*13*). Likewise, among welfare recipients, employment and car ownership show a positive correlation (*14, 15*), whereas transit service levels by residential location show little relationship with employment status (*16*).

Extensive literature has suggested that transportation limitations could contribute to social exclusion and isolation (*17*). Accessibility to services and activities beyond employment matter as well, as limited mobility and transportation challenges are barriers for accessing vital opportunities for many vulnerable populations. A lack of transportation can be a barrier to accessing important governmental or social services, such as support for injection drug users (*18*).

Thus, even as environmental concerns mount, private automobiles are an important transportation mode in the United States, including for many low-income and minority households. In 2015, in the United States, only 8.9% of households had no vehicle available, 33.5% had one vehicle available, and the majority had two or more available vehicles (*19*). Not surprisingly, the accessibility enabled by the private automobile is not experienced equally across groups. The majority of poor households have at least one vehicle, although 28% of U.S. households with earnings of less than US$20,000 did not have a vehicle in 2009 (*20*). Race shows a relationship with car ownership stability, as more Black and Latino households only sometimes (44% and 32%, respectively) owned a car relative to non-Latino, White households (18%) in a longitudinal study (*21*).

Given the ubiquity of car ownership and its association with some employments benefits for low-income populations, better understanding the distribution of automobile ownership burdens merits further study. For instance, obtaining financing to purchase a car can be challenging for welfare recipients (*22*), as can be covering the cost of insurance (*14*). In fact, the poor can be disadvantaged through reliance on the "fringe" economy for automobile financing and insurance (*23*).

In addition to these marketplace challenges, more specific study is needed on how vehicle registration requirements might cause disproportionate burdens for low-income and minority automobile owners. Public agencies that receive federal funds must comply with EJ rules (which protect low-income and minority populations from disproportionate harms) and Title VI of the Civil Rights Act. For instance, in an Alabama case, the United States Department of Transportation found that—in violation of Title VI—Black residents were disproportionally burdened by the limited locations of driver licensing services (*24*). Thus, this study considers how siting of vehicle emissions testing facilities could burden populations protected by EJ rules (minority and low-income groups), as well as the potential for negative environmental impacts, through additional VMT, due to the closure of testing sites.

Methods

Compliance costs associated with emissions testing were measured based on travel time and corresponding VMT to the nearest testing facility. 'Nearest' facility was based on shortest travel time (as opposed to distance) along the road network. This analysis was completed at the zip code level, and travel times were calculated using the distance from the geographic centroid of each zip code tabulation area (ZCTA) to the address of the nearest testing facility (in terms of time) both prior to and following the closure of four emissions testing facilities. Zip codes were used because that was the unit for which administrative data on vehicles tested was available. Thus, the measurements are an approximation by zip code, not a disaggregated door-to-door precise calculation. Furthermore, vehicle owners may travel to emissions testing facilities from other locations (e.g., work, as part of a tour of destinations).

The analysis was completed using ArcGIS Network Analyst and a network dataset built from ESRI's U.S. and Canada Detail Streets layer file (*25*). This layer file uses 2007 TomTom data, was created in 2012, and modified in 2014. Relative elevation data in the layer file allowed for identifying highway overpasses and ramps for a more accurate road network analysis. The network dataset modeled global turn delays using 2 s for a left or right turn (classified as having turn angles of 120°), 3 s for a reverse turn (60°), and 2 s for a straight turn (60°). Travel time along road segments was determined using speed limit data included in the layer file. Without information on the day of week and time of day for travel to testing, the model did not account for traffic delays and congestion; thus travel times represent an off-peak, best case scenario for travel time.

The analysis calculates a conservative estimate of additional Chicago-based VMT due to the closure of the two Chicago testing sites. Vehicle counts by zip code inspected at the two closed Chicago sites (2014–2015) were obtained from the Illinois EPA through a Freedom of Information Act request. Because the agency did not have data on the two closed suburban sites, this analysis considers only the increased travel time and distance for the vehicles from the city of Chicago. Thus, the analysis demonstrates the impact of just two of the four metropolitan area closures over a full two-year testing cycle for only vehicles registered in Chicago zip codes. To calculate the aggregate increase in travel time and distance, the number of vehicles tested (at the two sites) from each Chicago zip code was multiplied by the increased distance and time to the nearest facility for the zip code, as calculated through the ArcGIS analysis described above. The increased aggregate time and distance among the Chicago zip codes was then summed.

Pre- and post-closure (current) nearest facility distances were analyzed relative to demographic variables by ZCTA. As noted in the following section, 2014 five-year data (*26*) were used to align with the time period of administrative data. In addition, ZCTA was selected as the scale to align with administrative data discussed above. Because the scale of analysis in EJ analyses can yield different results and because impacts both for the city and region matter, the city and region were both considered. ZCTAs with a centroid in Cook, DuPage, and Lake Counties (where all zip codes are subject to testing), as well as those zip codes with a centroid in Kane, Kendall, McHenry, and Will counties and subject to testing were included in the regional analysis.

Average travel times were compared by race and poverty. To test average travel time and distance for concentrations of people of color, zip codes were categorized as having or not having a concentration of people of color, which was defined as greater than the average of affected zip codes at the regional scale (48.5%, which is greater among the affected zip codes than the metropolitan statistical area) or the city average (67.8%). A *t*-test of difference in means of time and distance was calculated using Microsoft Excel's statistical package. Another test of difference in means was calculated based on whether there was a concentration of individuals below the poverty line (above 14.1% for impacted regional zip codes or 22.7% for the city of Chicago).

The software program SPSS was used to calculate correlation coefficients between travel and demographic variables. Spearman's (rank order) correlation coefficients were used, as a non-linear relationship between the demographic variables (all of which have non-normal distributions) was anticipated. A positive correlation with pre- or post-closure distance indicates a relative disadvantage, as it indicates an increase in travel time or miles to the nearest facility. A negative correlation shows a relative advantage, a tendency toward fewer miles or minutes to the nearest facility. The next section provides more context on the location of the study before describing analysis findings.

Findings

Case Study Context

In 2016, the Illinois EPA announced closure of four emissions testing sites within metropolitan Chicago, including the only two located within city limits (See Figure 1). The Chicago metropolitan area fails to meet the EPA standards for ozone and is thus a nonattainment area, with a required biennial vehicle testing

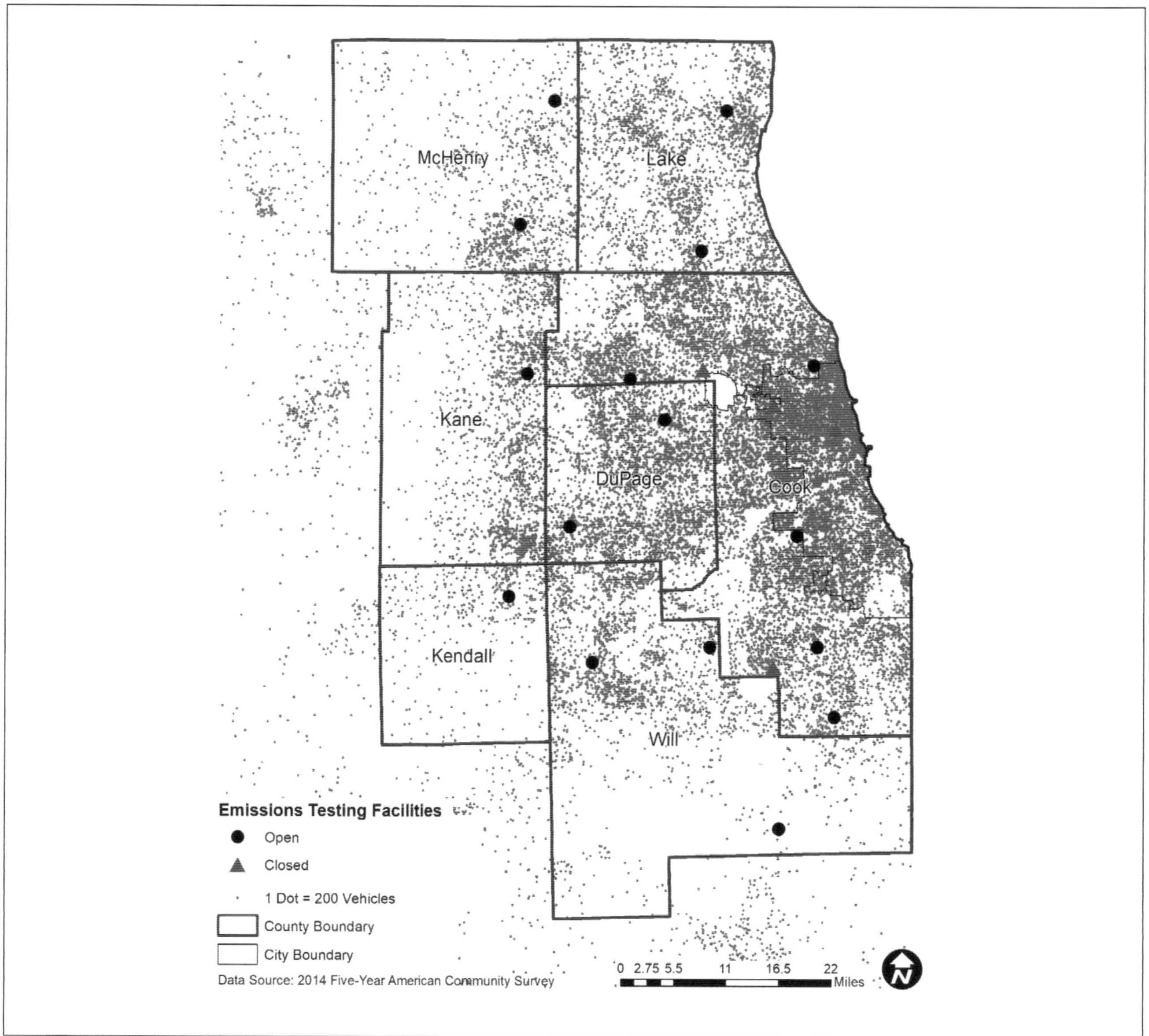

Figure 1. Distribution of emissions testing sites in metropolitan Chicago.

program for all vehicles that are four years old or older. The State of Illinois, through its EPA, contracts with a private vendor that operates vehicle emissions testing sites. As per state law, the contract specifies that all addresses with registered vehicles subject to mandatory testing must be within 12 miles of a testing facility. However, this is defined as a *straight line* distance, not distance on the roadway network, and does not reflect travel time. This worsening facilities scenario provided a context to consider the travel costs of emissions programs with specific attention to protected demographic groups that are relatively concentrated in the city of Chicago: people of color and those living below the poverty level.

The city of Chicago has a population of 2.7 million, and the metropolitan statistical area (Chicago-Naperville-Elgin, IL-IN-WI) has a population of 9.5 million (2014) (*27*). Because administrative data starting from 2014 were used, as described above in the Methods section, descriptive data from 2014 (five-year sampling 2010–2014) are used in the following subsection. Table 1 displays racial and ethnic demographic data for the metropolitan area and city, the latter which has a majority people of color population. The percent of individuals below the poverty level is 22.7% in the city and 14.1% in the region (*28*).

U.S. Census Bureau data show the aggregate number of available vehicles at 1.1 million in the city of Chicago and 5.6 million in the metropolitan area (*29*). The

Table 1. Population by Race and Ethnicity 2014 (27)

	Non-Latino			Latino
	White	Black	Asian	Any race
City of Chicago	32.2%	31.5%	5.7%	28.9%
Chicago metro area	54.3%	16.8%	5.9%	21.2%

Note: Percentages of each row will not total 100% due to rounding and omitted categories.

Table 2. Households with No Vehicle Available, 2006–2010 (*31*)

	By race/ethnicity		By poverty status	
	Householder is White alone, non-Latino	Other householder	Below poverty level	At or above poverty level
City of Chicago	22.8%	28.2%	52.7%	19.9%
Chicago metro area	8.3%	17.4%	36.9%	8.5%

Table 3. Unweighted Zip Code Average Travel Data

	Pre-closure time (minutes)	Pre-closure distance (miles)	Current time (minutes)	Current distance (miles)	Change in time (minutes)	Change in distance (miles)
Region	12.2	6.8	14.0	8.3	1.9	1.5
City	8.9	5.7	14.9	10.1	6.0	4.4

majority of households in both the city and metro have a least one vehicle available; only 12.2% of households in the metro area (26.9% in the city) have no vehicle available (*30*). Vehicles available by household vary by race, ethnicity, and poverty status as shown in the most recent Census Transportation Planning Products (2006–2010) tabulations (*31*). These data do show (Table 2), however, that the majority of households have at least one vehicle available, regardless of the householder's minority status. Within the city, a substantial share of households below the poverty line does have at least one vehicle available, even though a slight majority has no vehicle available. In the metro area, however, even a majority of households below the poverty line have at least one available vehicle.

Increased Vehicle Miles Traveled and Facility Siting

As expected, the closure of the four Chicago metropolitan area facilities led to an increase in average distance and travel time both for the region and the city. As shown in Table 3, prior to the closures, city ZCTAs on average (unweighted) had a shorter trip in time and distance to the nearest facility than ZCTAs across the region. ZCTAs within the city and across the region both saw an increase in average travel time, but after the closures, ZCTAs in the city had a longer trip in distance and time.

The analysis includes a conservative projection of the impact from the closure of two (of the four) facilities. Because of the lack of data for the suburban testing sites, as noted above, this analysis only considers the increase in travel time and distance for Chicago zip codes (weighted by the vehicles from each zip code inspected at the two closed city sites). During the 2014–2015 cycle of emissions inspections, the two Chicago sites tested 387,296 vehicles that were registered within zip codes in the city of Chicago. This number does not include vehicles registered outside of Chicago tested at each facility. It is lower than the total number of vehicles registered in Chicago, as it does not include vehicles registered in Chicago tested elsewhere or vehicles less than 4 years old that are not subject to testing. Furthermore, vehicles reflected in census data above could be legally registered elsewhere. When weighting the increase in travel to the nearest facility by the vehicles inspected per Chicago zip code, the average increase in time and distance is slightly longer than the unweighted average: 7.1 min and 5.0 mi. The aggregate increase in time to the nearest facility for Chicago-registered vehicles that were tested in Chicago is estimated at 45,672 h for a two-year cycle of testing. The aggregate estimated increase in VMT to the closest facility is 1.9 million mi for a two-year testing cycle, for just these vehicles.

Table 4. Spearman's Rho Correlation Coefficients between Travel Variables and Zip Code Demographic Variables, Chicago Region

	Pre-closure time	Pre-closure distance	Current time	Current distance
Non-Latino				
Percent below poverty line	−.127*	−.055	.051	.094
Percent White	.127*	.036	.066	−.004
Percent Black	−.046	.060	−.019	.040
Percent Asian	−.142*	−.162**	−.001	.016
Latino				
Any race	−.099	−.166**	−.103	−.116

Note: $n = 276$.
**Correlation is significant at the 0.01 level (2-tailed). *Correlation is significant at the 0.05 level.

Demographics and Travel to Testing Sites

The regional analysis showed some statistically significant travel variance by ZCTA demographics prior to the closures, as seen in Table 4, but none post-closure. Prior to the closure of testing sites, the demographic variables with statistically significant correlation coefficients with travel time were percent below the poverty line; percent non-Latino, White; and percent Asian. In this case, the population groups protected by EJ rules (Asian and those in poverty) actually had a pre-closure correlation showing a slight relative travel time advantage (demonstrated by the negative coefficient indicating less travel time). The percent non-Latino, White was actually associated with increased travel time. Likewise, among the statistically significant correlations for pre-closure distance, the percent Asian and percent Latino were also associated with a slight distance advantage. Following the closings, however, these groups lost their slight advantages, as no demographic group showed a statistically significant correlation with either current travel time or current distance.

At the regional scale, the average travel time and distance by zip codes with concentrations of people of color or those below the poverty line showed little statistical significance (and thus full results are not displayed). Prior to the closure, the only statistically significant difference (at the 99% level) was a travel time advantage for zip codes with a concentration of people below the poverty level: 10.9 min versus 12.7 min. After the closures, no statistically significant differences existed in average travel time or distance for concentrations of people of color or those below the poverty level at the regional scale.

The city analysis revealed more variability by demographic groups (see Table 5). Prior to the site closures, almost every demographic variable had a statistically significant correlation with travel time and distance to the nearest facility. Concentrations of two groups protected by EJ rules—those below the poverty line and Black residents—had statistically significant, positive correlations above .5 for both travel metrics, indicating higher concentrations of those groups were associated with increased travel time and distance. The Asian share had a negative coefficient, indicating a shorter travel time and distance, but the White, non-Latino share showed the largest negative coefficient for time and distance, indicating the largest pre-closure advantage. The Latino share had a negative coefficient for distance, indicating relatively close facilities, but the coefficient for travel time was not significant for the Latino share. Correlations after closures do not reflect improvements in travel time for any groups. Rather, any reduction in inequality is due to a concentration of a population group losing relative advantage. The Black and poor population shares maintained a statistically significant, positive correlation, indicating an association with more travel time. The coefficient is smaller post-closure, but again is due to other groups' degradation of proximity, not an improvement for ZCTAs with high shares of Black or poor residents. The non-Latino, White share no longer has a statistically significant negative correlation, indicating that because of increases in travel time and distance, ZCTAs with high shares of White residents no longer show a statistically significant relative advantage. The share of Asian residents in a ZCTA was no longer associated with a relative advantage in time or distance. The relative advantage for higher shares of Latino residents became stronger for both time and distance, showing the only remaining statistically significant correlation with current distance.

Similar disparities appear when analyzing average travel times by concentrations of people of color or those living below the poverty line at the city level. Zip codes with high shares of people of color had a statistically significant travel time and distance disadvantage prior to the facilities closures (see Table 6). Following the facilities closures, this disadvantage persisted, but the gap between zip codes with high shares of people of color and those without, as well as for high and low poverty zip codes, decreased as a result of lengthened travel times and distances especially for majority White and low poverty zip codes. The average travel time disadvantage for zip codes with concentrations of people of color or those

Table 5. Spearman's Rho Correlation Coefficients between Travel Variables and Zip Code Demographic Variables, City of Chicago

	Pre-closure time	Pre-closure distance	Current time	Current distance
Non-Latino				
Percent below poverty line	.580**	.539**	.350**	.081
Percent White	–.679**	–.607**	–.196	.056
Percent Black	.625**	.604**	.349**	.129
Percent Asian	–.566**	–.389**	–.088	.116
Non-Latino				
Any race	–.133	–.312*	–.346**	–.434**

Note: $n = 56$.
**Correlation is significant at the 0.01 level (2-tailed). *Correlation is significant at the 0.05 level.

Table 6. Average Travel Times by Zip Codes by Demographic Categories, City of Chicago

Zip code characteristics	*n*	Average pre-closure travel time (minutes)	Average pre-closure travel distance (miles)	Average current time (minutes)	Average current distance (miles)
By race/ethnicity					
Concentrated POC	22	12.1**	8.6**	16.6*	10.9
Not concentrated POC	34	6.8**	3.8**	13.8*	9.7
By income					
High poverty	23	11.6**	8.3**	16.8*	11.3
Not high poverty	33	7.0**	3.9**	13.5*	9.3

Note: POC = people of color.
**Difference is significant at the 0.01 level (2-tailed). *Difference is significant at the 0.05 level.

in poverty remained statistically significant, but after the closures the average distance difference was no longer statistically significant.

Implications and Conclusion

This analysis found that post-closure, current siting in terms of *distance* is not in conflict with federal EJ principles at the region or city scale, but that some protected populations within the city are at a statistically significant travel *time* disadvantage. Furthermore, a historic inequity within the city of Chicago has not been addressed by the reductions in access due to site closures.

The CAA has been transformational for air quality. Emissions testing is an important part of CAA implementation, but testing requires vehicle miles to be traveled to testing sites. Unfortunately, closure of the two Chicago testing sites will result in increased VMT and emissions. The partial impact—closures of two Chicago sites—is estimated as 1.9 million additional miles traveled over a two-year testing cycle by Chicago-registered vehicles. These additional VMT have environmental consequences. In 2014, the average age of vehicles was 11 years. Transportation statistics for model year 2003 indicate that light-duty vehicles on average emitted the criteria pollutant carbon monoxide at 13.2 g/mi, along with 1.2 g/mi of hydrocarbons and 1.1 g/mi of nitrogen oxide (*32*). Furthermore, 411 g/mi of carbon dioxide are emitted, on average (*33*).

The siting of these facilities in Chicago and other metropolitan areas with emissions testing programs should be reviewed and changed in order to minimize VMT, with attention to the distribution of EJ populations. To inform siting, future analyses could consider congestion levels if time of day and day of week data are available for when vehicles are tested. Confidentiality concerns may prevent a disaggregated model of travel and may not have racial and income data, but a smaller spatial unit of analysis would be desirable. In the case of Illinois, with its reliance on private vendors for emissions testing operations, new siting patterns may require administrative action in new contracts with vendors or legislative action that directs the administration to issue contracts with attention to VMT reduction and the pattern of travel time and distance by race and income.

Given the importance of automobiles for many within populations protected by EJ rules, the impacts of automobile registration requirements merit analysis for researchers and practitioners. This analysis shows historic and some continued inequity in the city of Chicago. The differing results at the city and region scale show the importance of analyzing multiple scales, when considering the EJ implications of public facilities related to automobile regulation and oversight. The difference in historic and current relationships also shows the need for analysis to be retrospective, as well as prospective. The relative time and distance advantage associated with a higher share of White, non-Latino residents went away in the post-closure, current siting scheme. However, from a historic perspective, this does not represent an EJ improvement, as rather than improving historically worse access for protected population groups, public facilities became less accessible. In other words, the public sector did not address historic and cumulative disadvantage for Black and poor residents by decreasing access for zip codes with higher shares of White, non-Latino residents. Although the analysis raises concern about access for concentrations of low-income and Black Chicagoans, more generally it also demonstrates the importance of a historic lens for EJ and multiple scales for EJ analysis.

Author Contributions

The authors confirm contribution to the paper as follows: study conception and design: KL, LN; data collection: na; analysis and interpretation of results: LN, KL; draft manuscript preparation: KL, LN. All authors reviewed the results and approved the final version of the manuscript.

References

1. Woodruff, T. J., J. D. Parker, A. D. Kyle, and K. C. Schoendorf. Disparities in Exposure to Air Pollution during Pregnancy. *Environmental Health Perspectives*, Vol. 111, No. 7, 2003, pp. 942–946.
2. Clougherty, J. E., J. I. Levy, L. D. Kubzansky, P. B. Ryan, S. Franco Suglia, M. Jacobson Canner, and R. J. Wright. Synergistic Effects of Traffic-Related Air Pollution and Exposure to Violence on Urban Asthma Etiology. *Environmental Health Perspectives*, Vol. 115, No. 8, 2007, pp. 1140–1146.
3. Houston, D., J. Wu, P. Ong, and A. Winer. Structural Disparities of Urban Traffic in Southern California: Implications for Vehicle-Related Air Pollution Exposure in Minority and High-Poverty Neighborhoods. *Journal of Urban Affairs*, Vol. 26, 2004, pp. 565–592.
4. Environmental Protection Agency. *EPA Report Underscores Clean Air Act's Successful Public Health Protections*. https://yosemite.epa.gov/opa/admpress.nsf/d0cf6618525a9efb85257359003fb69d/f8ad3485e788be5a8525784600540649%21OpenDocument. Accessed June 21, 2017.
5. Federal Highway Administration. *Transportation Conformity: A Basic Guide for State & Local Officials*. https://www.fhwa.dot.gov/environment/air_quality/conformity/2017_guide/fhwahep17034.pdf. Accessed June 21, 2017.
6. Makler, J., and A. M. Howitt. Regulating Transportation in New Nonattainment Areas Under the 8-hour Ozone Standard. *Transportation Research Record: Journal of the Transportation Research Board*, 2003. 1842: 39–46.
7. Environmental Protection Agency. *Regulatory Announcement: Proposed Rule for Amendments to Vehicle Inspection and Maintenance Program Requirements*. https://nepis.epa.gov/Exe/ZyPDF.cgi/P10006H1.PDF?Dockey=P10006H1.PDF. Accessed June 21, 2017.
8. Lachance, L. C., and E. Mierzejewski. Analysis of the Cost-Effectiveness of Motor Vehicle Inspection Programs for Reducing Air Pollution. *Transportation Research Record: Journal of the Transportation Research Board*, 1998. 1641: 105–111.
9. Blumenberg, E., and A. W. Agrawal. Getting Around When You're Just Getting By: Transportation Survival Strategies of the Poor. *Journal of Poverty*, Vol. 18, No. 4, 2014, pp. 355–378.
10. Grengs, J. Job Accessibility and the Modal Mismatch in Detroit. *Journal of Transport Geography*, Vol. 18, No. 1, 2010, pp. 42–54.
11. Kawabata, M., and Q. Shen. Commuting Inequality between Cars and Public Transit: The Case of the San Francisco Bay Area, 1990–2000. *Urban Studies*, Vol. 44, No. 9, 2007, pp. 1759–1780.
12. Ong, P. M., and D. Miller. Spatial and Transportation Mismatch in Los Angeles. *Journal of Planning Education and Research*, Vol. 25, No. 1, 2005, pp. 43–56.
13. Blumenberg, E., and G. Pierce. A Driving Factor in Mobility? Transportation's Role in Connecting Subsidized Housing and Employment Outcomes in the Moving to Opportunity (MTO) Program. *Journal of the American Planning Association*, Vol. 80, No. 1, 2014, pp. 52–66.
14. Ong, P. M. Car Ownership and Welfare-to-Work. *Journal of Policy Analysis and Management*, Vol. 21, No. 2, 2002, pp. 239–252.
15. Onésimo Sandoval, J. S., R. Cervero, and J. Landis. 2011. The Transition from Welfare-to-Work: How Cars and Human Capital Facilitate Employment for Welfare Recipients. *Applied Geography*, Vol. 31, No. 1, 2011, pp. 352–362.
16. Sanchez, T. W., Q. Shen, and Z.-R. Peng. Transit Mobility, Jobs Access and Low-Income Labour Participation in US Metropolitan Areas. *Urban Studies*, Vol. 41, No. 7, 2004, pp. 1313–1331.
17. Lucas, K. Transport and Social Exclusion: Where Are We Now? *Transport Policy*, Vol. 20, 2012, pp. 105–113.
18. Boeri, M. W., and B. D. Tyndall. A Contextual Comparison of Risk Behaviors among Older Adult Drug Users and Harm Reduction in Suburban versus Inner-city Social Environments. *Journal of Applied Social Science*, Vol. 6, No. 1, 2012, pp. 72–91.
19. United States Census Bureau. *American Community Survey One-Year Estimates, Table B08201. 2015*. https://factfinder.census.gov. Accessed June 27, 2017.
20. Blumenberg, E., and G. Pierce. Automobile Ownership and Travel by the Poor. *Transportation Research Record: Journal of the Transportation Research Board*, 2012. 2320: 28–36.
21. Klein, N. J., and M. J. Smart. Car Today, Gone Tomorrow: The Ephemeral Car in Low-Income, Immigrant and Minority Families. *Transportation*, Vol. 44, No. 3, 2017, pp. 495–510. https://doi.org/10.1007/s11116-015-9664-4
22. Holmes, J., J. Isham, and J. Wasilewski. Overcoming Information Asymmetries in Low-Income Lending: Lessons

from the 'Working Wheels' Program. *Southern Economic Journal*, Vol. 72, No. 2, 2005, pp. 329–351.
23. Karger, H. J. No Deals on Wheels: How and Why the Poor Pay More for Basic Transportation. *Journal of Poverty*, Vol. 7, No. 4, 2003, pp. 93–112.
24. United States Department of Transportation. *U.S. Department of Transportation Takes Action to Ensure Equitable Driver License Office Access for Alabama Residents*. https://www.transportation.gov/briefing-room/us-department-transportation-reaches-agreement-alabama-law-enforcement-agency-ensure. Accessed June 27, 2017.
25. Esri and TomTom. *U.S. and Canada Detailed Streets*. https://www.arcgis.com/home/item.html?id=f38b87cc295541fb88513d1ed7cec9fd. Accessed July 25, 2017.
26. United States Census Bureau. *American Community Survey, Five-Year Data*. Geodatabase, Zip Code Tabulation Area. https://www.census.gov/geo/maps-data/data/tiger-data.html. Accessed July 3, 2017.
27. United States Census Bureau. *American Community Survey, Five-Year Estimates*. https://factfinder.census.gov. Accessed June 26, 2017.
28. United States Census Bureau. *American Community Survey, Five-Year Estimates, Table S1701*. https://factfinder.census.gov. Accessed June 26, 2017.
29. United States Census Bureau. *American Community Survey, Five-Year Estimates, Table B25046*. https://factfinder.census.gov. Accessed June 26, 2017.
30. United States Census Bureau. *American Community Survey, Five-Year Estimates, Table B08201*. https://factfinder.census.gov. Accessed June 26, 2017.
31. United States Census Bureau. *American Community Survey, Five-Year Estimates, Special Tabulation: Census Transportation Planning, Tables A113201 and A112216*. 2006–2010. http://data5.ctpp.transportation.org/ctpp/Browse/browsetables.aspx. Accessed June 27, 2017.
32. Bureau of Transportation Statistics, United States Department of Transportation. *National Transportation Statistics*. https://www.rita.dot.gov/bts/sites/rita.dot.gov.bts/files/publications/national_transportation_statistics/index.html. Accessed July 11, 2017.
33. United States Environmental Protection Agency. *Greenhouse Gas Emissions from a Typical Passenger Vehicle*. https://www.epa.gov/greenvehicles/greenhouse-gas-emissions-typical-passenger-vehicle-0. Accessed July 27, 2017.

The Standing Committee on Environmental Justice in Transportation (ADD50) peer-reviewed this paper (18-03358).

Article

TRR
JOURNAL OF THE TRANSPORTATION RESEARCH BOARD

Transportation Research Record
2018, Vol. 2672(3) 194–205

DOI: 10.1177/0361198118790838
journals.sagepub.com/home/trr
SAGE

Does Location Matter? Performance Analysis of the Affordable Housing Programs with Respect to Transportation Affordability in Dallas Fort Worth (DFW) Metropolis

Shima Hamidi[1], Jinat Jahan[1], and Somayeh Moazzeni[1]

Abstract

Transportation costs are the second largest expenditure for a family, thus have a substantial influence on housing affordability. In an auto-oriented region like DFW, the situation is exacerbated for low-income families due to limited transportation options. This study seeks to evaluate the efficiency of major affordable housing programs for low-income people in terms of transportation affordability. This study uses a rigorous methodology that involves a solid transportation cost modeling with disaggregated data available at property level for housing assistance programs in DFW. Our findings show that about 69% of the assisted units in DFW are unaffordable in terms of transportation costs. The majority of them are spending about 17% to 20% of their income on transportation. The most affordable program is Low-Income Housing Tax Credit with 58% affordability rate and the least affordable program is the Continuum of Care with 9% affordability rate when accounting for transportation costs. We also found that almost all affordable units (regarding the transportation costs) are located in main economic hubs of the region such as Dallas and Fort Worth which have better access to jobs and public transit. In contrary, almost all housing properties in the areas between Dallas and Fort Worth are unaffordable. These are areas adjacent to the University of Texas at Arlington with a high number of transit dependent population and in Arlington, the biggest midsize city with no public transit. Our findings urge HUD to consider modifying these programs by incorporating the location-efficiency factors to ensure true affordability.

According to National Association of Home Builders, Dallas is the least affordable city in Texas and housing affordability has been declining in DFW since 2013 (*1*). As a result, the issue of housing affordability in DFW is receiving enormous attention in recent years. Such critical housing affordability issues are not confined at the local level but also nationwide. The federal government annually spends about 50 billion dollars to provide housing assistance specifically for low-income families (*2*). But housing costs burden still found to be critical for low-income households (*3*). This raises the question about effectiveness of housing assistance programs and the answer depends on how the definition of affordability is conceptualized.

Historically, the measure of housing affordability has been viewed as ratio of housing price and income (*4*). Since the United States National Housing Act of 1937, the income- housing costs ratio threshold was evolved and by 1981, the standard was raised to 30 percent of household income, which is still considered affordable for most housing assistance programs (*4*, *5*).

Transportation costs, on the other hand, are the second largest expenditure for an American family (*6*). Due to limited choices to transportation modes and in search of affordability, the vulnerable low-income families make critical trade-offs between housing and transportation costs. Previous literature explored that households may be willing to spend more for housing and less for commuting or may choose more affordable dwelling farther from jobs with higher transportation costs (*7*). This often leads low-income worker groups to suffer from spatial mismatch. The mismatch between individual needs and

[1]College of Architecture, Planning and Public Affairs, University of Texas at Arlington, Arlington, TX

Corresponding Author:
Address correspondence to Shima Hamidi: shima.hamidi@uta.edu

the location of daily services has been researched on the pertinent basis of transportation equity (*8–10*).

In an auto-oriented region like DFW, the situation might be exacerbated for low-income families with lack of private vehicles. It is not only the provision of affordable housing that matters, but also the access to daily needs and availability of a supportive system to enhance the well-being and livability of people and communities. This study seeks to identify these gaps by evaluating location efficiency for major affordable housing programs in DFW.

To address the gaps, this study uses an innovative and robust methodology with disaggregated data at property level and measures built environmental variables around each property. Then, compared to the methodology of Housing + Transportation Affordability Index and Location Affordability Index (LAI) that measure location affordability for typical households (*11*), this study estimates transportation costs specifically for low-income households to whom HUD is intended to provide housing support. Inclusion of address level disaggregated data for all available housing assistance programs in DFW and use of solid transportation costs modeling tailored for low income make this study unique than previous studies on affordable housing programs in DFW (*12*).

Transportation equity and location efficiency are growing concerns in DFW Metro area, so this study can provide an insight for current as well as long term performances of rental assisted properties. This study is conducted in an effort that can guide further federal initiatives to consider the combined effect of housing and transportation. It will tell HUD and other organizations whether the rental housing they subsidize is truly affordable and how, in the future, it can be made more so by directing subsidies to better (more compact, walkable, and transit-served) locations. For this, it is fundamental to understand the mechanism of major affordable housing programs of HUD which has been discussed as program overview in next section.

Literature Review

Overview of Affordable Housing Programs

HUD is responsible for the majority of the federal housing programs to support distressed economic group of people. In their recent report, HUD (*13*) have listed an overall 100 active and authorized housing assistance programs. Schwartz (*14*) in his book provides an overview of major national housing assistance programs under different categories. The major HUD programs primarily focus on availability of affordable housing units for the low and very low-income people as well as ensuring to protect people from housing discrimination. Table 1 presents a summary of major affordable programs covered in this study.

These programs have different funding mechanisms. The LIHTC program, for instance, allows investors to deduct their federal income taxes by $1 for every dollar of tax credit to cover the project costs. The rent of LIHTC assisted properties cannot exceed 30 percent of income limit for particular sized households and the property can be occupied for at least 15 years (*14*). Similarly, Housing Trust Funds with a minimum affordability period of 30 years, provides funding for construction and preservation of rental housings as well as support for homeownership to the households living below the poverty line (*13, 19*). On the other hand, the residents relocated due to HOPE VI receive vouchers to rent housing units in private market with the intention to allow them move into better neighborhoods (*15*).

Some of these assistance programs are designed for elderly people and persons with special needs like, Section 202 and Section 811. Under the Section 202 program, rental assistance is provided to the landlords to cover the difference between the renters' share toward rent and the HUD permitted expenses for senior citizens (*13*). The CoC is also aimed to provide housing facilities to immediately rehabilitate homeless people and provide them with long-term housing stability (*13*).

Another category of housing assistance initiatives are voucher programs like HOV, HCV and PBV programs. Among all the voucher programs, HCV is the largest federal rental subsidy program (*20*) which provides low-income families with vouchers to find their preferred housing unit (*13*). PBV, on the other hand, is location based and 25% of units in a project can be supported by the program to accommodate eligible low-income people. In this program, property owners initially can have 15 years of contract to receive financial security from the Public Housing Agencies (PHA) (*13, 21*).

In other assistance programs such as CDBG and HOME, HUD provides grants to the state and local governments to implement local housing strategies intended to increase homeownership for low-income people (*16, 22*).

Transportation Costs and Affordability Studies

Previous studies have criticized HUD's simple percentage (or ratio) of income-to-housing costs measure to define housing affordability. They indicated that the ratio fails to include other 'costs of living', like transportation and other living expenditures (*23*). In most cases, housing affordability studies disregarded the spatial dimensions of transportation costs despite the influence of travel distance on land price, housing value and household transportation costs (*24*).

Table 1. Summary of the Major Housing Assistance Programs Covered in This Study

Category	Housing programs	Established by	Eligibility	Location preference	Mechanism	Type
Public Housing	Housing Trust Fund	Article XVIII of the Private Housing Finance Law (PHFL)	Households $\leq$ 30 % Area Median Income (AMI) and $\leq$ 50 % AMI	Areas with distressed economic condition (*13*)	Fund through contributions from Fannie Mae and Freddie Mac	Rent and ownership
	HOPE VI	The United States Housing Act of 1937	Residents of severely distressed public housing	Neighborhoods with closer proximity to economic opportunities (*15*).	Capital cost to PHA for major reconstruction of housing. Vouchers to the recipients to rent housing in private market	Rent
	Section 32 Public Housing Homeownership Program	The United States Housing Act of 1937	Households $\leq$ 80 % AMI	–	Down payment assistance or subordinate mortgages, and/or below market financing to purchase home	Ownership
Nonprofit Sector and community development	HOME Investment Partnerships	The Cranston-Gonzalez National Affordable Housing Act	Households $\leq$ 60 % AMI and $\leq$ 80 % AMI	–	Fund to the state and local agencies to support rehabilitation and construction of rental housing.	Rent and ownership
	Community Development Block Grants (CDBG)	The Housing and Community Development Act of 1974	Households $\leq$ 50 % AMI; and $\leq$ 80 % AMI	Residential areas where at least 51 percent of residents are low and moderate income people (*16*)	Fund to state and local government to support acquisition, disposition or retention of property; rehabilitation of residential or non-residential buildings; social and economic development	Support community development
Multifamily or privately owned project-based	Section 8 Project-Based Rental Assistance Renewal Program	–	Households $\leq$ 30 %AMI and $\leq$ 50 %AMI; Or, $\leq$ 80 %AMI (depending on availability of unit)	–	Renewal of expiring contracts on units already receiving project-based Section 8 rental assistance.	Rent
	Rental Assistance Demonstration (RAD)	Consolidated and Further Continuing Appropriations Act, 2012	Owners of other any HUD assisted properties	–	Allow owners of other type HUD assisted properties convert units to project based Section 8 programs	Rent
	Section 236 or Below Market Interest Rate (BMIR)	National Housing Act, 1968	Households $\leq$ 80 %	–	Combined mortgage insurance and reduced interest rate for the mortgagee to develop low rental housings	Rent

(*continued*)

Table 1. (continued)

Category	Housing programs	Established by	Eligibility	Location preference	Mechanism	Type
Housing for people with special needs	Section 811 Project Rental Assistance (PRA) Program (PRAC811)	The National Affordable Housing Act of 1990	Households $\leq$ 30 % AMI or $\leq$ 50 % AMI and also have at least one adult member with a disability	–	Interest free capital advances to sponsors	Rent
	Section 202 Supportive Housing for the Elderly Program (202/PRAC)	The Housing Act of 1959	Households $\leq$ 30 % AMI and have at least one person 62 years of age or older	–	Capital grants and project rental assistance to developers	Rent
	Continuum of Care Program (CoC)	The Homeless Emergency Assistance and Rapid Transition to Housing Act of 2009 (HEARTH Act)	Homeless individuals or families	–	Fund to developers to provide housing facilities to accommodate homeless	Rent and ownership
Voucher programs	Housing Choice Vouchers (HCV)	The United States Housing Act of 1937	Households $\leq$ 30 % AMI	Participants' preference. Most of the voucher holders live in MSA neighborhoods (*17*)	Monthly rental subsidy to the recipients	Rent
Taxes and housing	Low-Income Housing Tax Credit (LIHTC)	The Tax Reform Act of 1986	AMI and $\leq$ 50 % AMI; Or, $\leq$ 60 %	Census tracts in which 50 percent or more of the households have incomes below 60% AMI (*18*)	Allocation of tax credits (an amount of money that can be offset against a tax liability) to developers to	Rent

Transportation, however, is the second largest spending sector of a household budget (*6*, *25*). In recent years, transportation costs and affordability has become a growing concern and is suggested to be factored in while measuring affordability for low-income families (*11*, *24*, *26*). Mattingly and Morrissey (*24*), for instance, found that there is a decrease in affordability in the peripheral areas of Auckland-Netherland if transportation costs are included in the measure. In some urban fringes, income spent for housing and transportation found to be more than 70%. Results of these studies suggest that when transportation costs are incorporated into the concept of affordability, they provide a more comprehensive picture of affordability than traditional measures of housing affordability.

The reason is simple: The residents living in location-efficient neighborhoods require less travel time and, in turn, lower transportation costs to access their daily necessities (*27*). Lipman (*28*) found that location is a major factor influencing the costs of housing and transportation, especially the distance between residential neighborhoods and work. A recent study by Center for Neighborhood Technology (CNT) (*29*) also found that lower income households pay more attention to the surrounding sites in search of transportation options to reduce VMT. Developing affordable housing for low-income families in location-efficient areas can reduce VMT at higher rates than for higher income households (*29*). Litman (*30*) thus considered "location efficient" or "Affordable-Accessible Developments" as those lower-priced residences in which households can reduce transportation costs due to higher accessibility, multimodal transportation choices, and affordable access to activities and services.

The big challenge in studying transportation affordability is how to measure transportation costs and how to incorporate that into the concept of affordability. In 2006, the CNT attempted to address this challenge by developing the Housing + Transportation Affordability Index. The H + T Index took into account not only the costs of housing but also the intrinsic value of location as quantified through transportation costs. According to CNT, housing is affordable if the combined costs of housing and transportation do not exceed 45 percent of a household's income. Later in 2013, in a joint project by the departments of Transportation and Housing Development, the index was upgraded to become the Location Affordability Index (LAI). The LAI is based on the same methodology as the H + T Affordability Index but uses the most recent and better quality data with more coverage (*11*). In the LAI Index, total housing costs are estimated based on the current housing sales prices and rents from the American Community Survey, and household transportation costs are estimated at the census block group level as three cost components: costs of auto ownership, auto use, and transit use. Auto use is modeled based on household VMT data from Chicago and St. Louis. Auto ownership is modeled based on vehicle ownership data from the American Community Survey. Transit use and associated costs are based on Google Transit Feeds. The LAI, however, has significant limitations in two of its three component models. The VMT model is based on data from two metropolitan areas which can hardly represent the national variation. Also the model used block group level aggregate data which may have problem of aggregation bias (11).

Most recently, Hamidi et al. (*11*), using disaggregate household travel data from 15 diverse regions in the U.S., estimated and summed automobile capital costs, automobile operating costs and transit fare costs for households at 8,857 HUD multifamily Section 8 rental assistance properties. Their models account for all of the so-called D variables found to affect travel and vehicle ownership in the peer-reviewed literature. Their analysis is based on disaggregate (household) travel and vehicle ownership data for tens of thousands of households in many diverse metropolitan regions of the U.S. All surveys provide XY coordinates for households and their trips. This allows travel to be modeled in terms of the precise built environment in which households reside and travel occurs. For travelers, individual age, employment status, driver's licensure and other variables are available from the survey data set. For households, household size, household income, vehicle ownership and other variables are available from the survey dataset. This allowed the authors to control for socio-demographic influences on travel at the household level. They found that the mean percentage of income spent on transportation is 15 percent for households at the eligible income scale. However, in highly sprawling metropolitan areas and in suburban areas of more compact metropolitan areas, a higher percentage of household budget is spent on transportation, which exceeds the 15 percent threshold.

This study is built on the work of Hamidi et al.'s (*11*) findings and modeling efforts. Our study, however, focuses on all major affordable housing programs in DFW. Focusing on various programs provides an opportunity to make comparison on the efficiency of these programs with respect to providing true affordability.

Methods

To calculate the housing and transportation affordability, this paper follows the work of Hamidi et al. (*11*). Inclusion of address level disaggregated data and use of transportation costs model designed to consider low-income people provides higher validity compared to the LAI methodology.

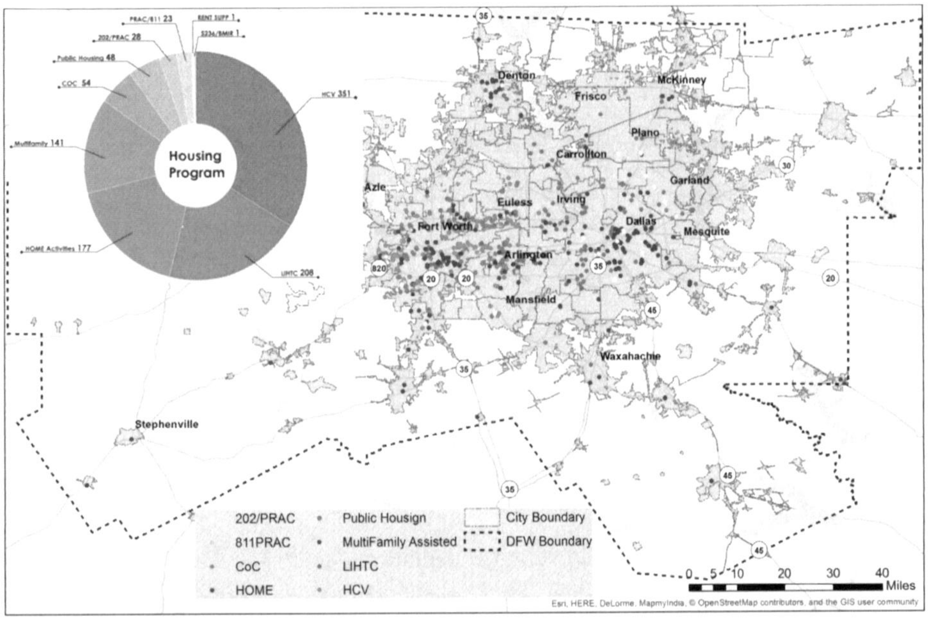

Figure 1. Location of assisted housing project in DFW.

Sample

We gathered address level data for properties in major affordable housing assistance programs in DFW through an extensive data collection process. We used three major data sources in our data collection process:

1) *"Picture of Subsidized Households" website:* This particular website of HUD contains the data of Public Housing, Housing Choice Vouchers (HCV), Moderate Rehabilitation, Section 8 Project Based Rental Assistance, Rent Supplement, Below market interest rate - section 236 (S236/BMIR), Section 202 Supportive Housing for the Elderly Program (S202 PRAC) and Section 811 Supportive Housing for Persons with Disabilities (S811 PRAC), HOME Investment Partnership Program and of Low-Income Housing Tax Credit (LIHTC).
2) *HUD eGIS Storefront:* This site provides HUD's geospatial datasets with web-based mapping tools. For our study, we extracted location data for Multifamily Assisted properties, Public Housing properties and HOME Investments Partnerships (HOME).
3) *Local Housing Agencies in DFW:* Contacted the local agencies in DFW, we acquired the data for Housing Choice Vouchers (HCV) and Continuum of Care (COC) for Tarrant County.

We successfully geocoded all 1,032 affordable housing assistance properties found in the sample. Figure 1 shows the number and spatial distribution of these properties in DFW.

In order to estimate transportation outcomes, we created network buffers of 0.25 mile, 0.50 mile and 1 mile around these locations using Business Analyst and measured D variables in these buffers. The D's that extensively are related to travel by literature are population and job *density* (actden), *diversity* measured in term of jobs–population balance and land-use entropy (entropy), *design* measured in terms of intersection density (intden) and street connectivity (int4way), *destination accessibility* measured in terms of jobs reachable within a given travel time by auto (emp10), and *distance to transit* measured as the frequency of transit service in the neighborhood (tfreq). Variables extracted from these datasets and used in subsequent transportation costs calculations are shown in Table 2. The table makes reference only to ½-mile buffers, but data for ¼-mile and 1-mile buffers are also available. This gave us a total of 21 built environment variables around each housing assistance property.

Transportation Models

In this study, we use the same methodology as CNT, LAI and Hamidi et al. (*11*) and estimate household transportation costs as the sum of three terms - auto ownership, auto use and public transit:

Table 2. Built Environmental Variables Used in the Household Transportation Costs Calculations

Symbol	Definition	Level
actden	Activity density within a half mile (sum of population and employment divided by gross land area in square miles)	Household
entropy	Land use mix within a half mile of a household (entropy index based on net acreage in different land use categories that ranges from 0, where all developed land is in one use, to 1, where developed land is evenly divided among uses)	Household
intden	Intersection density within a half mile (number of intersections divided by gross land area in square miles)	Household
int4way	Proportion of 4-way intersections with a half mile (4-or-more-way intersections divided by total intersections)	Household
emp10	Proportion of regional employment accessible within a 10-minute travel time via automobile	Household
sf	Single-family housing unit (dummy variable; yes = 1, no = 0)	
tfreq	Aggregate frequency of transit service within a quarter mile of block group boundary per hour during evening peak period	Block group

$$\text{Household T Costs} = [\text{C\_AO}*\text{F\_AO}(X)] + [\text{C\_AU}*\text{F\_AU}(X)] + [\text{C\_TU}*\text{F\_TU}(X)]$$

where,
C = cost factor (i.e. dollars per mile)
F = function of the independent variables (F_AO is auto ownership, F_AU is auto use, and F_TU is transit use)

For computing the F's, instead of developing transportation models, we borrow equations from Hamidi et al.'s (*11*) study since their models have all of the specifications we need for this study.

(1) **Household vmt** = oddsanyvmt*vmtpred
  a) oddsanyvmt = exp [1.7 + (0.2*hsize) + (0.3 *emp) + (0.03*inc) + (0.9*sf)–(0.02*emp10) –(0.7*entropy)–(0.003*intden)–(0.01*pct4wy) –(0.0009*tfreq)]
  b) vmtpred = exp [2.6 + (0.2*hsize) + (0.2 *emp) + (0.007*inc)–(0.008*emp10)–(0.005* actden/1000)–(0.3*entropy) –(0.002*intden)–(0.003*pct4wy)–(0.00009*tfreq)]

(2) **Household vehicle ownership** = exp [–0.1 + (0.1*hsize) + (0.1*emp) + (0.009*inc) + (0.3*sf) –(0.002*emp10)–(0.006*actden/1000)–(0.1*entropy) –(0.0009*intden)–(0.001*pct4wy)–(0.003*tfreq)]

(3) **Number of household transit trips** = oddsanytransit*transittrip
  a) oddsanytransit = exp [–2.8 + (0.2*hsize) + (0.3*emp)–(0.02*inc)–(0.8*sf) + (0.5*enthropy) + (0.003*intden) + (0.01*pct4wy) + (0.0009*tfreq)]
  b) transittrip = exp [0.9 + (0.1*hsize)–(0.006* inc) + (0.2*entropy)]

Finally, we used sociodemographic characteristics of a typical low-income household in all transportation equations since the focus of this paper is to evaluate affordability for a low-income household. As a result, in all five equations, sociodemographic characteristics (household size, household income and number of workers in the household) were kept constant so the variation in predicted transportation outcomes results from the built environmental variability and not sociodemographic characteristics. To obtain sociodemographic characteristics of a typical low-income household, we used the National Household Travel Survey (NHTS) database and extracted a subsample of households who qualify for HUD rental assistance, i.e. those with annual incomes of less than 80 percent of AMGI.

Transportation Costs Calculation

Transportation costs consist of vehicle costs (a household's expenses to own and use private vehicles) and public transit costs (transit fares). Vehicle costs are divided into fixed and variable costs. Fixed or ownership costs are not generally affected by the amount a vehicle is driven. Depreciation, insurance and registration fees are considered fixed. Variable costs are the incremental costs that increase with vehicle mileage. Fuel is a variable vehicle cost proportional to mileage (*31*).

We computed vehicles' fixed costs based on our household vehicle ownership model and the average cost of car ownership specific. Our average car ownership costs are based on a car ownership costs calculator called the True Cost to Own® pricing (TCO®) system developed by Edmunds Inc. The components of TCO® are depreciation, interest on financing, taxes and fees, insurance premiums, fuel, maintenance, repairs and any federal tax credit that may be available. In this paper, we used all categories except fuel because we treat fuel as a variable vehicle cost.

TCO® values are specific to each state as well as the vehicle's make, model and year. We were interested in costs for the most popular vehicles' model and make for low-income households. Therefore, we created a sample of low-income households from the NHTS database based on the HUD low-income standard and identified the 15 most popular vehicles owned by households in this sample. These vehicles account for more than 34 percent of vehicles owned by low-income households in the NHTS database. We acquired, for each state, the five-year average costs of car ownership for these 15 vehicles for the earliest year (2009) reported by the TCO® because, according to the NHTS database, low-income households tend to buy and own older cars. We then weighted the five-year average costs by the popularity of each make and model for low-income households in the NHTS database to obtain the average vehicle ownership costs for low-income households. We multiplied this by the predicted number of cars owned by a household to obtain the household's ownership or fixed vehicle costs.

Second, we computed auto operating costs based on our household VMT model. We acquired metropolitan-level average gasoline prices for 2010 from the Oil Price Information Service (OPIS), inflated them to 2014 dollars and then multiplied the fuel costs per mile by the predicated VMT to obtain the household's operating or variable vehicle costs.

Third, we computed transit costs based on our household transit trip model. Transit fare data comes from the National Transit Database. We computed average transit fare for each transit agency in the region by dividing the total transit revenue by the total number of unlinked passenger trips for the region. We multiplied the fare per transit trip by the predicted number of transit trips to obtain the household's public transit costs.

To estimate the overall household's transportation costs for each property in our sample, we added the three transportation cost components. Finally, we calculated the percentage of the household's income spent on transportation for a household with a size of three who qualifies for renting HUD assistance programs (average household size for a typical low-income household is 2.39).

Results and Discussion

Transportation is considered affordable if a household spends no more than 15% of its budget on transportation costs (*6*, *11*). Figure 2 shows the histogram distribution of transportation affordability for the 1032 properties in our sample. Interestingly, the distribution is not normal and is skewed toward less affordability. About 69% (718 properties out of 1032 properties) of the housing projects in DFW are spending more than 15% of their income for

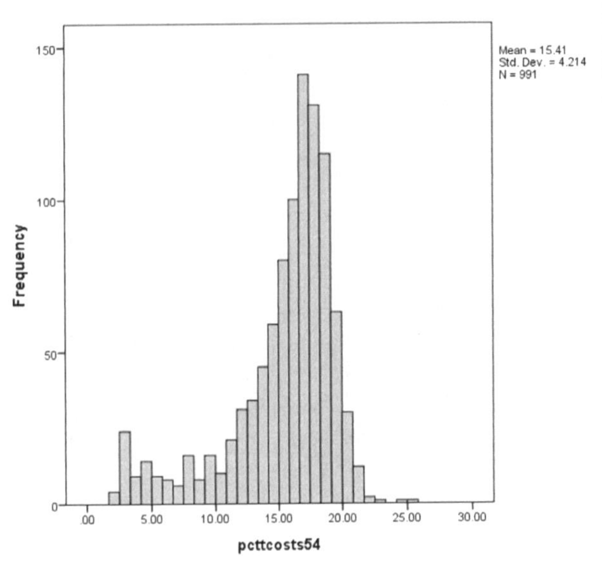

Figure 2. Frequency distribution of predicted transportation affordability (percentage of income spent for transportation costs).

transportation and thus unaffordable. The majority of them are spending about 17% to 20% of their income on transportation. The lowest expenditure found to be 2.14 % for a HOME property located in downtown Dallas. In contrast, a property supported by LIHTC program located in a suburban area, thirty five miles north from downtown Dallas, is spending the highest (25.76%) portion of their budget on transportation.

Figure 3 presents an overall picture of transportation affordability's spatial distribution for all housing assistance programs in our sample. The majority of the affordable units are concentrated in the main economic hubs of the region such as the City of Dallas and the City of Fort Worth. In addition to the sprawling northern and southern part of DFW, almost all housing properties in the areas between Dallas and Fort Worth are unaffordable. These are areas adjacent to the University of Texas at Arlington with a high number of transit dependent population and in Arlington, the biggest mid-size city with no public transit.

Regarding the performance analysis of individual programs, as evident from Figure 4, the LIHTC program is providing the highest number (121 samples) of affordable housing units among all programs in DFW and 58% of their supported housing is affordable in terms of transportation costs. As this study includes various programs, it would be interesting to make statistical comparison across the programs.

Table 3 provides descriptive statistics on the transportation affordability of various programs in DFW. The lowest average value is for the LIHTC program which is

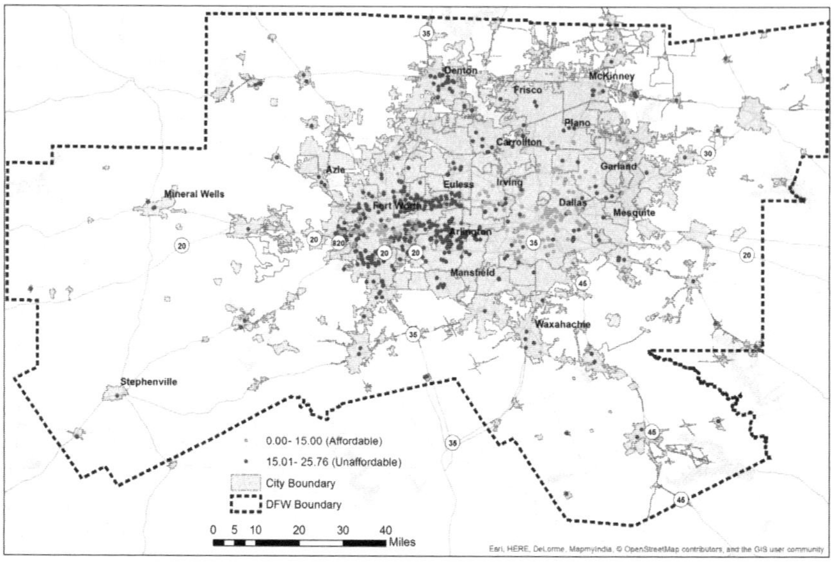

Figure 3. Transportation affordability of all the assisted housing properties in DFW.

Table 3. Percentage of Income Spent for Transportation under Different Programs

Housing program	Mean	Minimum	Maximum
202/PRAC	15.49	7.71	20.31
HOME	15.69	2.14	22.55
Multifamily assisted	16.17	4.14	21.61
PRAC811	17.00	12.14	20.02
Public housing	15.20	4.70	20.09
CoC	17.03	9.35	20.88
HCV	16.82	3.88	24.41
LIHTC	11.74	2.42	25.76

also found to provide maximum number of affordable units. On the other hand, the properties under CoC are spending the highest average percentage of transportation costs with only 9.26% affordable units which makes it the least affordable program in DFW. A follow up ANOVA analysis reveals that there is a statistically significant difference (with significance value of <0.001) in percentage of income spent on transportation across programs.

Figure 5 compares the spatial distribution of the most affordable (LIHTC) and the least affordable (CoC) programs. All the 121 affordable LIHTC housing units are found in six cities: Carrolton, Dallas, Fort Worth, Garland, Irving and Plano. These cities have public transit services provided by Dallas Area Rapid Transit (DART) and Trinity Rail Express (TRE) (*32, 33*) and hence are less car dependent. Those affordable units found to be located in areas close to the Medical District in Fort Worth, Ridgmar Mall, Downtown Dallas, and Downtown Fort Worth. Those areas are designed with a grid street pattern which provides increased accessibility, street connectivity, a vibrant mixture of land uses, and access to amenities and transit service.

On the contrary, only five housing units of the CoC program are located in areas of such characteristics. Most of the CoC (91%) units are found in areas with irregular street patterns that reduce the accessibility and street connectivity. Their units are located in sites which are distant from the central urban areas. Also, daily necessities and transit services are unavailable within a reachable distance. Previous studies explored that, development of housing for accommodating homeless individuals (or households) faces "not in my backyard" (NIMBY) syndrome of existing higher income residents which influences the placement of such programs (*18, 34*). Therefore, such properties tend to locate farther from the dense urban areas, which push residents to commute longer distance (*18, 24*). These factors lead to more auto dependency which results in higher

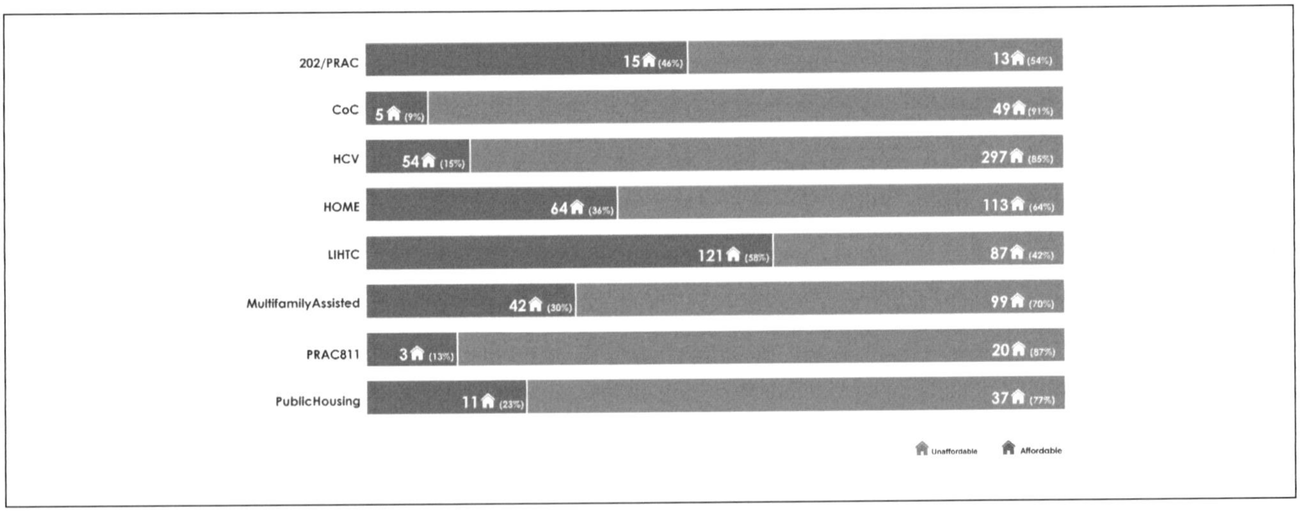

Figure 4. Number and percentage of affordable and unaffordable assisted housing properties in terms of transportation costs.

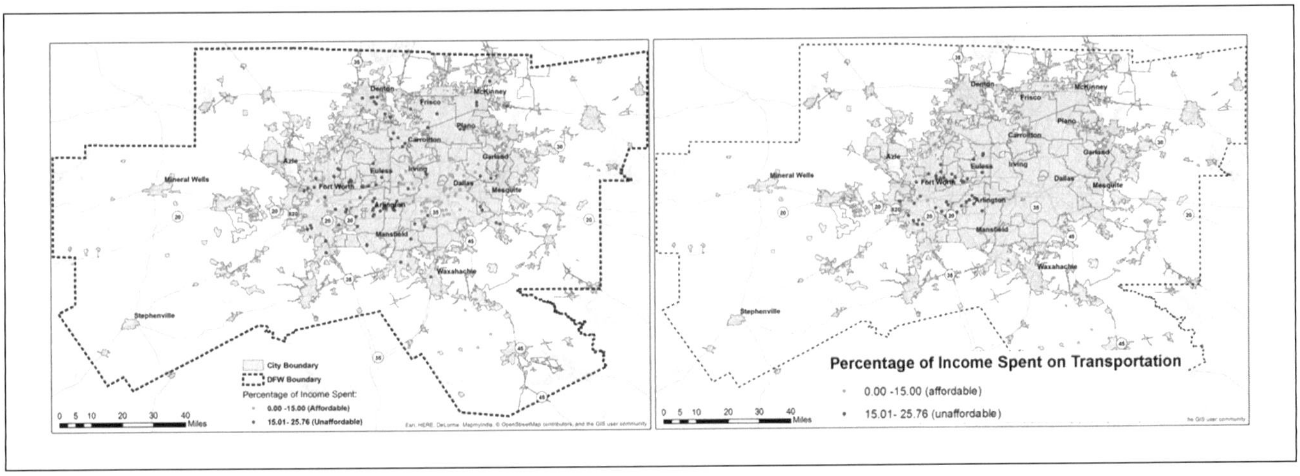

Figure 5. Transportation affordability for LIHTC (left) and CoC (right) assisted properties in DFW area.

transportation costs, making the HUD-designated properties less affordable in terms of transportation.

Even though the recipients of a rental assistance program like HCV have more flexibility to choose the location of their housing unit (*35*), most of the unaffordable HCV properties (84.62%) are interestingly found to be located in suburban neighborhoods which are designed with irregular street patterns (Figure 6). Literature suggests that HCV recipients tend to suburbanize themselves due to their preference for living in more spacious suburb neighborhoods (*36*).

For all cases, location and distance from daily destinations, availability of transit services, accessibility, street connectivity, and mixed use development are found to be influential for property's affordability in terms of transportation costs. It is clear from our findings that the "Location Matters" for assisted housing properties to be truly affordable in terms of both housing and transportation costs. The HUD and local housing authorities work with the mission of providing affordable housing units for low-income families. However according to our findings, living in inaccessible areas would cancel out the benefits of low housing costs.

Our findings urge HUD to consider modifying these programs by incorporating the location-efficiency factors to ensure true affordability. Government and city officials can control urban land market to incentivize low cost housing development in high accessible areas. Simultaneously, the existing unaffordable properties can be connected to the major destinations through improvement of transit services.

Our study has limitations. Data on transportation services like Uber and Lyft was not available for the assisted housing units and hence was not considered in the transportation costs calculation. This study is the first attempt to evaluate performance of assisted

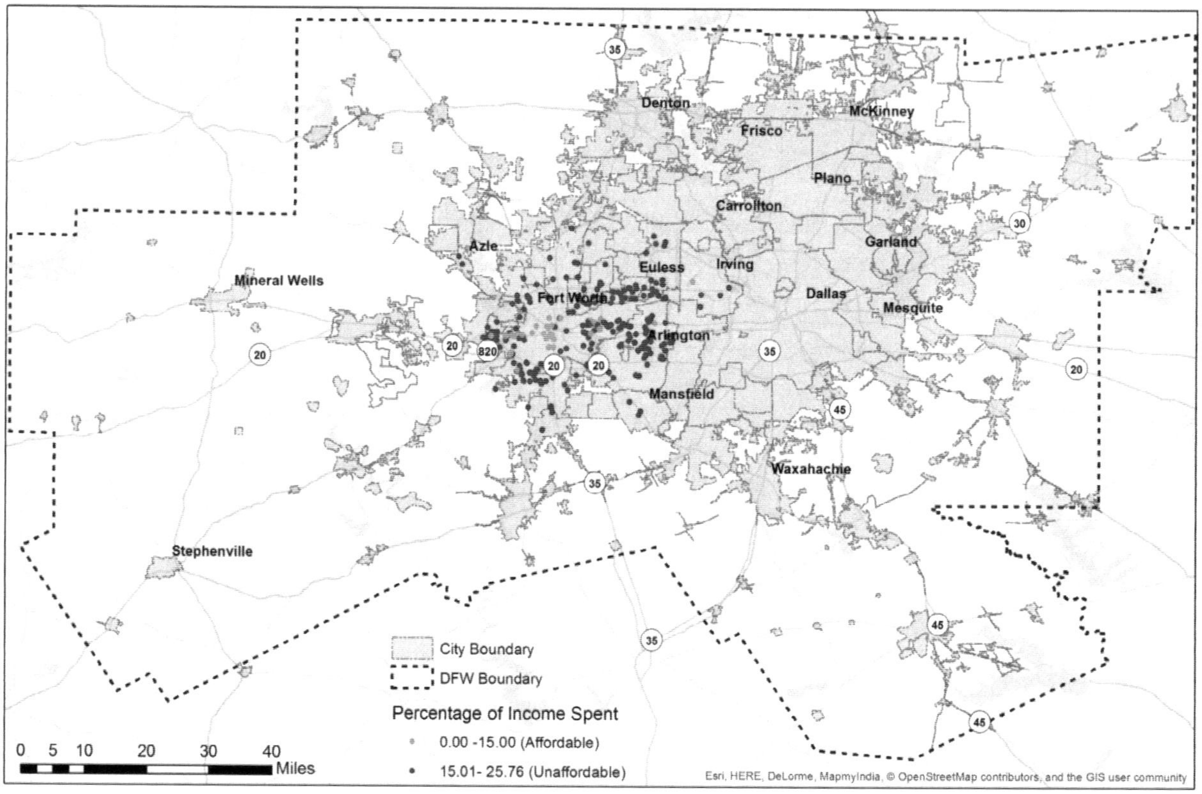

Figure 6. Transportation affordability for HCV assisted properties.

housing units in DFW area by incorporating the typical travel modes and related expenditure of low-income households.

Author Contributions

The authors confirm contribution to the paper as follows: study conception and design: Shima Hamidi; data collection: Somayeh Moazzeni, Jinat Jahan; analysis and interpretation of results: Shima Hamidi, Jinat Jahan; draft manuscript preparation: Shima Hamidi, Jinat Jahan. All authors reviewed the results and approved the final version of the manuscript.

References

1. Federal Reserve Bank of Dallas. *DFW Economic Indicators*. Dallas, TX, 2017.
2. Congressional Budget Office. *Federal Housing Assistance for Low-Income Households*. 2015.
3. Fischer, W., and B. Sard. *Chart Book: Federal Housing Spending is Poorly Matched to Need. Center on Budget and Policy Priorities*. http://www.cbpp.org/research/housing/chart-book-federal-housing-spending-is-poorly-matched-to-need. Accessed July 7, 2017.
4. Schwartz, M., and E. Wilson. *Who Can Afford To Live in a Home? A look at data from the 2006 American Community Survey*. The U.S. Census Bureau, Washington, D.C., 2008.
5. *Affordable Housing*. The U.S. Department of Housing and Urban Development. https://portal.hud.gov/hudportal/HUD?src=/program_offices/comm_planning/affordablehousing/. Accessed July 6, 2017.
6. Center for Transit-Oriented Development and Center for Neighborhood Technology. *The Affordability Index: A New Tool for Measuring the True Affordability of a Housing Choice. Metropolitan Policy Program*. The Brookings Institutions, Urban Markets Initiative, Washington, D.C., 2006.
7. Salvin, P. Factoring Transit Costs into Housing Affordability. *Urban Land: The Magazine of the Urban Land Institute*, 2014. https://urbanland.uli.org/news/factoring-transit-costs-housing-affordability/. Accessed July 8, 2017.
8. Welch, T. F. Equity in Transport: The Distribution of Transit Access and Connectivity Among Affordable Housing Units. *Transport Policy*, Vol. 30, 2013, pp. 283–293.
9. Grønbjerg, K. A., and L. Paarlberg. Community Variations in the Size and Scope of the Nonprofit Sector: Theory and Preliminary Findings. *Nonprofit and Voluntary Sector Quarterly*, Vol. 30, No.4, 2001, pp. 684–706.
10. Allard, S. W. *Out of Reach: Place, Poverty, and the New American Welfare State*. Yale University Press, New Haven, CT, 2008.

11. Hamidi, S., R. Ewing, and J. Renne. How Affordable is HUD Affordable Housing? *Housing Policy Debate*, Vol. 26, No. 3, 2016, pp. 437–455.
12. Zandt, S. V., and P. C. Mhatre. Growing Pains: Perpetuating Inequality through the Production of Low-Income Housing in the Dallas/Fort Worth Metroplex. *Urban Geography*, Vol. 30, No. 5, 2009, pp. 490–513. http://dx.doi.org/10.2747/0272-3638.30.5.490
13. U.S. Department of Housing and Urban Development. *Programs of HUD: Major Mortgage, Grant, Assistance, and Regulatory Programs*, 2016. https://portal.hud.gov/hudportal/documents/huddoc?id = HUDPrograms2016.pdf. Accessed 8 July, 2017.
14. Schwartz, A. F. *Housing Policy in the United States*. Routledge, New York, NY, 2015.
15. Goetz, E. G. Better Neighborhoods, Better Outcomes? Explaining Relocation Outcomes in HOPE VI. *Cityscape*, Vol. 12, No. 1, 2010, pp. 5–31.
16. HUD Exchange. HOME and CDBG. https://www.hudexchange.info/resources/documents/HOME-CDBGGuidebook.pdf. Accessed July 15, 2017.
17. Galvez, M. M. *What Do We Know About Housing Choice Voucher Program Location Outcomes? A Review of Recent Literature*, The Urban Institute, Washington, D.C., 2010.
18. Di, W., and J. C. Murdoch. *The Impact of LIHTC Program on Local Schools*. Research Department, Federal Reserve Bank of Dallas, Dallas, TX, 2010.
19. Homes and Community Renewal. Housing Trust Fund Program (HTF), March 2015. http://www.nyshcr.org/Programs/HousingTrustFund/. Accessed July 8, 2017.
20. Getsinger, L., L. Posey, G. MacDonald, and J. Leopold. *The Housing Affordability Gap for Extremely Low-Income Renters in 2014*. The Urban Institute, Washington, D.C., 2017.
21. Sard, B. *Vouchers: Project-Based Vouchers*. http://nlihc.org/sites/default/files/AG-2017/2017AG_Ch04-S12_Vouchers-Project-Based-Vouchers.pdf. Accessed July 12, 2017.
22. DeHaven, T. Community Development. *Downsizing the Federal Government*, 2009. https://www.downsizinggovernment.org/hud/community-development. Accessed July 10, 2017.
23. Jewkes, M. D., and L. M. Delgadillo. Weaknesses of Housing Affordability Indices Used by Practitioners. *Journal of Financial Counseling and Planning*, Vol. 21, 2010, pp. 43–52.
24. Mattingly, K., and J. Morrissey. Housing and Transport Expenditure: Socio-spatial Indicators of Affordability in Auckland. *Cities*, Vol. 38, pp. 69–83.
25. U.S. Bureau of Labor Statistics. *Quintiles of Income Before Taxes: Shares of Average Annual Expenditures and Sources of Income*. Consumer Expenditure Survey, Table 45, 2007. http://www.bls.gov/cex/2006/share/quintile.pdf. Accessed July 15, 2017.
26. Isalou, A. A., T. Litman, and B. Shahmoradi. Testing the Housing and Transportation Affordability Index in a Developing World Context: A Sustainability Comparison of Central and Suburban Districts in Qom, Iran. *Transport Policy*, Vol. 33, 2014, pp. 33–39.
27. Center for Neighborhood Technology. *Location Efficiency Recognizes the Inherent Efficiency of a Place*. locationefficiency.cnt.org/about/. Accessed July 10, 2017.
28. Lipman, B. J. *A Heavy Load: The Combined Housing and Transportation Burdens of Working Families*. Center for Housing Policy, Washington, D.C., 2006.
29. Center for Neighborhood Technology. *Income, Location Efficiency, and VMT: Affordable Housing as a Climate Strategy*. CNT Working Paper, 2015. http://www.cnt.org/sites/default/files/publications/CNT%20Working%20Paper%20revised%202015-12-18.pdf
30. Litman, T. *Transportation Affordability: Evaluation and Improvement Strategies*. Victoria Transport Policy Institute, 2017. http://www.vtpi.org/affordability.pdf
31. Litman, T. *Transportation Cost and Benefit Analysis*. Victoria Transportation Policy Institute, 2009. http://www.vtpi.org/tca/tca00.pdf
32. Dallas Area Rapid Transit. *DART Service Area*. https://www.dart.org/about/servicearea.asp. Accessed July 5, 2017.
33. Trinity Railway Express. http://www.trinityrailwayexpress.org/index.html. Accessed July 5, 2017.
34. Nguyen, M. Does Affordable Housing Detrimentally Affect Property Values? A Review of the Literature. *Journal of Planning Literature*, Vol. 20, No.1, 2005, pp. 15–26.
35. U.S. Department of Housing and Urban Development. Housing Choice Vouchers Fact Sheet. https://portal.hud.gov/hudportal/HUD?src = /program_offices/public_indian_housing/programs/hcv/about/fact_sheet. Accessed July 20, 2017.
36. Covington, K., L. Freeman, and M. A. Stoll. *The Suburbanization of Housing Choice Voucher Recipients*. Metropolitan Policy Program, Brookings, Washington, D.C., 2011.

The Standing Committee on Environmental Justice in Transportation (ADD50) peer-reviewed this paper (18-04466).

Sage Publications USA
sagepub.com

ISBN 978-1-5108-8175-4

CURRAN ASSOCIATES INC.
proceedings.com

Historic Ships 2023

Historic Vessels - Sustainable Futures

London, United Kingdom
28-29 November 2023

The Royal Institution of
Naval Architects